Bacterial Evasion of the Host Immune System

http://doi.org/10.21775/9781910190692

Edited by

Pedro Escoll

Institut Pasteur
Biologie des Bactéries Intracellulaires;
and Centre National de la Recherche Scientifique (CNRS) UMR 3525
Paris
France

Caister Academic Press

Caister Academic Press
Norfolk, UK

www.caister.com

British Library Cataloguing-in-Publication Data
A catalogue record for this book is available from the British Library

ISBN: 978-1-910190-69-2 (paperback)
ISBN: 978-1-910190-70-8 (ebook)

Cover design adapted from images provided by Pedro Escoll, showing *Legionella pneumophila* (red) replicating within a human primary macrophage.

Contents

Preface v

1 Secretion Systems Used by Bacteria to Subvert Host Functions 1
Chiara Rapisarda and Rémi Fronzes

2 Within-host Envelope Remodelling and its Impact in Bacterial Pathogen Recognition 43
M. Graciela Pucciarelli and Francisco García-del Portillo

3 Subversion of Macrophage Functions by Bacterial Protein Toxins and Effectors 61
Muyang Wan, Yan Zhou and Yongqun Zhu

4 Manipulation of Autophagy by Bacterial Pathogens Impacts Host Immunity 81
Tobias C. Kunz, Flávia Viana, Carmen Buchrieser and Pedro Escoll

5 Inflammasome-dependent Mechanisms Involved in Sensing and Restriction of Bacterial Replication 99
Warrison A. Andrade and Dario S. Zamboni

6 Molecular Mechanisms Used by *Salmonella* to Evade the Immune System 133
Joaquín Bernal-Bayard and Francisco Ramos-Morales

7 Immune-evasion Strategies of Mycobacteria and their Implications for the Protective Immune Response 169
Alexandra G. Fraga, Ana Margarida Barbosa, Catarina M. Ferreira, João Fevereiro, Jorge Pedrosa and Egídio Torrado

8 Role of Cyclic di-GMP in the Bacterial Virulence and Evasion of the Plant Immunity 199
Marta Martínez-Gil and Cayo Ramos

Index 223

Current Books of Interest

Illustrated Dictionary of Parasitology in the Post-Genomic Era	2017
Next-generation Sequencing and Bioinformatics for Plant Science	2017
The CRISPR/Cas System: Emerging Technology and Application	2017
Brewing Microbiology: Current Research, Omics and Microbial Ecology	2017
Metagenomics: Current Advances and Emerging Concepts	2017
Bacillus: Cellular and Molecular Biology (Third Edition)	2017
Cyanobacteria: Omics and Manipulation	2017
Foot-and-Mouth Disease Virus: Current Research and Emerging Trends	2017
Brain-eating Amoebae: Biology and Pathogenesis of *Naegleria fowleri*	2016
Staphylococcus: Genetics and Physiology	2016
Chloroplasts: Current Research and Future Trends	2016
Microbial Biodegradation: From Omics to Function and Application	2016
Influenza: Current Research	2016
MALDI-TOF Mass Spectrometry in Microbiology	2016
Aspergillus and *Penicillium* in the Post-genomic Era	2016
The Bacteriocins: Current Knowledge and Future Prospects	2016
Omics in Plant Disease Resistance	2016
Acidophiles: Life in Extremely Acidic Environments	2016
Climate Change and Microbial Ecology: Current Research and Future Trends	2016
Biofilms in Bioremediation: Current Research and Emerging Technologies	2016
Microalgae: Current Research and Applications	2016
Gas Plasma Sterilization in Microbiology: Theory, Applications, Pitfalls and New Perspectives	2016
Virus Evolution: Current Research and Future Directions	2016
Arboviruses: Molecular Biology, Evolution and Control	2016
Shigella: Molecular and Cellular Biology	2016
Aquatic Biofilms: Ecology, Water Quality and Wastewater Treatment	2016
Alphaviruses: Current Biology	2016
Thermophilic Microorganisms	2015
Flow Cytometry in Microbiology: Technology and Applications	2015
Probiotics and Prebiotics: Current Research and Future Trends	2015
Epigenetics: Current Research and Emerging Trends	2015
Corynebacterium glutamicum: From Systems Biology to Biotechnological Applications	2015
Advanced Vaccine Research Methods for the Decade of Vaccines	2015
Antifungals: From Genomics to Resistance and the Development of Novel Agents	2015

Full details at www.caister.com

Preface

Humans have an intimate relationship with bacteria as we harbour in our bodies a considerable amount of bacteria, known as our microbiota (Thaiss *et al.*, 2016). Indeed it has been demonstrated that a close interaction between our cells and the commensal microbes in our gut is essential for human health. However, humans also have contact with pathogenic bacteria and the infectious diseases that these cause continue to be a major threat to human health: predictions suggest that over the next few decades, these will account for one of five deaths globally (Mathers and Loncar, 2006). For example in 2015 the World Health Organization stated that tuberculosis (caused by *Mycobacterium tuberculosis*) was one of the top 10 causes of death worldwide (WHO, 2015). Another example is antibiotic-resistant *Salmonella*: it accounts for more than 100,000 infections in the USA per year (Medalla *et al.*, 2017). In a time when emergence of antibiotic resistance in bacteria is occurring worldwide, understanding how bacteria cause disease is the key to finding new therapeutic approaches to tackle infection. The study of host–pathogen interactions, in particular the interaction with the host immune system, is an essential area of research.

The immune system is thus composed of a combination of mechanisms and specialized cells (i.e. immune cells) that evolved in eukaryotes to constantly monitor the homeostasis of the host. Disruption of such homeostasis, for example during tissue damage, infection or tumour development activates the immune system (Goldszmid *et al.*, 2014). For bacterial pathogens, escaping the host immune system is thus essential for successful multiplication and the establishment of an infection. Indeed, failure of the immune system to deal with pathogenic bacteria and recover homeostasis has life-threatening consequences, as seen in immunocompromised individuals. In this volume we present a collection of timely reviews on the most current research in bacterial evasion of the host immune system.

Life is communication and all living systems communicate with their neighbours. The immune system is key in the establishment of this cross-talk between eukaryotes and bacteria. However, pathogenic bacteria are able to *hack* into this communication, causing *confusion* in the immune system and allowing them to evade defence mechanisms, subvert and exploit host functions in their benefit, and replicate within the host during infection, causing disease which can be fatal.

Pathogenic bacteria have evolved sophisticated mechanisms to exploit host functions and evade their recognition by immune cells during infection. One of these strategies is the injection of bacterial proteins (*effectors*) into the host cells during infection, a topic reviewed in Chapter 1. In Chapter 2, Pucciarelli and García-del Portillo review another interesting strategy; the *camouflage* of bacterial surface components during infection in order to alter the recognition of pathogenic signatures.

In the recent years, research on immunity against pathogenic bacteria has been focused on innate immunity, the first line of defence against infection. Among innate immune cells, macrophages have received special attention as these immune cells are professional phagocytes that engulf and analyse bacteria, recognizing pathogenic signatures and further orchestrating immune responses with the adaptive immune system (composed by B and T lymphocytes) to fight infection. Macrophages are therefore the first line guardians against infection, but many pathogenic bacteria *manipulate* these cells to prevent an effective immune response. In Chapter 3, Wan *et al.* review the mechanisms by which pathogenic bacteria subvert macrophage functions.

In addition to the innate and adaptive immune systems mentioned above, another essential part of any immune system, 'cell-autonomous immunity' (also called cellular self-defence), has been discovered. Cell-autonomous immunity operates in every cell of an organism and guards both individual immune and non-immune cells against the immediate threat of infection. It cooperates with conventional immunity (i.e. the innate and adaptive systems) and permits any cell to trigger the activation of an immune response. Not surprisingly, pathogenic bacteria also harbour mechanisms to disrupt these systems too. For instance, bacteria can subvert autophagy in an infected cell, thereby evading this catabolic process which removes undesirable material (including intracellular pathogens) within the cell. Chapter 4 analyses the consequences of bacterial modulation of autophagy on the immune response to infection. Furthermore, the recognition of bacterial components by cellular inflammasomes is another key self-defence mechanism against intracellular pathogens, a topic examined in Chapter 5.

Later in the volume, once the players, strategies and mechanisms have been presented, and in order to have a broad view of the mechanisms used by specific bacteria to evade the immune system, we provide an overview of *the whole dance* in two reviews examining the specific evasion strategies used by two important human pathogens during infection. The evasion strategies developed by *Salmonella* are discussed in Chapter 6, while the strategies utilized by *Mycobacterium* are reviewed in Chapter 7. Finally, we close this *volume* with a look into plant pathogens and plant immunity in Chapter 8.

In summary, this volume provides the reader with an overview of important current research on bacterial evasion of the immune system, a topic that is not only exciting but also essential for the elucidation of the mechanisms bacteria use to cause infection.

References

Goldszmid, R.S., Dzutsev, A., and Trinchieri, G. (2014). Host immune response to infection and cancer: unexpected commonalities. Cell Host Microbe *15*, 295–305. https://doi.org/10.1016/j.chom.2014.02.003.

Mathers, C.D., and Loncar, D. (2006). Projections of global mortality and burden of disease from 2002 to 2030. PLOS Med. *3*, e442.

Medalla, F., Gu, W., Mahon, B.E., Judd, M., Folster, J., Griffin, P.M., and Hoekstra, R.M. (2017). Estimated incidence of antimicrobial drug-resistant nontyphoidal Salmonella infections, United States, 2004-2012. Emerging Infect. Dis. *23*, 29–37.

Thaiss, C.A., Zmora, N., Levy, M., and Elinav, E. (2016). The microbiome and innate immunity. Nature *535*, 65–74. https://doi.org/10.1038/nature18847.

World Health Organization (2015). Global tuberculosis report 2015. 20th Edition. WHO/HTM/TB/2015.22

Pedro Escoll
Institut Pasteur, Biologie des Bactéries Intracellulaires; and Centre National de la Recherche Scientifique (CNRS) UMR 3525, Paris, France.

Secretion Systems Used by Bacteria to Subvert Host Functions

1

Chiara Rapisarda[1,2,3,4*] and Rémi Fronzes[1,2,3,4]

[1]G5 Biologie structurale de la sécrétion bactérienne, Institut Pasteur, Paris, France.
[2]Centre National de la Recherche Scientifique (CNRS) UMR 3528, Paris, France.
[3]Microbiologie fondamentale et pathogénicité, Institut Européen de Chimie et Biologie, Université de Bordeaux, Pessac, France.
[4]Centre National de la Recherche Scientifique (CNRS) UMR 5234, Pessac, France.

*Correspondence: c.rapisarda@iecb.u-bordeaux.fr

http://doi.org/10.21775/9781910190692.01

Abstract

In this chapter we examine the use of secretion systems by bacteria to subvert host functions. Bacteria have evolved multiple systems to interact with and overcome their eukaryotic host and other prokaryotes. Secretion systems are required for the release of several effectors through the bacterial membrane(s) into the extracellular space or directly into the cytoplasm of the host. We review the secretion systems of Gram-positive and Gram-negative bacteria and describe briefly the structural composition of the seven secretion systems that have been associated with increased virulence through subversion of host functions. Some of the effects of such systems on eukaryotic host processes have been studied extensively. We also describe the best-characterized effectors of each secretion system to give an overview of the molecular mechanisms employed by bacteria to hide from the immune system and convert eukaryotic cells into optimal ecological niches for their replication.

Introduction

Bacteria are the dominant form of life on earth. Around 50% of our cells are bacterial, and they have evolved to adapt to a multitude of ecological niches including multicellular eukaryotes (Sender *et al.*, 2016). Bacterial pathogenicity is greatly enhanced by their ability to evade and weaken the host defences, and to outcompete other microorganisms growing in the same milieu. To achieve that, bacteria can express and secrete several different macromolecules and toxins, called effectors, into the extracellular medium and in some cases directly into the cytoplasm of their target. Sometimes the macromolecules released will trigger a response from the host defences, in which case they are called virulence factors or elicitors (Jimenez *et al.*, 2016).

The machinery required to release such factors and the macromolecules themselves are usually encoded in pathogenicity islands in the bacterial genome or in plasmids that have been acquired through horizontal gene transfer, since the secretion machines are well conserved across species (Abby *et al.*, 2016).

Bacteria are contained in one or two membranes separated by a peptidoglycan layer. The passage of the effectors through these barriers cannot occur without expenditure of energy and is carried out by one of nine secretion systems (types I to IX). These complex multiprotein assemblies evolved from existing ones such as the bacterial phage tail (type VI) from the machinery required for shuffling macromolecules through organelles (type I), for piliation (type II), flagellar or gliding motion (types III and IX), or conjugation (type IV). Each secretion system has its own unique mechanism; some release the effector in one stage, others require an additional secretion step carried out by another membrane protein. Some of them are encoded in operons; each species might express none, one or more of the systems and when present they are normally associated with increased virulence and pathogenicity. Very few secretion systems (types I and III) are expressed in commensal organisms and mutations that lead to loss of function in one of their components results in phenotypically avirulent population. For this reason the presence of secretion systems on the surface makes them a likely target of host defence mechanisms to discriminate highly pathogenic from commensal species (Vance *et al.*, 2009).

Gram-positive and Gram-negative bacteria share some of the components of type IV and type VII secretion systems, but each has evolved their own unique systems to export proteins through one or two membranes (Schneewind and Missiakas, 2012).

The type VIII secretion system is encoded by two csg operons (Hammar *et al.*, 1995) and is involved in the secretion of curli fibres composed of the protein CsgA through a pore formed by CsgG monomers (Goyal *et al.*, 2014). Curli are involved in biofilm formation and surface attachment (Olsén *et al.*, 1989; Pawar *et al.*, 2005) and will not be discussed further in this review.

In this review we provide a brief overview of the eight bacterial secretion systems involved in the subversion of the host system and their mechanisms of action, with a focus on some of the best-characterized secreted effectors and how they function. The review is divided in three main sections: (i) Mechanism of protein translocation by secretion systems; (ii) mechanism of protein secretion by Gram-positive bacteria; and (iii) secreted virulence factors that interact with the host cell.

Mechanism of protein translocation by bacterial secretion systems

Bacteria communicate with their host and other organisms by secreting macromolecules into the external environment (Waters and Bassler, 2005). Bacterial secretion systems that span one or two membranes modulate such function through direct or indirect host interaction.

In Gram-negative bacteria, where the effectors have to be secreted through two membranes, this could be done in one step through the type I, III, IV, VI and VII, or in two steps by secreting first the effector in the periplasm through a common Sec/TAT pathway and finally secreting it into the extracellular space thought the type II, type V or type IX secretion system (see *Secretion of effectors to the periplasm*). Gram-positive bacteria only have the outer

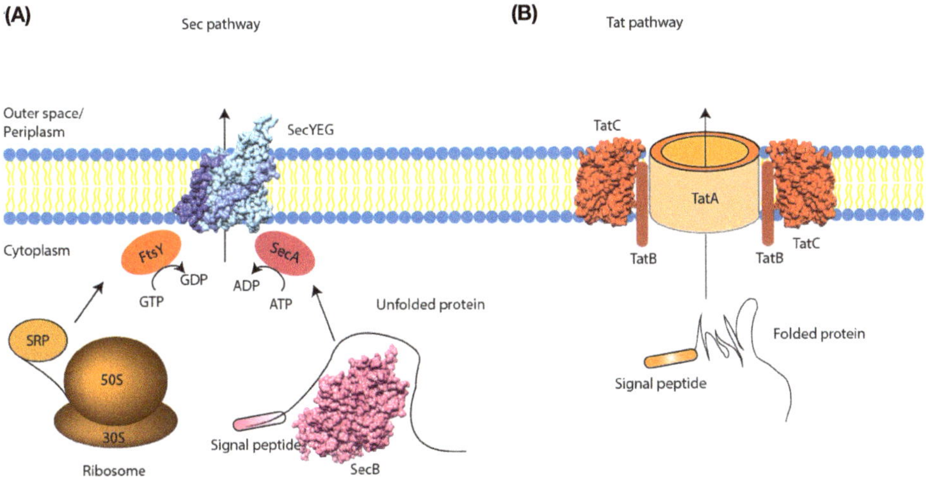

Figure 1.1 Sec and Tat pathway of protein secretion. (A) Diagram of the Sec secretion pathways. SecY in light blue, SecG in dark blue and SecE in cornflower blue (PDB: 5AWW). The nascent peptide binds to either SRP or SecB (PDB: 1OZB) and is secreted by SecYEG in an unfolded state with the assistance of FtsY or SecA respectively. (B) Diagram of the TAT secretion system. Folded proteins with an appropriate signal peptide are secreted through a pore formed by TatA, B and C (PDB: 4HTS).

portion of the two-step mechanism as they only have one membrane. Mycobacteria are an exception, since that have a membrane surrounding the cytoplasm and a unique mycomembrane, and thus require a one-step secretion system (Brennan and Nikaido, 1995). We will describe the first Sec/TAT systems required for the two-step mechanism used by the type II, V and IX secretion systems, which will then be described. We will continue this section by describing the rest of secretion systems (type I, III, IV, VI and VII), which perform the secretion of effectors in only one step.

Secretion of effectors to the periplasm: the two-step mechanism requires the Sec translocase complex or the TAT system

The effectors that require secretion through the type II and V systems are first exported to the periplasm through the inner membrane. The Sec translocase is used if effectors are secreted in an unfolded conformation, but if they require cytoplasmic-assisted folding the twin-arginine-dependent translocation (TAT) system is used (Voulhoux, 2001).

The Sec translocase complex consists of a pore made of three membrane proteins, SecYEG, and a membrane associated ATPase (SecA) that provides the energy for translocation. Once the nascent protein emerges from the ribosome and the signal peptide becomes available for binding, two possible pathways of Sec-dependent secretion are possible (Fig. 1.1):

1 If the N-terminal signal sequence is hydrophobic it slows down translation and allows the signal recognition particle (SRP) to bind to it (Pechmann *et al.*, 2014). SRP then associates to its receptor FtsY, located on the membrane and transfers the nascent polypeptide chain to the SecYEG channel (Gilmore *et al.*, 1982).

2 For non-hydrophobic signal sequences, the trigger factor (TF) chaperone binds and transfers the nascent protein to SecB, another chaperone that keeps it into an unfolded state and subsequently to SecA, an ATPase that powers protein transport across SecYEG (Randall *et al.*, 1997).

The twin-arginine-dependent translocation (TAT) apparatus can transport across the plasma membrane fully folded proteins that contain the specific signal sequence SRRXFLK (Weiner *et al.*, 1998). It is composed of three essential transmembrane proteins: two single-pass proteins (TatA and TatB) and a protein with six transmembrane helices (TatC). The mechanism by which the protein is transported fully folded is not yet clear. It is thought that TatA might form oligomers that destabilize transiently the plasma membrane and allow the protein to pass through its pore (Westermann *et al.*, 2006) (Fig. 1.1).

Once in the periplasm the effector protein can be secreted through the type II, V or IX secretion systems.

Type II secretion systems are homologous to type IV pili

The type II secretion system (T2SS) is widely expressed in pathogenic and non-pathogenic bacteria such as *Klebsiella pneumoniae* (encoded by *pul* genes) (d'Enfert *et al.*, 1987), *Vibrio cholerae* (*eps*) (Sandkvist *et al.*, 1997), *Pseudomonas aeruginosa* (*xcp* and *hxc*) (Bally *et al.*, 1992; Nampiaparampil and Shmerling, 2004), *Yersinia enterocolitica* (*yts*) (Iwobi *et al.*, 2003), *Erwinia chrysanthemi* (*out*) (He *et al.*, 1991) and in *E. coli* where it is encoded in an operon consisting of 12–15 general secretion pathway proteins (*gsp*) (Francetic and Pugsley, 1996). Some proteins are not conserved across all species that express the T2SS; the core genes essential for secretion are GspC–O. GspA and GspB are accessory single-pass membrane proteins that, when present, assemble into a large heteromultimeric complexes capable of rearranging the peptidoglycan layer for better assembly of the core complex (Strozen *et al.*, 2011). Mutations in any of the core genes abolish secretion. Fig. 1.2 shows the arrangement of the operon in *V. cholerae* and *E. coli* K12 coloured on the basis of the subcomplex composition and their location in the membrane. The nomenclature of the proteins across species is not well conserved, for clarity in this review we will be using only the gsp nomenclature used in *E. coli*, other reviews have a comprehensive list of all the different nomenclatures used in other species (Douzi *et al.*, 2012; Korotkov *et al.*, 2012).

The complex is divided into four subcomplexes: the pseudopilus (in red the major pseudopilin GspG and in yellow the minor pseudopilins GspH, I, J K), the inner-membrane platform (in green, GspC, E, F, L, M), the outer membrane complex (in blue, GspD) and the secretion ATPase (in orange, GspE).

The secretion ATPase GspE, which is anchored to the membrane through GspL (Sandkvist *et al.*, 1995), provides the energy through ATP hydrolysis for GspF to assemble the pseudopilus, a filament composed of major and minor pseudopilins. The pseudopilus is thought to act as a piston to push the proteins through the outer membrane secretin GspD channel, a dodecameric bell-shaped structure with a hollow centre and a gate at the periplasmic edge (Fig. 1.2) (Reichow *et al.*, 2010). Secretins are well conserved components of the type III (Koster *et al.*, 1997) and type IV (Collins *et al.*, 2001) secretion systems, and in type II secretion systems they interacts with GspC, a transperiplasmic protein that is the least conserved of the core proteins (Korotkov *et al.*, 2011). The outer membrane complex aids in the formation of the inner-membrane complex (Lybarger *et al.*, 2009). The substrates

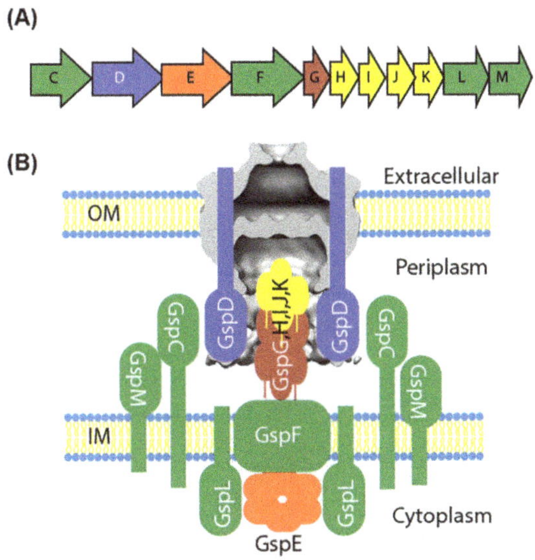

Figure 1.2 The type II secretion system structural organization. (A) Operon composition of the type II secretion system. In blue is the secretin component, in red and yellow the major and minor pilins, in green the inner-membrane components and in orange the ATPase. (B) Structure of the secretin of type II secretion system (EMDB 1763) and a diagram of protein–protein interaction of all the other components. The colouring scheme is the same as in (A).

that are transported across the type II secretion system are usually folded in the periplasm, they are exclusively targeted for excretion to the type II secretion system and they contain a specific three-dimensional signal that has not yet been identified (Filloux, 2010).

Several proteins are released by the type II secretion systems, most notably the cholera toxin and other enzymes such as phosphatases, lipases, proteases that function mostly extracellularly and are required for the bacterial motility and its virulence. None of these proteins is involved in evasion of host defences (Parsot, 2009; Sandkvist, 2001).

Type V secretion systems and the autotransport mechanism

Type V secretion systems were identified in several pathogenic bacteria such as *Neisseria gonorrhoeae* (Pohlner *et al.*, 1986), *Neisseria meningitidis* (van Ulsen *et al.*, 2003), *Pseudomonas aeruginosa* (Salacha *et al.*, 2010) and *Bordetella pertussis* (Leininger *et al.*, 1991). Several variants of this system exist, broadly classified into five types based on their composition and insertion mechanism into the membrane (Leo *et al.*, 2012). They all share the common feature of being a Sec-dependent transport system that requires neither ATP nor a proton gradient. The only energy available for insertion into the membrane and transport is the free-energy of protein folding (Junker *et al.*, 2009).

The secretion process involves the following steps (Fig. 1.3):

1 All proteins involved in secretion are expressed with an N-terminal signal that targets them for transport to the periplasm through the Sec pathway (Brandon *et al.*, 2003; Sijbrandi *et al.*, 2003). The proteins are unfolded at the time of transport and are likely kept in a translocation-competent state by chaperones (Adams *et al.*, 2005; Ruiz-Perez *et al.*, 2009).

2 The β-barrel or translocator domain of one of the proteins inserts in the outer membrane (OM) independently or through the Bam (β-barrel assembly) complex (Rossiter *et al.*, 2011).
3 The secreted domain or protein, termed passenger, binds to at least one periplasmic polypeptide transport-associated (POTRA) domain on BamA or the translocase protein. It then translocates through the pore of the β-barrel and is either secreted directly, cleaved by a serine protease domain and secreted, or it is secreted to the extracellular space while remaining anchored to the translocator domain (e.g. adherins) (Leo *et al.*, 2012).

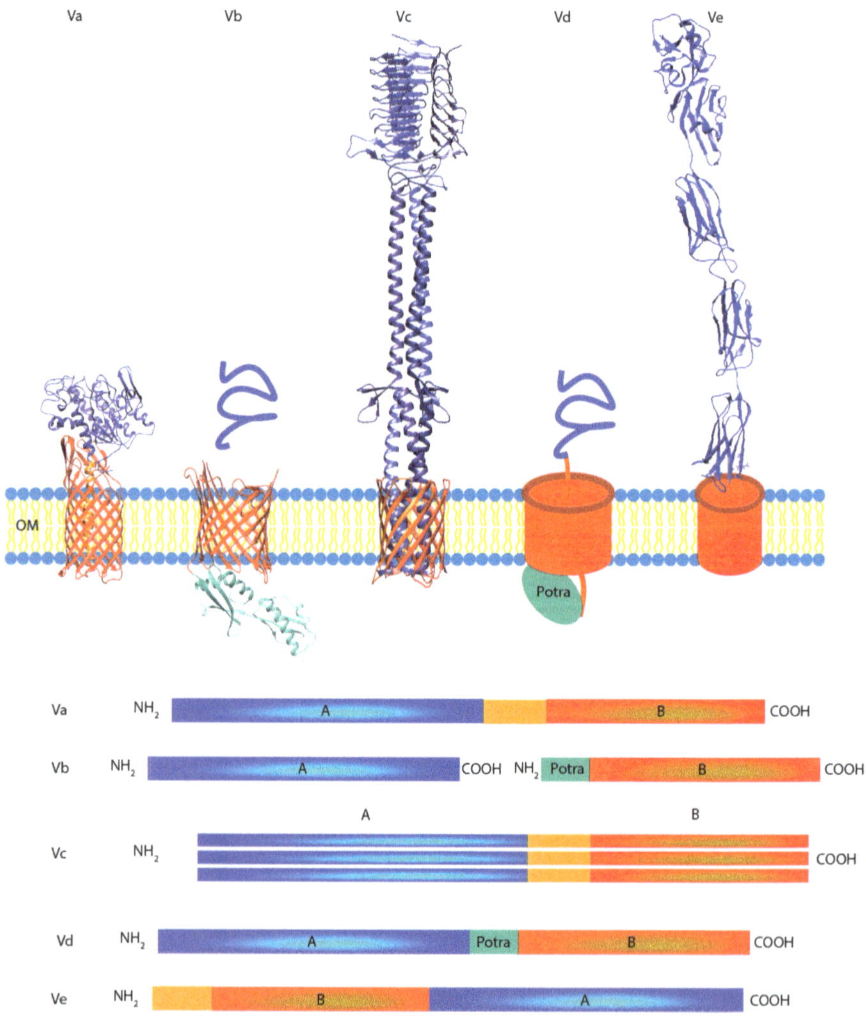

Figure 1.3 Summary of type V secretion mechanisms. There are five known mechanisms for secretion of type V (Va–e). The difference between the mechanisms involves the direction of insertion and the presence of one or more polypeptides. Several structures have been solved by crystallography: 3KVN for the type Va, 4QKY for type Vb, 2XQH and 2GR7 for type Vc, and 1CWV.

Table 1.1 Summary of type V secretion systems classes

Class	Alternative name	Translocator/ passenger	Bam-assisted	POTRA location	Direction of secretion
Va	Autotransporter	1 protein	Yes	BamA	N to C
Vb	Two-partner	2 proteins (TpsA and TpsB)	No	TpsB	N to C
Vc	Trimeric autotransporter	3 copies of 1 protein	Yes	BamA	N to C
Vd	Fused two-partner	1 proteins	No	Translocase	N to C
Ve	Inverted autotransport	1 protein	Yes	None	C to N

More than one class of type V secretion systems can co-exist in one organism. Their characteristics are summarized in Table 1.1.

Several proteins are secreted through the type V secretion system, most are involved in pathogenicity through adhesion (AIDA-I, Pertactin, shdA), biofilm formation (Ag43, BcpA), serum resistance (BrkA), cytotoxicity (EspP) and cell communication (van Ulsen *et al.*, 2014). Type V effectors will be discussed further in the following section.

Type IX secretion system is also a gliding motility apparatus

In *Porphyromonas gingivalis*, a periodontopathic pathogen that lacks type I to VIII secretion systems, a new mechanism for protein secretion and gliding motility was recently discovered (Sato *et al.*, 2010a). The type IX secretion system (T9SS) is composed of 11 proteins (porK-X and sov) scattered along the genome. PorK, L, M and N form a large complex of 1.4 MDa resistant to SDS PAGE electrophoresis (Sato *et al.*, 2010b) and smaller subcomplexes of porK and porN assemble into multimeric rings of 32–26 subunits in the outermembrane (Gorasia *et al.*, 2016). PorL and porM are inner-membrane proteins that provide the energy required for secretion (Hammar *et al.*, 1995).

Proteins are targeted for secretion mainly by the presence of an N-terminal signal that targets them for translocation across the IM through the Sec system (Pavloff *et al.*, 1995), and C-terminal signal sequence (Glew *et al.*, 2012) that is cleaved by the peptidase PorU (Gorasia *et al.*, 2015). The secretion can then take place across the OM through a pore likely made of porK and porN (Hammar *et al.*, 1995). The secreted effectors will be discussed in section 3.

With the exception of the type II, V and IX secretion systems, all other systems require no periplasmic intermediary step and the secretion occurs in one step.

The type I secretion system is a bacterial efflux system

The type I secretion system (T1SS) consists of three proteins that are essential to its function: an ABC (ATP-binding cassette) transporter fused to an inner membrane-spanning domain, a membrane fusion protein (MFP) in the periplasm and an outer membrane factor (OMF) homologous to TolC receptors (Fig. 1.4).

The ABC transporter most likely works as a dimer (Dawson and Locher, 2006) and contains a nucleotide binding domain (NBD) that shuttles between a cytoplasm-facing to periplasm-facing conformations upon nucleotide binding (Fig. 1.4; Yang and Rees, 2015).

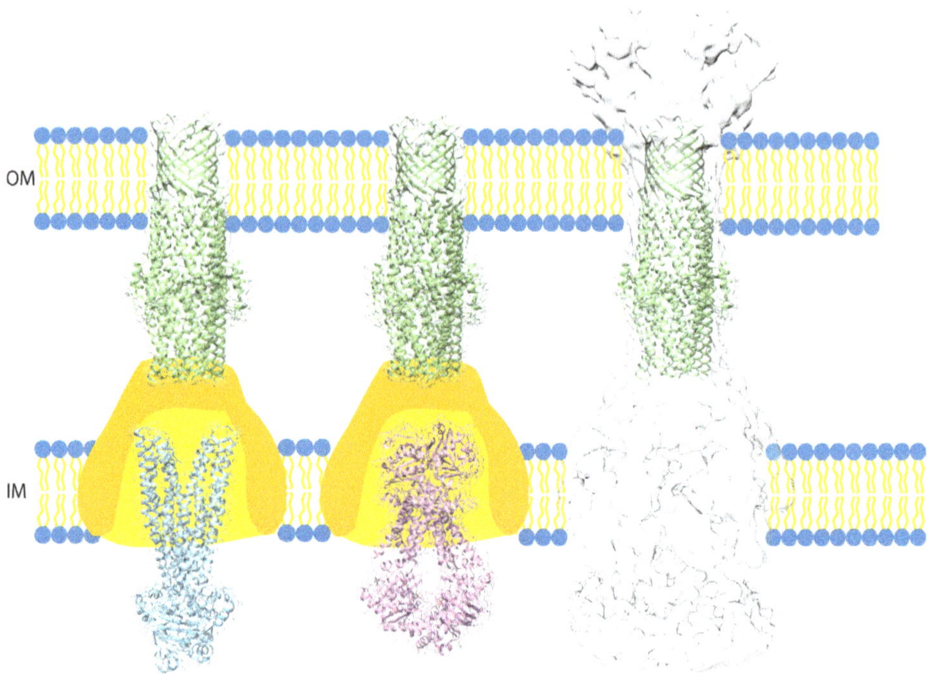

Figure 1.4 Type I secretion system proposed structure. In green is the TolC receptor (1EK9), in orange is the membrane fusion protein (MFP) whose structure is unknown and in blue (2ONJ) is the closed state of the ABC (ATP-binding cassette) transporter and in magenta is the open state (2ONK) modelled on a multi drug transporter from *S. aureus*. The structure at high resolution of a related drug transporter shows the tolC in an open conformation depicting a potential mechanism of secretion (EMD 3330).

The OMF consists of a short cytoplasmic domain, a membrane-spanning domain and a periplasmic region that anchors it to the ABC transporter. It is associated to the ABC transporter even in the absence of the secreted substrate (Uhlén *et al.*, 2000a) and it is thought to form hexamers as its analogue AcrA, a member of the resistance-nodulation-division (RND) family of multidrug efflux pumps (Du *et al.*, 2014; Jeong *et al.*, 2016).

The TolC protein trimerizes and forms a long tube made up of a large β-barrel with a very long but narrow periplasmic domain of several α-helices (Koronakis *et al.*, 2000). The α-helices form a narrow channel in the unbound tolC (Koronakis *et al.*, 2000) but open up upon binding to the OMF and the ABC transporter (Du *et al.*, 2014; Jeong *et al.*, 2016). The assembly of the complex is initiated by the substrate binding to the ABC transporter (Balakrishnan *et al.*, 2001). All the substrates transported across T1SS are unfolded during transport and contain a glycine rich motif (GGxGxDxxx) at their C-terminus that targets them for secretion (Nicaud *et al.*, 1986). The glycines form calcium-binding domains that promote the protein folding once outside the cell at a calcium concentration above its K_d (Gangola and Rosen, 1987).

The substrates secreted by T1SS vary in size between 19 kDa, e.g. the haemophore HasA (Létoffé *et al.*, 1994), to 900 kDa, e.g. LapA, surface layer protein (Hinsa *et al.*, 2003), and most notably include pore-forming haemolysin A (Uhlén *et al.*, 2000b), toxins, lipases, adenylate cyclases and proteases, all proteins associated with increased virulence. T1SS effectors will be further discussed in *Virulence factors in host cell interaction*.

Type III secretion systems inject the effectors directly into the cytoplasm of the host

Type III secretion systems (T3SS) have been often associated with pathogenicity in bacteria, though they can also be involved in symbiosis (Dale *et al.*, 2002). They have been identified mostly through genome sequencing in *Salmonella* (Ochman and Groisman, 1996), *Shigella*, *Vibrio* (Park *et al.*, 2004), *Yersinia* (Cornelis, 2002), *Pseudomonas* species (Yahr *et al.*, 1996), some *E. coli* strains (Ren *et al.*, 2004), and many more species (Pallen *et al.*, 2005). There are eight subclasses of T3SS identified, each with a different target (animals, plants or other pathogens) and effect on the host. More than one T3SS can be encoded in the genome of the same species, though usually of different subclasses. Examples are the SPI-1 and SPI-2 systems found in *Salmonella*, required respectively for entry into the host, and survival and replication (SPI-2) at the intracellular stage of infection (Diepold and Armitage, 2015). T3SS is derived from the flagellar export system in bacteria and the two systems share a strong homology (Abby and Rocha, 2012). While the flagellar genes are encoded in the genome of the bacterium and co-evolved with it, the T3SS is encoded in virulence plasmids or pathogenicity islands and thus is more well-conserved across species as a result of horizontal gene transfer (Gophna *et al.*, 2003). The expression of the genes on the pathogenic islands is not constitutive but is induced by contact with the host (Pettersson *et al.*, 1996) since the complex formation requires a large energy expenditure to express and assemble more than 20 proteins that form a needle spanning the two membranes (Fig. 1.5). In this review we will be using the Sct (secretion and cellular translocation) nomenclature except for proteins that are exclusively expressed in certain species (Hueck, 1998).

The secretion complex can be divided into several substructures, described below.

The basal body

Stacks of concentric rings, referred to as the basal body, connect the inner and outer membrane leaflets. The outer membrane (OM) ring is a member of the secretin family analogous to that of T2SS (Koster *et al.*, 1997; Reichow *et al.*, 2010). It is composed of 12–15 copies of the SctC protein giving the overall basal body structure a 12- and a 3-fold symmetry respectively (Hodgkinson *et al.*, 2009; Schraidt and Marlovits, 2011). The inner-membrane (IM) ring is composed of two concentric circles of different diameters, the inner ring is made up of 24 molecules of the lipoprotein SctJ and the outer ring of 24 molecules of SctD, both single-pass membrane-spanning proteins (Hodgkinson *et al.*, 2009).

Export apparatus

The export apparatus is essential to secretion and is composed of five membrane-spanning proteins: SctR, SctS and SctT being mostly periplasmic (Berger *et al.*, 2010), and SctU and SctV with mostly cytoplasmic domains (Abrusci *et al.*, 2013; Zarivach *et al.*, 2008). The export apparatus is thought to recognize and bind to the substrate and proceeds to its secretion. It is visible in the EM structure as a central density at the height of the IM (Fig. 1.5) (Hodgkinson *et al.*, 2009).

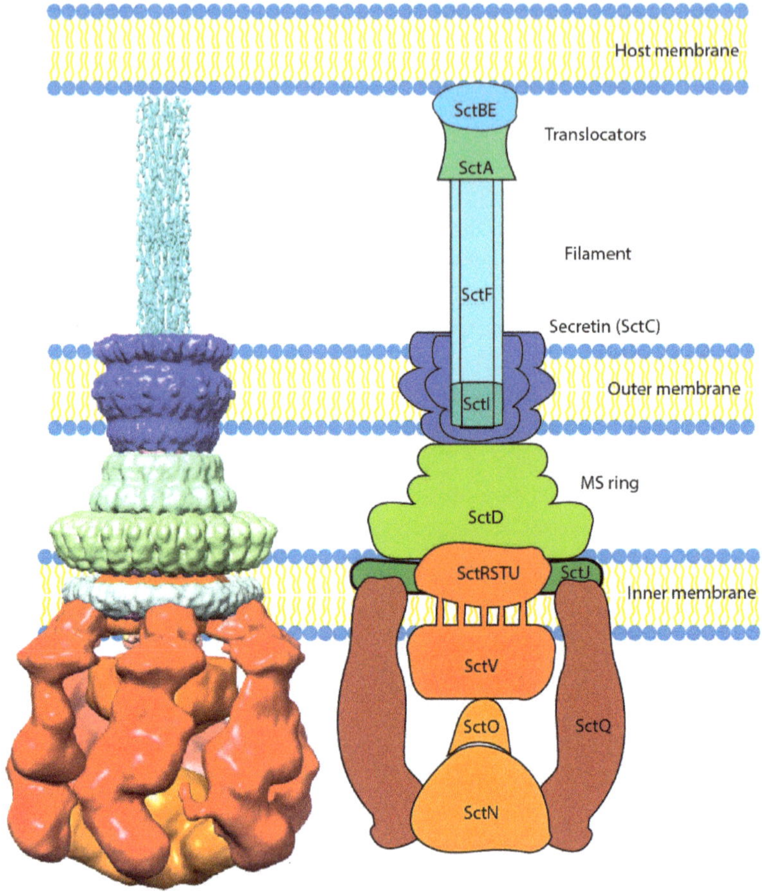

Figure 1.5 Type III secretion system architecture. Left. The EM structure of type III secretion system obtained from the superimposition of tomographic, single particle and helical reconstruction cryoEM studies (EMD 2667 and EMD 1875, EMD-5352 respectively). Right. A summary of the different components of the secretion system coloured according to their position in the bacterial membrane.

Cytosolic components

The cytosolic components make up a very dynamic complex that stabilizes upon contact with the host (Nans *et al.*, 2015). This complex, homologous to V/F-type ATPases (Pallen *et al.*, 2006), is composed of five proteins: sctN, sctO, sctQ, sctL and sctK. sctN, the hexameric ATPase, upon contact with the substrate–chaperone complex, disassembles it and unfolds the substrate through ATP hydrolysis to prepare it for secretion (Akeda and Galán, 2005). The other proteins anchor the ATPase to the membrane and assist in the secretion process (Lara-Tejero *et al.*, 2011).

The needle

The needle is a helical oligomer of SctF (Loquet *et al.*, 2012) that probably self assembles (Fujii *et al.*, 2012) and whose length is controlled by SctP (Kubori *et al.*, 2000) and SctI

(Lefebre and Galán, 2014). It projects out of the bacterial envelope towards the target's membrane (Kubori, 1998). It is hollow inside and the unfolded proteins are secreted through the lumen (Radics *et al.*, 2014). The tip of the needle contains a pentameric hydrophilic protein (SctA-IpaD) that provides a scaffold for two translocator proteins (SctB-IpaB and E-IpaC) to insert into the host membrane, thus connecting the two cytosols (Cheung *et al.*, 2015).

The T3SS complex is assembled independently of any stimuli, but substrate export is activated upon contact with the host cell (Ménard *et al.*, 1994). Following a change in pH of the cytosol, a 'plug' protein complex (SctW) is released from the cytosolic component and the translocator proteins can be successfully exported to the tip of the needle, leading to the formation of a pore on the target membrane (Yu *et al.*, 2010).

Type IV is a secretion and conjugation apparatus that spans two membranes and the membrane of the host

The type IV secretion systems are found in both Gram-negative and Gram-positive bacteria such as *Helicobacter pylori* (Hofreuter *et al.*, 2001; Kersulyte *et al.*, 2003; Odenbreit, 2000), *Brucella* (O'Callaghan, 1999), *Bartonella* (Schmiederer and Anderson, 2000), and *Streptococcus* species (Zhang *et al.*, 2012). They can be broadly classified into three subfamilies based either on their function, the coding region of the genomic DNA or the plasmid that encodes them. Type IV secretion systems possible functions are conjugation (Lawley *et al.*, 2003), DNA uptake and release (Hamilton *et al.*, 2005), and effector secretion.

Based on the coding region of their DNA, they are divided into type IVa, b and c (Fig. 1.6). Type IVa is composed of proteins homologous to the virB and VirD operons in *Agrobacterium tumefacens* (Ghai and Das, 1989; Ward *et al.*, 1988). Type IVb has no sequence similarity but is functionally and structurally related to the type IVa proteins (Ghosal *et al.*, 2016) and is composed of 25 subunits encoded in *Legionella pneumophila* by the Dot/Icm (defective in organelle 45 trafficking/intracellular multiplication) operon (Segal and Shuman, 1997). Finally, the type IVc system, mainly found in Gram-positive bacteria, is a complex composed of 5 subunits derived from the type IVa system (Zhang *et al.*, 2012) (Fig. 1.6). These classifications are not rigid since the same secretion systems can have more than one function and proteins can be exchanged between the different systems.

The secretion complex of type IVa can be divided into several substructures:

Type 4 coupling protein (T4CP)

The type IV coupling protein (T4CP) VirD4 is an ATPase that mediates the interaction between cytoplasmic proteins (chaperones, substrates or accessory proteins) and the inner-membrane complex (IMC) (Atmakuri *et al.*, 2004). VirD4 and its homologues DotL and TrwB, are integral membrane proteins that form hexamers and have a large cytoplasmic domain (Gomis-Rüth, 2001). They interact with the IMC through the transmembrane portion of the VirB10 protein (Gomis-Rüth, 2001).

Inner-membrane complex (IMC)

The bipartite inner-membrane complex is made of VirB3, VirB4, VirB5, VirB6, VirB8, and VirB11, together forming the stalk and the arch in the structure of the whole system (Low *et al.*, 2014). VirB8 is the nucleating protein around which the other proteins assemble (Kumar *et al.*, 2000). VirB3 interacts with VirB4 and VirB5 (Shamaei-Tousi *et al.*, 2004), and has shares some homology with the pilins (Hapfelmeier *et al.*, 2000), suggesting a possible

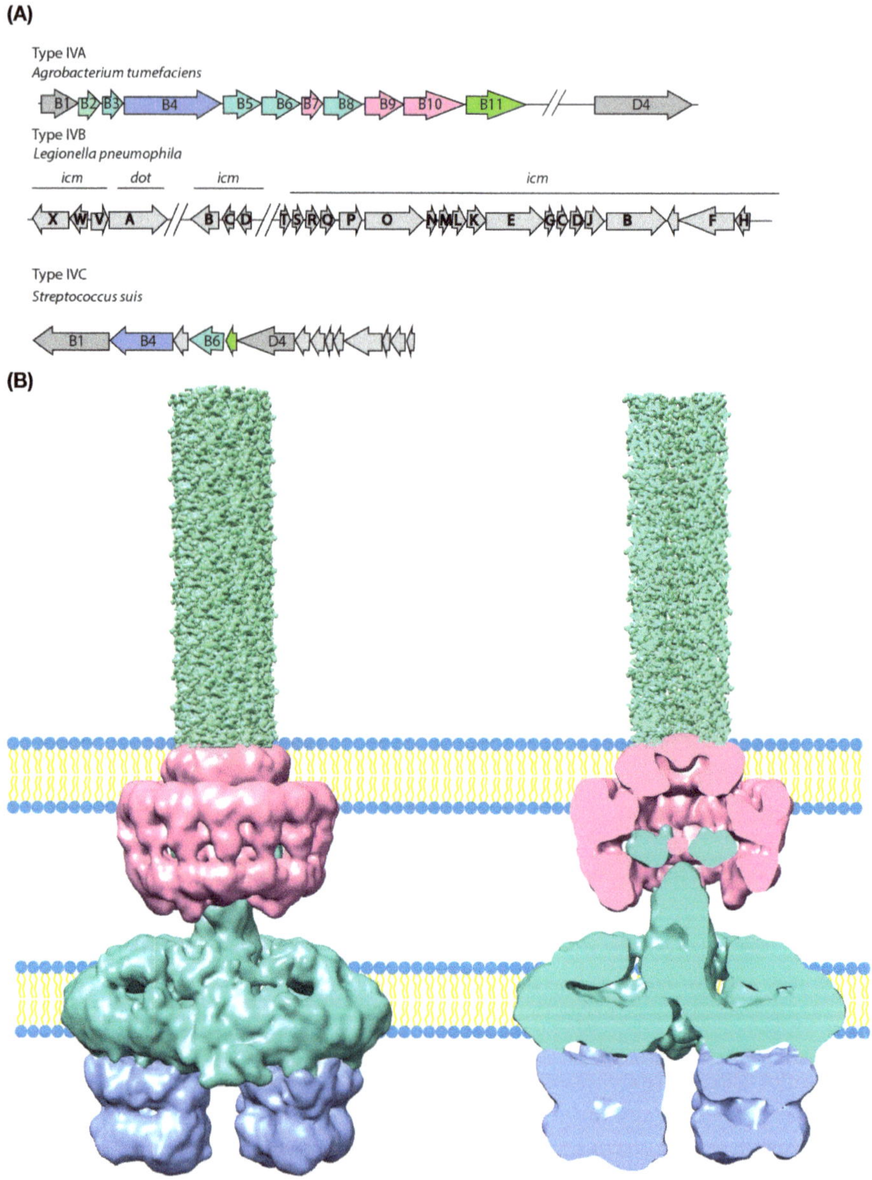

Figure 1.6 Type IV secretion systems arrangement and structure. (A) Genomic arrangement of the three different types of Secretion system IV. The proteins are coloured according to their special arrangement in the EM structure. (B) Structure of the type IV secretion system as solved by negative stain EM (EMDB 2567) with its pilus (EMD-4042). The colour scheme is maintained from section (A).

role in the assembly of the pilus. The VirB4 ATPase forms two hexameric barrels on either side of the complex (Low *et al.*, 2014) and likely mediates the dislocation of the virB2 pilin from the membrane (Kerr and Christie, 2010). VirB6 is a polytopic membrane protein that interacts with the substrate and the virB9 component of the outer membrane core complex

(Jakubowski *et al.*, 2004). Two ATPases, DotB and DotO, belong to the same class at the VirB11 and VirB4 ATPases (Segal *et al.*, 2005).

Outer membrane 'core' complex (OMC)

The outer membrane core complex is a large barrel that extends between the two membranes and it is composed of 14 subunits of VirB7, VirB9, and VirB10. It forms a central cavity through which the pili can extend (Fronzes *et al.*, 2009). In the Dot/Icm complex, DotF and DotG are located in the periplasm (REF) DotC and DotD are two lipoproteins like VirB7 and DotH is an outer membrane protein that is inserted by the two lipoproteins (Vincent *et al.*, 2006).

Cell surface pili or adhesins

The conjugative or secretion pilus is composed of polymerized VirB2 protein–phospholipid complex. It mediates the interaction of the secretion system with its target and provides a channel, 28Å large, through which macromolecules can be transported across the bacterial membranes directly to the cytoplasm of the target cell (Costa *et al.*, 2016).

The type IVb system is more complex than the type IVa and is composed of at least 25 proteins (Segal and Shuman, 1997). Only one of its components, DotG, shares sequence homology limited to its C-terminus with part of VirB10, the OMC component (Kubori and Nagai, 2016).

The secretion process differs between proteins and DNA, and begins with the assembly of the T4SS complex and the pilus at the cell poles (Judd *et al.*, 2005). For the process of conjugation, several proteins bind to the origin-of-transfer (OriT), a specific region of the DNA to be secreted (Fu *et al.*, 1991), create a nick and unwind the double-stranded DNA (Byrd and Matson, 1997). This complex, called a relaxosome, is then specifically recruited to the T4CP by a chaperone (Disqué-Kochem and Dreiseikelmann, 1997). On the contrary, proteins are directly recruited to VirD4 through a well conserved signalling sequence at the N- or C-terminus and are maintained in a secretion-ready state by chaperones (Alvarez-Martinez and Christie, 2009). Some species such as *B. pertussis* do not have a TC4P, but they exploit the Sec-dependent translocation system for the movement of the substrate from the IM to the periplasm (Farizo *et al.*, 2002). The effectors are likely transported in an unfolded manner directly to the cytoplasm of the target cell where they exert their function. The macromolecules secreted by T4SS increase the fitness of the bacterium by allowing spreading of antibiotics resistance, switching of surface antigens, the killing of neighbouring bacteria (Souza *et al.*, 2015) and the subversion of host cell response.

The type VI secretion system evolved as a killing machine

The type VI secretion system (T6SS) is present only in Gram-negative bacteria and consists of a two membrane–spanning complex that requires the assembly of at least 13 proteins encoded in one genomic locus (Boyer *et al.*, 2009) (Fig. 1.7). Two of the genes, *tssL* and *tssM* are homologous to *icmH* and *icmF* of the type Vb secretion systems and several proteins share structural, but not sequence, homology with bacteriophage contractile tail machinery (Cascales and Cambillau, 2012). The T6SS system is used as a defensive and offensive weapon against competitors and is activated upon contact with another bacterium or a host cell. The bacterium secretes toxins indiscriminately of its target and is protected from self-harm by the presence of immunity proteins that are not expressed in other species (Hood

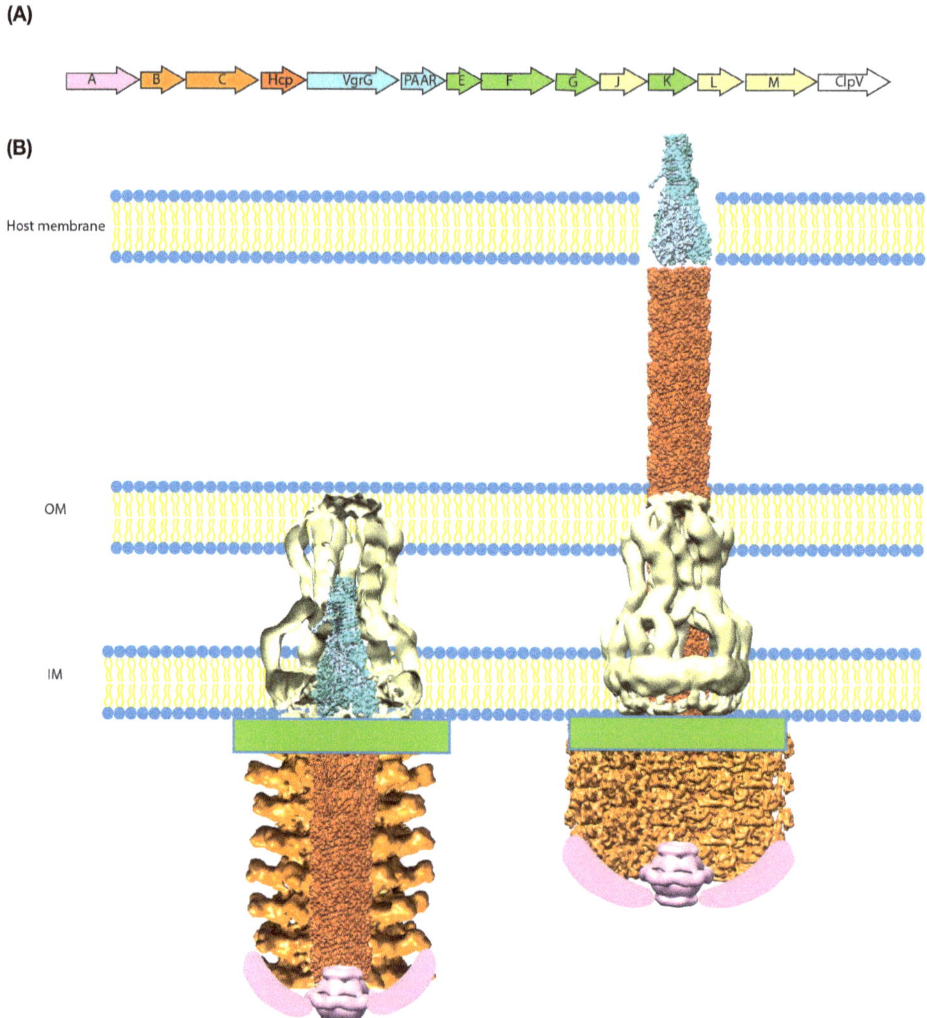

Figure 1.7 Mode of action of the type VI secretion. (A) Type VI secretion operon coloured on the basis of its structure. (B) Structure of the extended (left) and contracted sheath (right) showing the predicted mechanism of secretion. The type VI complex in yellow (EMD-2927), the VgrG trimer in blue (4UHV), the tail tube in dark orange, (hcp analogue, 1Y12), the extended and the contracted sheath of a homologue of VirB/C in orange (EMD-1126 and EMD-2524 respectively) and TssA in pink (EMD-3282).

et al., 2010). The T6SS is required both for virulence against eukaryotic hosts (Mougous *et al.*, 2006) and for intra-bacterial competition in the reproductive niche (Hood *et al.*, 2010).

The secretion apparatus can be divided into two substructures (Fig. 1.7).

The membrane complex (MC)

The MC is needed to anchor the baseplate complex to the membrane. It is composed of three proteins, TssJ, TssL and TssM that form a 'rocket-shape' structure of 1.7 MDa that spans both the outer and the inner membrane (Durand *et al.*, 2015). TssJ is the first one to be recruited to the outer membrane, followed by TssM, an elongated protein that spans the periplasm and the IM. TssL is the last protein to be added to the complex and is a single pass inner-membrane protein that interacts with TssK and other cytoplasmic components (Durand *et al.*, 2015; Zoued *et al.*, 2013).

The baseplate complex (BC)

The BC assembles independently of the MC and one of its main components is VgrG a protein that forms a spike needed to break the membrane of the target cells (Brunet *et al.*, 2014). VgrG recruits HCP1, a protein that oligomerizes into hollow tubes formed by homohexamers stacked head-to-tail as in the phage tail (Leiman *et al.*, 2009). The tail tube is surrounded by a sheath formed by oligomers of TssB (VipA) and TssC (VipB), an assembly that is attached to the baseplate through interaction with TssE. The sheath forms an extended micrometre-long contractile structure that contracts to power secretion (Basler *et al.*, 2012; Kube *et al.*, 2014). Once contracted the sheath is disassembled by ClpV through ATP hydrolysis (Kudryashev *et al.*, 2015). Not much is know about the other components of the BC. TssE is homologous to the gp25 protein in the phage base plate and co-purifies with Hcp1 (Leiman *et al.*, 2009). TssF and TssG were demonstrated to interact with each other and with TssE *in vitro* (Brunet *et al.*, 2015). TssA is recruited to the MC and mediates the priming and assembly of the base plate complex and tail (Zoued *et al.*, 2016).

Once the type VI secretion system assembles, the contraction of the sheath causes the injection of the effector in the cytoplasm of the target cell. The protein VgrG forms a trimeric spike, sharpened by the PAAR (proline-alanine-alanine-arginine repeat) protein, and is capable of piercing the membrane of the target cell. The effectors are bound to either protein and are released into the cytoplasm by a mechanism not yet clear (Shneider *et al.*, 2013).

Gram-positive mechanism of secretion

In Gram-positive bacteria some toxins can be secreted through the Sec pathway and they enter the cell through the SLO (haemolysin streptolysin O), a cytolysin that forms pores in the host membrane and allows the targeting of toxins, of even large size, to the membrane of the host (Madden *et al.*, 2001). In some Gram-positive bacteria and in mycobacteria some effectors are secreted through the type VII secretion system.

Type VII is the unique secretion system in Gram-positive bacteria

The type VII secretion system (T7SS) is exclusively found in Gram-positive bacteria. Initially identified in a virulence island in *Mycobacterium tuberculosis* (Gey Van Pittius *et al.*, 2001), homologues were later found in *Firmicutes* such as *Bacillus* and *Staphylococcus* species (Pallen, 2002). There are seven conserved genes that are required for secretion: two are in the cytoplasm (EspG and EccA) and the remaining five are in the membrane (EccB, EccC, EccD, EccE, and MycP). All the proteins that are not conserved and that are specific to each system are most likely ATPases, substrates or specific chaperones. In *Bacillus* and *Staphylococcus* species only homologues of EccC and of the secreted substrates are present, and for this reason it is considered as an alternative type VIIb system (Fig. 1.8). The type VIIa system is divided into five classes (ESX-1 to 5), depending on the conservation of the subunits.

In the case of cytoplasmic proteins, EspG does not interact with any of the other

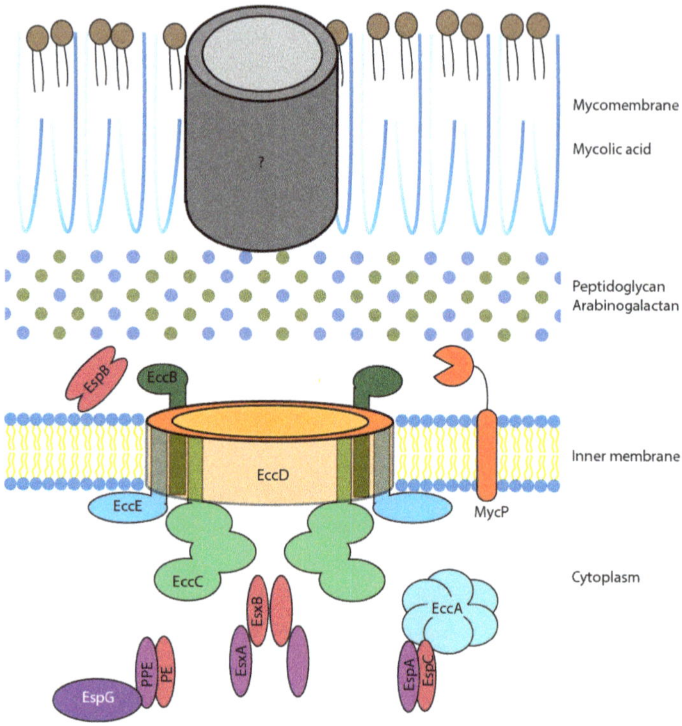

Figure 1.8 Type VII secretion system structure. Proposed assembly of the type VII secretion system across the inner membrane of Gram-positive bacteria. So far no transporter protein has been identified that would be active in the mycomembrane.

components of the system except the secreted substrates PE and PPE (Daleke *et al.*, 2012). EspG is necessary to prevent the aggregation of the substrate and is not present in the ESX-4 system probably because of the solubility of the secreted proteins (Korotkova *et al.*, 2014). EccA is the hexameric AAA+ ATPase that provides the energy for secretion. It has two distinct domains: the C-terminal domain that exhibits ATPase activity (Luthra *et al.*, 2008) and the N-terminal domain, homologous to the pilF of type IV pili, which contains several tetratricopeptide repeats that are involved in protein–protein interaction (Wagner *et al.*, 2014).

The membrane proteins, which form a 1.5 MDa complex, have large cytosolic or periplasmic domains and one single transmembrane helix (Houben *et al.*, 2012). EccD is the exception; it is a polytopic membrane protein expected to have 11 transmembrane helices and is thought to be the pore required for secretion (Wagner *et al.*, 2016). It contains a ubiquitin-like domain in its N-terminal cytoplasmic domain and might form dimers (Wagner *et al.*, 2016). EccB, whose function is yet unknown, has a small cytoplasmic domain and contains a functional ATP binding site in its large periplasmic C-terminal domain (Wagner *et al.*, 2016; Zhang *et al.*, 2015). EccC is a member of the tsK/SpoIIIE-like ATPase family to which T4CP also belongs. It has three distinct nucleotide binding domains (NBD1–3) each with a specific function, NBD3 is required for substrate binding and NBD1 hydrolyses

ATP (Rosenberg *et al.*, 2015). EccE is a protein not present in all secretion systems and thus is thought to be peripheral to the membrane complex (Houben *et al.*, 2012). Mycosin P (MycP) is a membrane protein with a large periplasmic, subtilisin-like, protease domain (Brown *et al.*, 2000). Its presence is essential to secretion but its function as a protease is not (Ohol *et al.*, 2010). Nevertheless, the cleavage of the C-terminus of EspB by MycP causes a change in the quaternary structure of the secreted substrate and might be important to its function (Solomonson *et al.*, 2015).

The mechanism by which all these proteins can act in concert to modulate substrate secretion remains to be uncovered. Moreover, mycobacteria are contained in an outer membrane composed of mycolic acids linked to an arabinogalactan-peptidoglycan matrix embedded in unique lipids, proteins, polysaccharides and lipoglycans that differ from species to species (Daffé *et al.*, 2014). The transporter necessary to guide the substrate across the mycomembrane remains to be identified (Fig. 1.8).

Virulence factors in host cell interaction

The effectors released into the cytoplasm of the target host, have a profound effect on the physiology of the cell, mostly on cellular processes that respond to bacterial infections: cytoskeletal reshuffling, actin-dependent motility, autophagy evasion and phagosomal escape, adhesion, inhibition of signalling, obstruction of the immune response, (de)-ubiquitination, induction of apoptosis and formation of multinucleated cells.

Many of the effectors have one or more of these functions. Here we will separate the effector based on the secretion system responsible for their release into the extracellular space or the target's cytoplasm. Most of the best-characterized effectors are secreted by type III and type IV secretion systems, but a few are present in all other systems. The number and nature of effectors known is very large and a comprehensive list of all of them is beyond the scope of this review. We will be mentioning a few of the best-characterized effectors based on their function, hoping that it will give an overview of the diversity of mechanisms that bacteria have evolved to maintain their supremacy over other forms of life.

Type I secretion system secretes unfolded effectors

The T1SS secrete, in a single step through the double membrane, proteins of the repeats-in-toxin (RTX) family. RTX proteins vary in size (40 to more than 600 kDa) and in function (Linhartová *et al.*, 2010). They are characterized by the presence of several repeats of the nonapeptide X(L/I/F)XGGXG(N/D)D, a calcium-binding motif that forms a parallel β-roll (Baumann *et al.*, 1993). The proteins are secreted unfolded and are able to fold thanks to the high concentration of calcium ions present in the extracellular space (Thomas *et al.*, 2014).

The best-characterized function of RTX toxins is the pore forming and membrane disrupting activity of leukotoxins such as HlyA from *E. coli* (Welch *et al.*, 1992), LktA from *Mannheimia* haemolytica (Chanter, 1990), and CyaA from *Bordetella species* (Benz *et al.*, 1994). To be active these proteins require fatty acylation before secretion (Stanley *et al.*, 1994). Once in the extracellular space they interact with the host target membrane non-specifically, or through β_2 integrin surface receptors in leucocytes (Lally *et al.*, 1997), and glycophorin in erythrocytes (Cortajarena *et al.*, 2003). RTX toxins form pores though which calcium and potassium can enter and through which important metabolites, such as ATP, can exit the cells (Bhakdi *et al.*, 1989). Depending on the amount of toxin released by the bacteria, the host cells will undergo necrosis or apoptosis (Russo, 2005).

Additionally to the pore-forming function the *Vibrio cholerae* MARTX toxin also contains an actin cross-linking domain (Sheahan *et al.*, 2004) that prevents actin polymerization (Kudryashov *et al.*, 2008) and a Rho GTPase-inactivation domain that causes cell rounding through the inhibition of the Rho GTPase family (Sheahan and Fullner Satchell, 2007). The cyaA toxin in *Bordetella* spp. also contains an additional domain with an adenylate cyclase function. Once inside the host leucocytes, cyaA is activated by calmodulin and produces excess cAMP that abolishes immune functions such as phagocytosis, chemotaxis and interleukins release (Vojtova *et al.*, 2006).

Type II effectors are mostly housekeeping proteins

The proteins secreted through the T2SS vary in terms of function and are mostly required for survival in the environment. The presence of T2SS nevertheless was shown to reduce the response of macrophages to infection through a yet unidentified mechanism (McCoy-Simandle *et al.*, 2011). HapA in *Vibrio cholerae* processes the HlyA secreted from type I secretion system, thus enhancing the effect of the toxin on the host (Nagamune *et al.*, 1996). The chitinase domain-containing exoprotein chiA has also been linked with persistence of *Legionella* in the lung (DebRoy *et al.*, 2006). The T2SS has also been associated with intracellular growth in amoebas of *Legionella penumophila* (Hales and Shuman, 1999). Recently a yet unidentified effector secreted by T2SS was shown to be required for retention of a Rab1B GTPase to the *Legionella*-containing vacuoles (LCV), a process required for the successful intracellular growth of *L. pneumophila* (White and Cianciotto, 2016). Another secreted protein from T2SS was also found to affect cytokine secretion during infection in macrophages and inhibit IL-6 production through interference with the myeloid differentiation primary response (MyD88), TANK binding kinase 1 (TBK1) and the Toll-like receptor 2 (TLR2) pathways (Mallama *et al.*, 2017). More work remains to be done to identify such T2SS effectors and to better our understanding of the influence of these exoproteins on the host interaction and evasion.

Type III secreted effectors affect several host functions

Type III secretion systems target effectors directly to the cytoplasm of the host cell. Most share a common secretion signal, a chaperone-binding, and an active domain. We can classify the exoproteins on the basis of the changes elicited on the host metabolism.

Focal adhesion/tight junctions modulators

YopH in *Yersinia* species exhibits a protein tyrosine phosphatase (PTP) activity that, once inside the cytoplasm, targets proteins in focal adhesion complexes and blocks integrin-mediated phagocytosis in epithelial cells and macrophages (Rosqvist *et al.*, 1988). Its targets include paxilin, a protein involved in Fc receptor-mediated phagocytosis in macrophages (Greenberg *et al.*, 1993), $p130^{Cas}$, a focal adhesion-associated protein (Black and Bliska, 1997), the focal adhesion kinase (FAK) (Persson *et al.*, 1997), Cas and Fyn-binding protein (Hamid *et al.*, 1999), all of which when dephosphorylated disrupt focal adhesion complexes.

To colonize the epithelium effectively, *Shigella flexneri* secretes OspE1/2, an effector that binds to the integrin-linked kinase (ILK) leading to an increase in β1 integrin concentration on the host plasma membrane and a decreased phosphorylation of paxilin and FAK. The resulting decrease in focal adhesion turnover prevents cell detachment and promotes cell-to-cell dissemination of the bacterium (Kim *et al.*, 2009).

The gut of the mammalian host has an epithelial cell barrier that specifically selects the contents to absorb and those to keep out. The *E. coli* effectors NleA/EspI (Kim *et al.*, 2007), EspF (McNamara *et al.*, 2001) and Map (Dean and Kenny, 2004) affect the tight junctions between intestinal epithelial cells to allow the bacterium to enter through the gaps between the cells. NleA inhibits, upon binding, COPII, the protein required for ER vesicles trafficking (Kim *et al.*, 2007) while EspF and Map action is dependent on the interaction with the bacterial surface receptor intimin in a yet unidentified mechanism (Dean and Kenny, 2004).

In *Salmonella enterica* SipA, SopB, SopE and SopE2 decrease the expression of ZO and other proteins involved in tight junction formation, leading to an increased permeability of the epithelial layer (Boyle *et al.*, 2006).

Cytoskeleton modulators: actin microfilaments

Intracellular facultative and obligate pathogens release effectors that affect the stability of the actin cytoskeleton. YopE (Black and Bliska, 2000; Evdokimov *et al.*, 2002) in *Yersinia* spp., ExoS (Goehring *et al.*, 1999) and ExoT (Krall *et al.*, 2000) in *Pseudomonas aeruginosa*, SptP in *Salmonella* spp. (Fu and Galán, 1999) and AexT in *Aeromonas salmonicida* (Burr *et al.*, 2003; Litvak and Selinger, 2007) are all proteins that target Rho GTPases and mimic the guanidine exchange activation protein (GAP). Rho GTPases regulate actin polymerization and cytoskeleton dynamics (Hall, 1998). Their activity is dependent on their nucleotide state. Nucleotide exchange factors activate the protein by exchanging GDP for GTP. GAPs promote nucleotide hydrolysis by providing an arginine finger required for positioning of GTP in the binding site (Vetter and Wittinghofer, 2001). The activation of Rho GTPases causes the ruffling of the membranes and the internalization of the bacterium by micropinocytosis. The inactivation of Rho GTPases inhibits phagocytosis, providing another mechanism of host defences evasion. All the effectors mentioned above except YopE, contain an ADP-ribosyltransferase domain that targets Ras GTPase proteins for ADP ribosylation leading to a decrease in downstream signalling and alteration of several cellular processes (Vincent *et al.*, 1999). SopE (Hardt *et al.*, 1998), SopE2 (Stender *et al.*, 2000) in *Salmonella* spp. are Guanidine exchange factors that interact with Rho GTPases and induce actin filament formation and stabilization. This induces cell entry into the host through membrane ruffling and intracellular replication through a mechanism not yet clear (Vonaesch *et al.*, 2014). SipA (Kenny, 1999) and SipC modulate actin directly through binding. SipA stimulates actin bundling by increasing the activity of T-plastin (Zhou *et al.*, 1999) and inhibiting the depolymerizing cofactor ADF/cofilin (McGhie *et al.*, 2004), while SipC directly induces actin nucleation (Chang *et al.*, 2005) and bundling (Myeni and Zhou, 2010). The translocated intimin receptor (Tir) is one of the most studied effectors to hijack host systems. It is secreted by enterohaemorrhagic and enteropathogenic *E. coli* (EHEC and EPEC) and it inserts into the host membrane and acts as a receptor for intimin to lead to actin rearrangements (Kenny, 1999). Another EPEC/EHEC effector that alters actin recycling leading to the formation of filipodia is the mitochondrial associated protein (Map). Map contains a GEF and a PDZ domain that polarizes the actin control protein cdc42 to the membrane (Orchard *et al.*, 2012).

Cytoskeleton modulators: microtubules

EspG from EHEC/EPEC, and VirA in *Shigella* destabilize microtubules (Shaw *et al.*, 2005; Yoshida *et al.*, 2002) leading to the release of several factors, one of them being GEF-H1 the activator of RhoA. Such activity is thought to contribute to actin stress fibre formation and ruffling of the host membrane to facilitate bacterial entry (Matsuzawa *et al.*, 2004). EspG also localizes to the Golgi, where it binds to the Golgi proteins GM130, ADP-ribosylation factor GTPase (ARF) and p21-activated kinases (PAK), leading to the inhibition of protein secretion by the host (Clements *et al.*, 2011; Selyunin *et al.*, 2011).

Immune response modulators

YopJ is an acetyl transferase in *Yersinia* spp. that inhibits the innate immune response by reducing the expression of pro-inflammatory signals. It acetylates and thus inhibits IKKβ (Zhou *et al.*, 2004), Mitogen-activated protein kinase (MAPK) signalling (Mittal *et al.*, 2006) and TGFβ-activated kinase TAK (Paquette *et al.*, 2012), and deubiquitinates key proteins in the NF-κB signalling cascade (Zhou *et al.*, 2005).

NleE and NleB from EPEC and OspZ from *Shigella* also affect the same signalling cascade by inhibiting the nuclear translocation and the activation of NF-κB. This leads to the inhibition of IL-8 production and of the host inflammatory response (Newton *et al.*, 2010).

YopM (Cullinane *et al.*, 2008) inhibits platelet aggregation and the production of interleukin 10 by binding to α-thrombin, α1 anti-trypsin, Rsk1 and Prk2 (Leung *et al.*, 1990; McPhee *et al.*, 2010). YopJ is not essential for virulence, rather it activates the caspase-1 cascade and induces the release of IL-1β/IL18, balancing the effect of YopM in the host cells (Ratner *et al.*, 2016).

SptP in *Salmonella* spp. has a similar function to YopH in Mast Cells (MC). Through its phosphatase domain it dephosphorylates the protein responsible for vesicle fusion N-ethylmalemide-sensitive factor (NSF) and Syk. This results in the blockage of MC degranulation and the blockage of neutrophils recruitment to clear the bacteria (Choi *et al.*, 2013).

Caspase inhibitors

Cells respond to infection by activating the caspase cascade that leads to pyropotosis: cell death and release of cytokines that direct the host defences to the site of infection (Ashida *et al.*, 2011). YopM in *Yersinia* spp. binds to caspase-1 and inhibits the pyroptosis, leading to an increase in virulence (Larock and Cookson, 2012).

Vacuole reorganization/autophagy evasion

Inc proteins from *Chlamydia* are secreted by the type III secretion system. IncA is homologous to SNAREs, vesicular scaffold proteins that mediate vesicle fusion, and induce vacuole re-organization by recruiting them (Delevoye *et al.*, 2008). IcsB in concert with virA in *Shigella* and BopA of *Burkholderia pseudomallei* (Cullinane *et al.*, 2008) are linked to autophagy escape through a yet unidentified mechanism that might involve cholesterol binding and the inhibition of the VirG effector (Kayath *et al.*, 2010; Ogawa *et al.*, 2005).

Type IV effectors vary depending on the replication niche of the bacterium

The effectors secreted by the type IV secretion system aim at transforming a hostile environment into an optimal ecological niche for the bacteria. The effectors secreted by T4SS are similar in function to the T3SS effectors but in this section we will, rather, characterize them based on the pathogen's reproductive cycle: extracellular, obligate or facultative intracellular, since the requirements for each niche differ.

Extracellular pathogens

In *B. pertussis* the T4SS secretes exclusively the pertussis toxin (PT) through a Sec-mediated mechanism. Once secreted, it is able to enter the cells through receptor-mediated endocytosis (Plaut and Carbonetti, 2008). Once inside the cell, PT ADP-ribosylates a trimeric signal-transducing G protein, which locks it into an inactive state (Katada *et al.*, 1983). This results in an increase of cyclic AMP (cAMP) intracellular concentration and subsequently in the disruption of several signalling cascades required for function and recruitment of immune cells (Carbonetti, 2010).

Another well-characterized toxin of extracellular pathogens is the cytotoxin-associated antigen A (cagA) from *Helicobacter pylori*. cagA is phosphorylated (Selbach *et al.*, 2002) and then secreted through the T4SS directly into the cytoplasm of the host cell through the integrin α5β1 (Kwok, 2007). Once inside the cell, cagA and other effectors secreted by the T4SS mimic eukaryotic proteins leading to alterations in cell shape, signalling cascades and immune responses. They might affect phosphorylation, protein–protein interaction, ubiquitination, guanidine nucleotide exchange and other cellular processes (Gonzalez-Rivera *et al.*, 2016).

Facultative intracellular pathogens

Several bacterial species can evade host detection by hiding inside cells temporarily. One such species, *Bartonella* spp., infects erythrocytes and once inside the cells secretes up to seven *Bartonella* effector proteins (Beps) (Schulein *et al.*, 2005). Beps contain a Bep intracellular delivery (BID) domain required for secretion and function (Schulein *et al.*, 2005), a filamentation induced by cAMP (FIC) domain that post translationally modifies proteins through addition of a metabolite and an EPIYA motif, both of unknown function in *Bartonella* effectors. BepC and BepF interfere with Rho GTPases (Truttmann *et al.*, 2011) and BepG interrupts endocytosis by promoting F-actin rearrangements (Rhomberg *et al.*, 2009). BepE prevents apoptosis of its host by binding to the adenylyl cyclase and inducing cAMP production (Pulliainen *et al.*, 2012) and targets the Rho pathway, thus promoting normal migration of dendritic cells which facilitates dissemination of the infection (Okujava *et al.*, 2014).

During infection, *Brucella* species enters the host cells in an unknown site and is incorporated in *Brucella*-containing vacuoles (BCVs) that fuse with lysosomes. The fused vacuoles reach the endoplasmic reticulum (ER) where the bacteria are able to replicate and exit the cell through autophagy-like vacuoles (Pizarro-Cerdá *et al.*, 1998; Starr *et al.*, 2012). To induce toxicity in macrophages, the type IV effector VceC secreted by the T4SS induces ER stress via the unfolded protein response (UPR) and induction of inflammation possibly by binding and inactivating the ER chaperone Bip (de Jong *et al.*, 2013). Two effectors, BtpA and BtpB, translocate into the cytoplasm of host cells and contain a toll/interleukin-1 receptor (TIR) that mediates the NF-κB signalling pathway and both cause the failed maturation of dendritic cells (Salcedo *et al.*, 2013). TIR domains are also present in the effector proteins of other species such as TlpA of *Salmonella enterica* (Newman *et al.*, 2006), Tdp in *Yersinia pestis* (Rana *et al.*, 2011), TcpC in *Escherichia coli* (Yadav *et al.*, 2010), PdTLP in *Paracoccus denitrificans* (Low *et al.*, 2007) and TirS in *Staphylococcus aureus* (Askarian *et al.*, 2014). Recently identified effectors BspA, BspB and BspF prevent membrane trafficking and thus inhibit host protein secretion through mechanisms not yet elucidated (Myeni *et al.*, 2013).

Legionella pneumophila infects phagocytic cells such as macrophages and amoebas, and enters the cells through phagocytosis in *Legionella*-containing vacuoles (LCVs). Over 300 effectors are secreted through its type IVb secretion system (Hubber and Roy, 2010). The phosphatase SidP and the phospholipase VipD promote inactivation or removal of phosphatidylinositol-3-phosphate (PI3P) from the surface of endosomes thus blocking their fusion with the LCVs (Gaspar and Machner, 2014; Toulabi *et al.*, 2013). DrrA/SidM and RalF are GEFs that modulate the fusion of LCV with ER vesicles by activating the GTPases Rab1 and Arf1 respectively (Murata *et al.*, 2006). SidE effector family can also ubiquitinate without ATP Rab1 GTPase thus altering its function (Qiu *et al.*, 2016). LseA, YlfA and YlfB mimic the soluble NSF attachment protein receptor (SNARE) and mediate membrane fusion of LCVs to Golgi-associated vesicles (Campodonico *et al.*, 2016; King *et al.*, 2015). LepB has a Rab modulating function as a Guanidine activating protein (GAP) that disrupts vesicles trafficking in the cells (Mishra *et al.*, 2013). To escape killing by autophagy, *Legionella* releases RavZ, a protease that cleaves Atg8, a protein essential for the correct autophagosome maturation (Choy *et al.*, 2012). The effector sphingosine-1 phosphate lyase (LpSpl) also limits autophagy by blocking sphingosine biosynthesis (Rolando *et al.*, 2016). Organelles and vesicles trafficking can also be mediated by actin remodelling effectors to prevent lysosomes from fusing with the LCVs; LegK2 is a kinase that phosphorylates the ARP2/3 complex that recruits nucleation promoting factors (Michard *et al.*, 2015), and VipA accelerates actin polymerization and binds to endosomal organelles (Franco *et al.*, 2012). LubX is an E3 ligase that ubiquitinates Clk1, a protein required for optimal growth of the host (Kubori *et al.*, 2008) and AnkB (Lpp2082) and mimics the ubiquitin-binding site of parvin B, stabilizing the protein and leading to caspase-3 activation and apoptosis (Lomma *et al.*, 2010). *L. pneumophila* is unique in that it can also controls the expression of host innate immune genes through methylation of chromain by the effector RomA (Rolando *et al.*, 2013).

Obligate intracellular pathogens

Obligate intracellular pathogens require the host to be remain viable throughout the infection process, thus they contain several effectors that inhibit apoptosis. In *Coxiella* species, AnkG, CaeA and CaeB all have anti-apoptotic effects through interaction with the mitochondria or the nucleus and interfering with several signalling cascades (Eckart *et al.*, 2014; Klingenbeck *et al.*, 2013).

CvpA is thought to recruit clathrin from the host and to allow its assembly in the *Coxiella*-containing vacuoles (CCV) (Larson *et al.*, 2013). Vesicular fusion leads to the formation of large CCVs where the bacterium can modulate and it is modulated by two effectors: cirA which stimulates the activity of the Rho GTPase (Weber *et al.*, 2016) and CvpB that prevents the degradation of PI3P leading to membrane aggregation and fusion (Martinez *et al.*, 2016).

Type V effectors have multiple functions

Type V secreted exoproteins are mainly involved in adherence, but a few are also involved in pathogenicity. The IgA protease in *Neisseria* spp. is an effector that cleaves the region between the Fab and the Fc domains of the IgA antibody thus preventing clearance of the bacterium by the immune cells (Plaut *et al.*, 1975). Moreover the Fab fragment is still capable of binding to the antigen on the surface of the bacterium, thus masking it from other immunoglobulins (Mansa and Kilian, 1986).

Another important autotransporter (type Va) required for host manipulation by *Shigella* is IcsA (VirG), an effector that mediates actin-based motility (ABM) (Lett *et al.*, 1989; Suzuki *et al.*, 1995). ABM is the mechanism by which the bacterium, having escaped from the vacuole onto the cytosol of the host, rearranges actin filaments to form a comet-like tail and promote bacterial movement (Goldberg, 2001). IcsA passenger domain is released from the surface by the protease IcsP (Shere *et al.*, 1997) and it is then free to bind N-WASP, an actin regulator protein. Arp2/3 complex is activated by N-WASP and leads to actin polymerization that allows the bacterium to gain propulsion (Egile *et al.*, 1999). *Burkholderia* species require BimA, a trimeric autotransporter for ABM, but the mechanism of action differ between species. *B. thailandensis* BimA activates Arp2/3 complex that leads to branched actin filaments as with the IcsA of *Shigella*. BimA in *B. pseudomallei* and B. mallei instead acts independently of Arp2/3, rather it acts like a human Ena/VASP protein leading to the nucleation, elongation and formation of bundles of F-actin, rather than branching as in the case of BtBimA (Benanti *et al.*, 2015). LepA from *Pseudomonas aeruginosa* is a two-partner secreted protease that is capable of activating through cleavage PAR-1, -2 and -4 leading to increased NF-κB and thus modulation of the immune response (Kida *et al.*, 2008).

Type VI effectors are mainly anti-bacterial but some have an effect on the eukaryotic hosts

The anti-bacterial function of the T6SS has been very well characterized and has been the focus of most research in the field. Little is known about its function in host–pathogen interactions. The presence of T6SS on the surface of pathogens has been linked to increased virulence, but the effect might be secondary due to its ability to outcompete other species. Nevertheless several effects in the eukaryotic host have been linked to the surface expression of T6SS and a few effectors have been recently identified (Hachani *et al.*, 2016).

The spike protein required for piercing the membrane of the host cell is very often also a toxin that contributes to pathogenicity. The VgrG5 in *Burkholderia* spp. mediates membrane fusion of neighbouring macrophages and the creation of multinucleated giant cells, a process that is essential for cell-to-cell spread and ultimately pathogenicity (Schwarz *et al.*, 2014; Toesca *et al.*, 2014). *Aeromonas hydrophila* secretes, in the eukaryotic host, VgrG1, the spike protein that contains an ADP ribosyltransferase site. VgrG1 ribosylates actin, preventing filament formation and leading to cell rounding and death of the host (Suarez *et al.*, 2010). *Pseudomonas aeruginosa* is a facultative intracellular pathogen that secretes VgrG2b into non-phagocytic cells to promote invasion. VgrG2 interacts with the γ-tubulin ring complex (γTuRC) and precedes microtubule- and actin-dependent phagocytosis by the host cell (Sana *et al.*, 2015). *Vibrio cholerae* secretes into the cytoplasm of the host cell the spike protein VgrG1, which not only serves as a puncturing device, but also has an actin cross-linking domain similar to the MARTX toxin of type I secretion systems (Pukatzki *et al.*, 2007). The cross-linking action causes the permanent polymerization of actin and eventually leads to cell death (Pukatzki *et al.*, 2007)

Vibrio cholerae also secretes into the extracellular space VasX, an effector with a pleckstrin homology (PH) domain. VasX binds to phosphoinositides on the eukaryotic host membrane and disrupts signalling (Miyata *et al.*, 2011). The two effectors PldA and PldB, secreted by two different TYPEs of T6SS in *Pseudomonas aeruginosa*, exhibit a phospholipase D (PLD) activity. PldA and PldB target the lipids of competing bacteria and when injected into the cytosol of a eukaryotic host they bind to Akt1 and Akt2 to induce PI3K activation, a common mechanism for pathogen host invasion (Jiang *et al.*, 2014). *Yersinia pseudotuberculosis* secretes the effector YezP that is unique in its capacity to bind Zn^{2+} and thus provides the bacterium at the same time with a potential source of Zn^{2+} and protection against reactive oxygen species that are released by macrophages (Wang *et al.*, 2015).

Many effectors are known, but many more need to be identified to understand better the influence of T6SS in host–pathogen interactions. The identification of a secretion signal common to all the effectors would in all likelihood increase the identification of the effectors that have shown unique mechanisms of action compared to other secretion systems.

Type VII secreted proteins elicit a plethora of effects on the host

The expression of type VII secretion systems is not always associated with pathogenicity; rather they can sometimes be associated with development (Fyans *et al.*, 2013), DNA transfer (Coros *et al.*, 2008), and have a general role in physiology (Serafini *et al.*, 2009). The ESX-1 system has been mainly associated with phagosomal rupture as a means to escape phagocytosis by macrophages (van der Wel *et al.*, 2007), inflammation (Majlessi *et al.*, 2005), apoptosis and bacterial dissemination (Simeone *et al.*, 2012). EsxA/EsxB, a heterodimer of secreted effectors, creates pores in the lipid bilayers in red blood cells and thus allow the bacterium to escape phagocytosis (Smith *et al.*, 2008). In *Staphylococcus aureus* EsxA acts as an inhibitor of apoptosis and EsxB prevents the exit of the bacterium from the host cells to evade detection by the immune system (Korea *et al.*, 2014). The ESX-5 system secretes mainly substrates of the PE and PPE families that modulate the inflammatory response in the host (Bottai *et al.*, 2012). EspB, released by ESX-1, is composed of two domains homologous to PE and PPE and is involved in the dissemination of *M. tuberculosis* in mice (Ohol *et al.*, 2010).

The complex of EsxG and EsxH is secreted by ESX-3 in *Mycobacterium tuberculosis* and it targets the host factor Hrs, a component of the ESCRT complex. Secretion of the complex thus prevents the fusion of the mycobacteria-containing vacuoles with the lysosomes through the impairment of the ESCRT complex (Mehra *et al.*, 2013).

A better understanding of the molecular details of the effectors action on the host and the discovery of new effectors will greatly aid in the development of new drugs targeted at the elusive mycobacteria and other Gram-negative pathogens.

Gingipains are the effectors secreted by type IX secretion systems

Gingipains are the main effectors secreted by type IX secretion systems and they are Arg-X and Lys-X specific proteases that accumulate around the cell surface (Narita *et al.*, 2014).

Their targets include cytokines and complements that hinder the normal inflammatory response and cell surface molecules to prevent clotting (Fitzpatrick *et al.*, 2009).

Recent bioinformatics analysis of the genome of the *Bacteroidetes* phylum to which *Porphyromonas gingivalis* belongs, showed that the CTD is present in hundreds of proteins with different functions: proteases, glycosidases, adhesins, haemagglutinins and internalins (Veith *et al.*, 2013). Future work will be needed to understand better the effectors secreted by type IX secretion systems and their mechanism of pathogenesis.

Conclusion

Secretion systems are fascinating machines that have finely evolved to target eukaryotic and prokaryotic adversaries. The study of secretion systems has been accelerating in the last few years, thanks to advances in the understanding of the structural assembly of these well-evolved and efficient machineries of protein translocation. A lot is now known about how they assemble and the mechanism by which they are capable of secreting a plethora of effectors.

In recent years the emergence of antibiotic resistance in highly pathogenic bacteria has led to the exploration of new targets for the development of antimicrobial agents. Secretion systems would be an ideal target since they are mostly expressed in pathogenic bacteria and would leave the healthy commensal microbiota intact. Much work has been done to understand these systems, but more has to be done to be able to fully exploit them as a means to control the most dominant form of life.

References

Abby, S.S., and Rocha, E.P. (2012). The non-flagellar type III secretion system evolved from the bacterial flagellum and diversified into host-cell adapted systems. PLOS Genet. *8*, e1002983. https://doi.org/10.1371/journal.pgen.1002983.

Abby, S.S., Cury, J., Guglielmini, J., Néron, B., Touchon, M., and Rocha, E.P. (2016). Identification of protein secretion systems in bacterial genomes. Sci. Rep. *6*, 23080. https://doi.org/10.1038/srep23080.

Abrusci, P., Vergara-Irigaray, M., Johnson, S., Beeby, M.D., Hendrixson, D.R., Roversi, P., Friede, M.E., Deane, J.E., Jensen, G.J., Tang, C.M., *et al.* (2013). Architecture of the major component of the type III secretion system export apparatus. Nat. Struct. Mol. Biol. *20*, 99–104. https://doi.org/10.1038/nsmb.2452.

Adams, T.M., Wentzel, A., and Kolmar, H. (2005). Intimin-mediated export of passenger proteins requires maintenance of a translocation-competent conformation. J. Bacteriol. *187*, 522–533. https://doi.org/10.1128/JB.187.2.522-533.2005.

Akeda, Y., and Galán, J.E. (2005). Chaperone release and unfolding of substrates in type III secretion. Nature *437*, 911–915. https://doi.org/10.1038/nature03992.

Alvarez-Martinez, C.E., and Christie, P.J. (2009). Biological diversity of prokaryotic type IV secretion systems. Microbiol. Mol. Biol. Rev. *73*, 775–808. https://doi.org/10.1128/MMBR.00023-09.

Ashida, H., Mimuro, H., Ogawa, M., Kobayashi, T., Sanada, T., Kim, M., and Sasakawa, C. (2011). Cell death and infection: a double-edged sword for host and pathogen survival. J. Cell Biol. *195*, 931–942. https://doi.org/10.1083/jcb.201108081.

Askarian, F., van Sorge, N.M., Sangvik, M., Beasley, F.C., Henriksen, J.R., Sollid, J.U., van Strijp, J.A., Nizet, V., and Johannessen, M. (2014). A *Staphylococcus aureus* TIR domain protein virulence factor blocks TLR2-mediated NF-κB signaling. J. Innate Immun. *6*, 485–498. https://doi.org/10.1159/000357618.

Atmakuri, K., Cascales, E., and Christie, P.J. (2004). Energetic components VirD4, VirB11 and VirB4 mediate early DNA transfer reactions required for bacterial type IV secretion. Mol. Microbiol. *54*, 1199–1211.

Balakrishnan, L., Hughes, C., and Koronakis, V. (2001). Substrate-triggered recruitment of the TolC channel-tunnel during type I export of hemolysin by *Escherichia coli*. J. Mol. Biol. *313*, 501–510. https://doi.org/10.1006/jmbi.2001.5038.

Bally, M., Filloux, A., Akrim, M., Ball, G., Lazdunski, A., and Tommassen, J. (1992). Protein secretion in *Pseudomonas aeruginosa*: Characterization of seven xcp genes and processing of secretory apparatus components by prepilin peptidase. Mol. Microbiol. *6*, 1121–1131. https://doi.org/10.1111/j.1365-2958.1992.tb01550.x.

Basler, M., Pilhofer, M., Henderson, G.P., Jensen, G.J., and Mekalanos, J.J. (2012). Type VI secretion requires a dynamic contractile phage tail-like structure. Nature *483*, 182–186. https://doi.org/10.1038/nature10846.

Baumann, U., Wu, S., Flaherty, K.M., and McKay, D.B. (1993). Three-dimensional structure of the alkaline protease of *Pseudomonas aeruginosa*: a two-domain protein with a calcium binding parallel beta roll motif. EMBO J. *12*, 3357–3364.

Benanti, E.L., Nguyen, C.M., and Welch, M.D. (2015). Virulent *Burkholderia* species mimic host actin polymerases to drive actin-based motility. Cell *161*, 348–360. https://doi.org/10.1016/j.cell.2015.02.044.

Benz, R., Maier, E., Ladant, D., Ullmann, A., and Sebo, P. (1994). Adenylate cyclase toxin (CyaA) of Bordetella pertussis. Evidence for the formation of small ion-permeable channels and comparison with HlyA of *Escherichia coli*. J. Biol. Chem. *269*, 27231–27239.

Berger, C., Robin, G.P., Bonas, U., and Koebnik, R. (2010). Membrane topology of conserved components of the type III secretion system from the plant pathogen *Xanthomonas campestris* pv. *vesicatoria*. Microbiology *156*, 1963–1974. https://doi.org/10.1099/mic.0.039248-0.

Bhakdi, S., Greulich, S., Muhly, M., Eberspächer, B., Becker, H., Thiele, A., and Hugo, F. (1989). Potent leukocidal action of *Escherichia coli* hemolysin mediated by permeabilization of target cell membranes. J. Exp. Med. *169*, 737–754. https://doi.org/10.1084/jem.169.3.737.

Black, D.S., and Bliska, J.B. (1997). Identification of p130Cas as a substrate of Yersinia YopH (Yop51), a bacterial protein tyrosine phosphatase that translocates into mammalian cells and targets focal adhesions. EMBO J. *16*, 2730–2744. https://doi.org/10.1093/emboj/16.10.2730.

Black, D.S., and Bliska, J.B. (2000). The RhoGAP activity of the *Yersinia pseudotuberculosis* cytotoxin YopE is required for antiphagocytic function and virulence. Mol. Microbiol. *37*, 515–527. https://doi.org/10.1046/j.1365-2958.2000.02021.x.

Bottai, D., di Luca, M., Majlessi, L., Frigui, W., Simeone, R., Sayes, F., Bitter, W., Brennan, M.J., Leclerc, C., Batoni, G., et al. (2012). Disruption of the ESX-5 system of *Mycobacterium tuberculosis* causes loss of PPE protein secretion, reduction of cell wall integrity and strong attenuation. Mol. Microbiol. *83*, 1195–1209. https://doi.org/10.1111/j.1365-2958.2012.08001.x.

Boyer, F., Fichant, G., Berthod, J., Vandenbrouck, Y., and Attree, I. (2009). Dissecting the bacterial type VI secretion system by a genome wide in silico analysis: what can be learned from available microbial genomic resources? BMC Genomics. *10*, 104. https://doi.org/10.1186/1471-2164-10-104.

Boyle, E.C., Brown, N.F., and Finlay, B.B. (2006). *Salmonella enterica* serovar Typhimurium effectors SopB, SopE, SopE2 and SipA disrupt tight junction structure and function. Cell. Microbiol. *8*, 1946–1957. https://doi.org/10.1111/j.1462-5822.2006.00762.x.

Brandon, L.D., Goehring, N., Janakiraman, A., Yan, A.W., Wu, T., Beckwith, J., and Goldberg, M.B. (2003). IcsA, a polarly localized autotransporter with an atypical signal peptide, uses the Sec apparatus for secretion, although the Sec apparatus is circumferentially distributed. Mol. Microbiol. *50*, 45–60. https://doi.org/10.1046/j.1365-2958.2003.03674.x.

Brennan, P.J., and Nikaido, H. (1995). The envelope of mycobacteria. Annu. Rev. Biochem. *64*, 29–63. https://doi.org/10.1146/annurev.bi.64.070195.000333.

Brown, G.D., Dave, J.A., Gey van Pittius, N.C., Stevens, L., Ehlers, M.R., and Beyers, A.D. (2000). The mycosins of *Mycobacterium tuberculosis* H37Rv: a family of subtilisin-like serine proteases. Gene *254*, 147–155. https://doi.org/10.1016/S0378-1119(00)00277-8.

Brunet, Y.R., Hénin, J., Celia, H., and Cascales, E. (2014). Type VI secretion and bacteriophage tail tubes share a common assembly pathway. EMBO Rep. *15*, 315–321. https://doi.org/10.1002/embr.201337936.

Brunet, Y.R., Zoued, A., Boyer, F., Douzi, B., and Cascales, E. (2015). The Type VI secretion TssEFGK-VgrG phage-like baseplate is recruited to the TssJLM membrane complex via multiple contacts and serves as assembly platform for tail tube/sheath polymerization. PLOS Genet. *11*, e1005545. https://doi.org/10.1371/journal.pgen.1005545.

Burr, S.E., Stuber, K., and Frey, J. (2003). The ADP-ribosylating toxin, AexT, from *Aeromonas salmonicida* subsp. *salmonicida* is translocated via a type III secretion pathway. J. Bacteriol. *185*, 6583–6591. https://doi.org/10.1128/JB.185.22.6583-6591.2003.

Byrd, D.R., and Matson, S.W. (1997). Nicking by transesterification: the reaction catalysed by a relaxase. Mol. Microbiol. *25*, 1011–1022. https://doi.org/10.1046/j.1365-2958.1997.5241885.x.

Campodonico, E.M., Roy, C.R., and Ninio, S. (2016). *Legionella pneumophila* type IV effectors YlfA and YlfB are SNARE-like Proteins that form homo- and heteromeric complexes and enhance the efficiency of vacuole remodeling. PLOS ONE *11*, e0159698. https://doi.org/10.1371/journal.pone.0159698.

Carbonetti, N.H. (2010). Pertussis toxin and adenylate cyclase toxin: key virulence factors of *Bordetella pertussis* and cell biology tools. Future Microbiol. *5*, 455–469. https://doi.org/10.2217/fmb.09.133.

Cascales, E., and Cambillau, C. (2012). Structural biology of type VI secretion systems. Phil. Trans. R. Soc. Lond. B. Biol. Sci. *367*, 1102–1111. https://doi.org/10.1098/rstb.2011.0209.

Chang, J., Chen, J., and Zhou, D. (2005). Delineation and characterization of the actin nucleation and effector translocation activities of Salmonella SipC. Mol. Microbiol. *55*, 1379–1389. https://doi.org/10.1111/j.1365-2958.2004.04480.x.

Chanter, N. (1990). Molecular aspects of the virulence of Pasteurella multocida. Can. J. Vet. Res. *54*, S45–7.

Cheung, M., Shen, D.K., Makino, F., Kato, T., Roehrich, A.D., Martinez-Argudo, I., Walker, M.L., Murillo, I., Liu, X., Pain, M., et al. (2015). Three-dimensional electron microscopy reconstruction and cysteine-mediated cross-linking provide a model of the type III secretion system needle tip complex. Mol. Microbiol. *95*, 31–50. https://doi.org/10.1111/mmi.12843.

Choi, H.W., Brooking-Dixon, R., Neupane, S., Lee, C.J., Miao, E.A., Staats, H.F., and Abraham, S.N. (2013). Salmonella Typhimurium impedes innate immunity with a mast-cell-suppressing protein tyrosine phosphatase, SptP. Immunity *39*, 1108–1120. https://doi.org/10.1016/j.immuni.2013.11.009.

Choy, A., Dancourt, J., Mugo, B., O'Connor, T.J., Isberg, R.R., Melia, T.J., and Roy, C.R. (2012). The Legionella effector RavZ inhibits host autophagy through irreversible Atg8 deconjugation. Science *338*, 1072–1076. https://doi.org/10.1126/science.1227026.

Gonzalez-Rivera, C., Bhatty, M., and Christie, P.J. (2016). Mechanism and function of Type IV secretion during infection of the human host. Microbiol. Spectr. *4*. https://doi.org/10.1128/microbiolspec.VMBF-0024-2015.

Clements, A., Smollett, K., Lee, S.F., Hartland, E.L., Lowe, M., and Frankel, G. (2011). EspG of enteropathogenic and enterohemorrhagic E. coli binds the Golgi matrix protein GM130 and disrupts the Golgi structure and function. Cell. Microbiol. *13*, 1429–1439. https://doi.org/10.1111/j.1462-5822.2011.01631.x.

Collins, R.F., Davidsen, L., Derrick, J.P., Ford, R.C., and Tønjum, T. (2001). Analysis of the PilQ secretin from *Neisseria meningitidis* by transmission electron microscopy reveals a dodecameric quaternary structure. J. Bacteriol. *183*, 3825–3832. https://doi.org/10.1128/JB.183.13.3825-3832.2001.

Cornelis, G.R. (2002). The Yersinia Ysc-Yop 'type III' weaponry. Nat. Rev. Mol. Cell Biol. *3*, 742–752. https://doi.org/10.1038/nrm932.

Coros, A., Callahan, B., Battaglioli, E., and Derbyshire, K.M. (2008). The specialized secretory apparatus ESX-1 is essential for DNA transfer in *Mycobacterium smegmatis*. Mol. Microbiol. *69*, 794–808. https://doi.org/10.1111/j.1365-2958.2008.06299.x.

Cortajarena, A.L., Goni, F.M., and Ostolaza, H. (2003). A receptor-binding region in *Escherichia coli* alpha-haemolysin. J. Biol. Chem. *278*, 19159–19163. https://doi.org/10.1074/jbc.M208552200.

Costa, T.R., Ilangovan, A., Ukleja, M., Redzej, A., Santini, J.M., Smith, T.K., Egelman, E.H., and Waksman, G. (2016). Structure of the Bacterial Sex F Pilus Reveals an Assembly of a Stoichiometric Protein-Phospholipid Complex. Cell *166*, 1436–1444.e10. https://doi.org/10.1016/j.cell.2016.08.025.

Cullinane, M., Gong, L., Li, X., Lazar-Adler, N., Tra, T., Wolvetang, E., Prescott, M., Boyce, J.D., Devenish, R.J., and Adler, B. (2008). Stimulation of autophagy suppresses the intracellular survival of Burkholderia pseudomallei in mammalian cell lines. Autophagy *4*, 744–753.

d'Enfert, C., Ryter, A., and Pugsley, A.P. (1987). Cloning and expression in *Escherichia coli* of the Klebsiella pneumoniae genes for production, surface localization and secretion of the lipoprotein pullulanase. EMBO J. *6*, 3531–3538.

Daffé, M., Crick, D.C., and Jackson, M. (2014). Genetics of Capsular Polysaccharides and Cell Envelope (Glyco)lipids. Microbiol. Spectr. *2*, MGM2-0021-2013. https://doi.org/10.1128/microbiolspec.MGM2-0021-2013.

Dale, C., Plague, G.R., Wang, B., Ochman, H., and Moran, N.A. (2002). Type III secretion systems and the evolution of mutualistic endosymbiosis. Proc. Natl. Acad. Sci. U.S.A. *99*, 12397–12402. https://doi.org/10.1073/pnas.182213299.

Daleke, M.H., van der Woude, A.D., Parret, A.H., Ummels, R., de Groot, A.M., Watson, D., Piersma, S.R., Jiménez, C.R., Luirink, J., Bitter, W., *et al.* (2012). Specific chaperones for the type VII protein secretion pathway. J. Biol. Chem. *287*, 31939–31947. https://doi.org/10.1074/jbc.M112.397596.

Dawson, R.J., and Locher, K.P. (2006). Structure of a bacterial multidrug ABC transporter. Nature *443*, 180–185. https://doi.org/10.1038/nature05155.
de Jong, M.F., Starr, T., Winter, M.G., den Hartigh, A.B., Child, R., Knodler, L.A., van Dijl, J.M., Celli, J., and Tsolis, R.M. (2013). Sensing of bacterial type IV secretion via the unfolded protein response. MBio *4*, e00418–12. https://doi.org/10.1128/mBio.00418-12.
Dean, P., and Kenny, B. (2004). Intestinal barrier dysfunction by enteropathogenic *Escherichia coli* is mediated by two effector molecules and a bacterial surface protein. Mol. Microbiol. *54*, 665–675. https://doi.org/10.1111/j.1365-2958.2004.04308.x.
DebRoy, S., Dao, J., Söderberg, M., Rossier, O., and Cianciotto, N.P. (2006). *Legionella pneumophila* type II secretome reveals unique exoproteins and a chitinase that promotes bacterial persistence in the lung. Proc. Natl. Acad. Sci. U.S.A. *103*, 19146–19151. https://doi.org/10.1073/pnas.0608279103.
Delevoye, C., Nilges, M., Dehoux, P., Paumet, F., Perrinet, S., Dautry-Varsat, A., and Subtil, A. (2008). SNARE protein mimicry by an intracellular bacterium. PLOS Pathog. *4*, e1000022. https://doi.org/10.1371/journal.ppat.1000022.
Diepold, A., and Armitage, J.P. (2015). Type III secretion systems: the bacterial flagellum and the injectisome. Phil. Trans. R. Soc. Lond. B. Biol. Sci. *370*, 20150020. https://doi.org/10.1098/rstb.2015.0020.
Disqué-Kochem, C., and Dreiseikelmann, B. (1997). The cytoplasmic DNA-binding protein TraM binds to the inner membrane protein TraD in vitro. J. Bacteriol. *179*, 6133–6137.
Douzi, B., Filloux, A., and Voulhoux, R. (2012). On the path to uncover the bacterial type II secretion system. Phil. Trans. R. Soc. Lond. B. Biol. Sci. *367*, 1059–1072. https://doi.org/10.1098/rstb.2011.0204.
Du, D., Wang, Z., James, N.R., Voss, J.E., Klimont, E., Ohene-Agyei, T., Venter, H., Chiu, W., and Luisi, B.F. (2014). Structure of the AcrAB-TolC multidrug efflux pump. Nature *509*, 512–515. https://doi.org/10.1038/nature13205.
Durand, E., Nguyen, V.S., Zoued, A., Logger, L., Péhau-Arnaudet, G., Aschtgen, M.S., Spinelli, S., Desmyter, A., Bardiaux, B., Dujeancourt, A., *et al.* (2015). Biogenesis and structure of a type VI secretion membrane core complex. Nature *523*, 555–560. https://doi.org/10.1038/nature14667.
Eckart, R.A., Bisle, S., Schulze-Luehrmann, J., Wittmann, I., Jantsch, J., Schmid, B., Berens, C., and Lührmann, A. (2014). Antiapoptotic activity of *Coxiella burnetii* effector protein AnkG is controlled by p32-dependent trafficking. Infect. Immun. *82*, 2763–2771. https://doi.org/10.1128/IAI.01204-13.
Egile, C., Loisel, T.P., Laurent, V., Li, R., Pantaloni, D., Sansonetti, P.J., and Carlier, M.F. (1999). Activation of the CDC42 effector N-WASP by the *Shigella flexneri* IcsA protein promotes actin nucleation by Arp2/3 complex and bacterial actin-based motility. J. Cell Biol. *146*, 1319–1332. https://doi.org/10.1083/jcb.146.6.1319.
Evdokimov, A.G., Tropea, J.E., Routzahn, K.M., and Waugh, D.S. (2002). Crystal structure of the *Yersinia pestis* GTPase activator YopE. Protein Sci. *11*, 401–408. https://doi.org/10.1110/ps.34102.
Farizo, K.M., Fiddner, S., Cheung, A.M., and Burns, D.L. (2002). Membrane localization of the S1 subunit of pertussis toxin in Bordetella pertussis and implications for pertussis toxin secretion. Infect. Immun. *70*, 1193–1201. https://doi.org/10.1128/IAI.70.3.1193-1201.2002.
Filloux, A. (2010). Secretion signal and protein targeting in bacteria: a biological puzzle. J. Bacteriol. *192*, 3847–3849. https://doi.org/10.1128/JB.00565-10.
Fitzpatrick, R.E., Wijeyewickrema, L.C., and Pike, R.N. (2009). The gingipains: scissors and glue of the periodontal pathogen, Porphyromonas gingivalis. Future Microbiol. *4*, 471–487. https://doi.org/10.2217/fmb.09.18.
Francetic, O., and Pugsley, A.P. (1996). The cryptic general secretory pathway (gsp) operon of *Escherichia coli* K-12 encodes functional proteins. J. Bacteriol. *178*, 3544–3549.
Franco, I.S., Shohdy, N., and Shuman, H.A. (2012). The *Legionella pneumophila* effector VipA is an actin nucleator that alters host cell organelle trafficking. PLOS Pathog. *8*, e1002546. https://doi.org/10.1371/journal.ppat.1002546.
Fronzes, R., Schäfer, E., Wang, L., Saibil, H.R., Orlova, E.V., and Waksman, G. (2009). Structure of a type IV secretion system core complex. Science *323*, 266–268. https://doi.org/10.1126/science.1166101.
Fu, Y., and Galán, J.E. (1999). A salmonella protein antagonizes Rac-1 and Cdc42 to mediate host-cell recovery after bacterial invasion. Nature *401*, 293–297. https://doi.org/10.1038/45829.
Fu, Y.H., Tsai, M.M., Luo, Y.N., and Deonier, R.C. (1991). Deletion analysis of the F plasmid oriT locus. J. Bacteriol. *173*, 1012–1020.
Fujii, T., Cheung, M., Blanco, A., Kato, T., Blocker, A.J., and Namba, K. (2012). Structure of a type III secretion needle at 7-Å resolution provides insights into its assembly and signaling mechanisms. Proc. Natl. Acad. Sci. U.S.A. *109*, 4461–4466. https://doi.org/10.1073/pnas.1116126109.

Fyans, J.K., Bignell, D., Loria, R., Toth, I., and Palmer, T. (2013). The ESX/type VII secretion system modulates development, but not virulence, of the plant pathogen Streptomyces scabies. Mol. Plant Pathol. *14*, 119–130. https://doi.org/10.1111/j.1364-3703.2012.00835.x.

Gangola, P., and Rosen, B.P. (1987). Maintenance of intracellular calcium in *Escherichia coli*. J. Biol. Chem. *262*, 12570–12574.

Gaspar, A.H., and Machner, M.P. (2014). VipD is a Rab5-activated phospholipase A1 that protects *Legionella pneumophila* from endosomal fusion. Proc. Natl. Acad. Sci. U.S.A. *111*, 4560–4565. https://doi.org/10.1073/pnas.1316376111.

Gey Van Pittius, N.C., Gamieldien, J., Hide, W., Brown, G.D., Siezen, R.J., and Beyers, A.D. (2001). The ESAT-6 gene cluster of *Mycobacterium tuberculosis* and other high G+C Gram-positive bacteria. Genome Biol. *2*, RESEARCH0044. https://doi.org/10.1186/gb-2001-2-10-research0044.

Ghai, J., and Das, A. (1989). The virD operon of *Agrobacterium tumefaciens* Ti plasmid encodes a DNA-relaxing enzyme. Proc. Natl. Acad. Sci. U.S.A. *86*, 3109–3113.

Ghosal, D., Chang, Y.-W., Jeong, K.C., Vogel, J.P., and Jensen, G.J. (2016). Structure of the Legionella Dot/Icm type IV secretion system in situ by electron cryotomography. bioRxiv. https://doi.org/10.1101/085977.

Gilmore, R., Blobel, G., and Walter, P. (1982). Protein translocation across the endoplasmic reticulum. I. Detection in the microsomal membrane of a receptor for the signal recognition particle. J. Cell Biol. *95*, 463–469. https://doi.org/10.1083/jcb.95.2.463.

Glew, M.D., Veith, P.D., Peng, B., Chen, Y.Y., Gorasia, D.G., Yang, Q., Slakeski, N., Chen, D., Moore, C., Crawford, S., *et al.* (2012). PG0026 is the C-terminal signal peptidase of a novel secretion system of Porphyromonas gingivalis. J. Biol. Chem. *287*, 24605–24617. https://doi.org/10.1074/jbc.M112.369223.

Goehring, U.M., Schmidt, G., Pederson, K.J., Aktories, K., and Barbieri, J.T. (1999). The N-terminal domain of *Pseudomonas aeruginosa* exoenzyme S is a GTPase-activating protein for Rho GTPases. J. Biol. Chem. *274*, 36369–36372. https://doi.org/10.1074/jbc.274.51.36369.

Goldberg, M.B. (2001). Actin-based motility of intracellular microbial pathogens. Microbiol. Mol. Biol. Rev. *65*, 595–626, table of contents. https://doi.org/10.1128/MMBR.65.4.595-626.2001.

Gomis-Rüth, F.X., Moncalián, G., Pérez-Luque, R., González, A., Cabezón, E., de la Cruz, F., and Coll, M. (2001). The bacterial conjugation protein TrwB resembles ring helicases and F1-ATPase. Nature *409*, 637–641. https://doi.org/10.1038/35054586.

Gophna, U., Ron, E.Z., and Graur, D. (2003). Bacterial type III secretion systems are ancient and evolved by multiple horizontal-transfer events. Gene *312*, 151–163. https://doi.org/10.1016/S0378-1119(03)00612-7.

Gorasia, D.G., Veith, P.D., Chen, D., Seers, C.A., Mitchell, H.A., Chen, Y.Y., Glew, M.D., Dashper, S.G., and Reynolds, E.C. (2015). Porphyromonas gingivalis Type IX Secretion Substrates Are Cleaved and Modified by a Sortase-Like Mechanism. PLOS Pathog. *11*, e1005152. https://doi.org/10.1371/journal.ppat.1005152.

Gorasia, D.G., Veith, P.D., Hanssen, E.G., Glew, M.D., Sato, K., Yukitake, H., Nakayama, K., and Reynolds, E.C. (2016). Structural Insights into the PorK and PorN Components of the Porphyromonas gingivalis Type IX Secretion System. PLOS Pathog. *12*, e1005820. https://doi.org/10.1371/journal.ppat.1005820.

Goyal, P., Krasteva, P.V., Van Gerven, N., Gubellini, F., Van den Broeck, I., Troupiotis-Tsaïlaki, A., Jonckheere, W., Péhau-Arnaudet, G., Pinkner, J.S., Chapman, M.R., *et al.* (2014). Structural and mechanistic insights into the bacterial amyloid secretion channel CsgG. Nature *516*, 250–253. https://doi.org/10.1038/nature13768.

Greenberg, S., Chang, P., and Silverstein, S.C. (1993). Tyrosine phosphorylation is required for Fc receptor-mediated phagocytosis in mouse macrophages. J. Exp. Med. *177*, 529–534.

Hachani, A., Wood, T.E., and Filloux, A. (2016). Type VI secretion and anti-host effectors. Curr. Opin. Microbiol. *29*, 81–93. https://doi.org/10.1016/j.mib.2015.11.006.

Hales, L.M., and Shuman, H.A. (1999). *Legionella pneumophila* contains a type II general secretion pathway required for growth in amoebae as well as for secretion of the Msp protease. Infect. Immun. *67*, 3662–3666.

Hall, A. (1998). Rho GTPases and the actin cytoskeleton. Science *279*, 509–514. https://doi.org/10.1126/science.279.5350.509.

Hamid, N., Gustavsson, A., Andersson, K., McGee, K., Persson, C., Rudd, C.E., and Fällman, M. (1999). YopH dephosphorylates Cas and Fyn-binding protein in macrophages. Microb. Pathog. *27*, 231–242. https://doi.org/10.1006/mpat.1999.0301.

Hamilton, H.L., Domínguez, N.M., Schwartz, K.J., Hackett, K.T., and Dillard, J.P. (2005). *Neisseria gonorrhoeae* secretes chromosomal DNA via a novel type IV secretion system. Mol. Microbiol. *55*, 1704–1721. https://doi.org/10.1111/j.1365-2958.2005.04521.x.
Hammar, M., Arnqvist, A., Bian, Z., Olsén, A., and Normark, S. (1995). Expression of two csg operons is required for production of fibronectin- and congo red-binding curli polymers in *Escherichia coli* K-12. Mol. Microbiol. *18*, 661–670. https://doi.org/10.1111/j.1365-2958.1995.mmi_18040661.x.
Hapfelmeier, S., Domke, N., Zambryski, P.C., and Baron, C. (2000). VirB6 is required for stabilization of VirB5 and VirB3 and formation of VirB7 homodimers in *Agrobacterium tumefaciens*. J. Bacteriol. *182*, 4505–4511. https://doi.org/10.1128/JB.182.16.4505-4511.2000.
Hardt, W.D., Chen, L.M., Schuebel, K.E., Bustelo, X.R., and Galán, J.E. (1998). *S. typhimurium* encodes an activator of Rho GTPases that induces membrane ruffling and nuclear responses in host cells. Cell 93, 815–826. https://doi.org/10.1016/S0092-8674(00)81442-7.
He, S.Y., Lindeberg, M., Chatterjee, A.K., and Collmer, A. (1991). Cloned Erwinia chrysanthemi out genes enable *Escherichia coli* to selectively secrete a diverse family of heterologous proteins to its milieu. Proc. Natl. Acad. Sci. U. S. A. *88*, 1079–1083.
Hinsa, S.M., Espinosa-Urgel, M., Ramos, J.L., and O'Toole, G.A. (2003). Transition from reversible to irreversible attachment during biofilm formation by *Pseudomonas fluorescens* WCS365 requires an ABC transporter and a large secreted protein. Mol. Microbiol. *49*, 905–918. https://doi.org/10.1046/j.1365-2958.2003.03615.x.
Hodgkinson, J.L., Horsley, A., Stabat, D., Simon, M., Johnson, S., da Fonseca, P.C. a, Morris, E.P., Wall, J.S., Lea, S.M., and Blocker, A.J. (2009). Three-dimensional reconstruction of the Shigella T3SS transmembrane regions reveals 12-fold symmetry and novel features throughout. Nat. Struct. Mol. Biol. *16*, 477–485. https://doi.org/10.1038/nsmb0809-897b.
Hofreuter, D., Odenbreit, S., and Haas, R. (2001). Natural transformation competence in *Helicobacter pylori* is mediated by the basic components of a type IV secretion system. Mol. Microbiol. *41*, 379–391. https://doi.org/10.1046/j.1365-2958.2001.02502.x.
Hood, R.D., Singh, P., Hsu, F., Güvener, T., Carl, M.A., Trinidad, R.R., Silverman, J.M., Ohlson, B.B., Hicks, K.G., Plemel, R.L., *et al.* (2010). A type VI secretion system of *Pseudomonas aeruginosa* targets a toxin to bacteria. Cell Host Microbe 7, 25–37. https://doi.org/10.1016/j.chom.2009.12.007.
Houben, E.N., Bestebroer, J., Ummels, R., Wilson, L., Piersma, S.R., Jiménez, C.R., Ottenhoff, T.H., Luirink, J., and Bitter, W. (2012). Composition of the type VII secretion system membrane complex. Mol. Microbiol. *86*, 472–484. https://doi.org/10.1111/j.1365-2958.2012.08206.x.
Hubber, A., and Roy, C.R. (2010). Modulation of host cell function by *Legionella pneumophila* type IV effectors. Annu. Rev. Cell Dev. Biol. *26*, 261–283. https://doi.org/10.1146/annurev-cellbio-100109-104034.
Hueck, C.J. (1998). Type III protein secretion systems in bacterial pathogens of animals and plants. Microbiol. Mol. Biol. Rev. *62*, 379–433. https://doi.org/10.1111/j.1365-2958.2006.05301.x.
Iwobi, A., Heesemann, J., Garcia, E., Igwe, E., Noelting, C., and Rakin, A. (2003). Novel virulence-associated type II secretion system unique to high-pathogenicity Yersinia enterocolitica. Infect. Immun. *71*, 1872–1879. https://doi.org/10.1128/IAI.71.4.1872-1879.2003.
Jakubowski, S.J., Krishnamoorthy, V., Cascales, E., and Christie, P.J. (2004). *Agrobacterium tumefaciens* VirB6 domains direct the ordered export of a DNA substrate through a type IV secretion System. J. Mol. Biol. *341*, 961–977. https://doi.org/10.1016/j.jmb.2004.06.052.
Jeong, H., Kim, J.S., Song, S., Shigematsu, H., Yokoyama, T., Hyun, J., and Ha, N.C. (2016). Pseudoatomic Structure of the Tripartite Multidrug Efflux Pump AcrAB-TolC Reveals the Intermeshing Cogwheel-like Interaction between AcrA and TolC. Structure *24*, 272–276. https://doi.org/10.1016/j.str.2015.12.007.
Jiang, F., Waterfield, N.R., Yang, J., Yang, G., and Jin, Q. (2014). A *Pseudomonas aeruginosa* type VI secretion phospholipase D effector targets both prokaryotic and eukaryotic cells. Cell Host Microbe *15*, 600–610. https://doi.org/10.1016/j.chom.2014.04.010.
Jimenez, A., Chen, D., and Alto, N.M. (2016). How Bacteria Subvert Animal Cell Structure and Function. Annu. Rev. Cell Dev. Biol. *32*, 373–397. https://doi.org/10.1146/annurev-cellbio-100814-125227.
Judd, P.K., Kumar, R.B., and Das, A. (2005). Spatial location and requirements for the assembly of the *Agrobacterium tumefaciens* type IV secretion apparatus. Proc. Natl. Acad. Sci. U.S.A. *102*, 11498–11503.
Junker, M., Besingi, R.N., and Clark, P.L. (2009). Vectorial transport and folding of an autotransporter virulence protein during outer membrane secretion. Mol. Microbiol. *71*, 1323–1332. https://doi.org/10.1111/j.1365-2958.2009.06607.x.

Katada, T., Tamura, M., and Ui, M. (1983). The A protomer of islet-activating protein, pertussis toxin, as an active peptide catalyzing ADP-ribosylation of a membrane protein. Arch. Biochem. Biophys. *224*, 290–298. https://doi.org/10.1016/0003-9861(83)90212-6.

Kayath, C.A., Hussey, S., El hajjami, N., Nagra, K., Philpott, D., and Allaoui, A. (2010). Escape of intracellular Shigella from autophagy requires binding to cholesterol through the type III effector, IcsB. Microbes Infect. *12*, 956–966. https://doi.org/10.1016/j.micinf.2010.06.006.

Kenny, B. (1999). Phosphorylation of tyrosine 474 of the enteropathogenic *Escherichia coli* (EPEC) Tir receptor molecule is essential for actin nucleating activity and is preceded by additional host modifications. Mol. Microbiol. *31*, 1229–1241. https://doi.org/10.1046/j.1365-2958.1999.01265.x.

Kerr, J.E., and Christie, P.J. (2010). Evidence for VirB4-mediated dislocation of membrane-integrated VirB2 pilin during biogenesis of the Agrobacterium VirB/VirD4 type IV secretion system. J. Bacteriol. *192*, 4923–4934. https://doi.org/10.1128/JB.00557-10.

Kersulyte, D., Velapatiño, B., Mukhopadhyay, A.K., Cahuayme, L., Bussalleu, A., Combe, J., Gilman, R.H., and Berg, D.E. (2003). Cluster of type IV secretion genes in *Helicobacter pylori*'s plasticity zone. J. Bacteriol. *185*, 3764–3772. https://doi.org/10.1128/JB.185.13.3764-3772.2003.

Kida, Y., Higashimoto, Y., Inoue, H., Shimizu, T., and Kuwano, K. (2008). A novel secreted protease from *Pseudomonas aeruginosa* activates NF-κB through protease-activated receptors. Cell. Microbiol. *10*, 1491–1504. https://doi.org/10.1111/j.1462-5822.2008.01142.x.

Kim, J., Thanabalasuriar, A., Chaworth-Musters, T., Fromme, J.C., Frey, E.A., Lario, P.I., Metalnikov, P., Rizg, K., Thomas, N.A., Lee, S.F., et al. (2007). The Bacterial Virulence Factor NleA Inhibits Cellular Protein Secretion by Disrupting Mammalian COPII Function. Cell Host Microbe 2, 160–171. https://doi.org/10.1016/j.chom.2007.07.010.

Kim, M., Ogawa, M., Fujita, Y., Yoshikawa, Y., Nagai, T., Koyama, T., Nagai, S., Lange, A., Fässler, R., and Sasakawa, C. (2009). Bacteria hijack integrin-linked kinase to stabilize focal adhesions and block cell detachment. Nature *459*, 578–582. https://doi.org/10.1038/nature07952.

King, N.P., Newton, P., Schuelein, R., Brown, D.L., Petru, M., Zarsky, V., Dolezal, P., Luo, L., Bugarcic, A., Stanley, A.C., *et al.* (2015). Soluble NSF attachment protein receptor molecular mimicry by a *Legionella pneumophila* Dot/Icm effector. Cell. Microbiol. *17*, 767–784. https://doi.org/10.1111/cmi.12405.

Klingenbeck, L., Eckart, R.A., Berens, C., and Lührmann, A. (2013). The *Coxiella burnetii* type IV secretion system substrate CaeB inhibits intrinsic apoptosis at the mitochondrial level. Cell. Microbiol. *15*, 675–687. https://doi.org/10.1111/cmi.12066.

Korea, C.G., Balsamo, G., Pezzicoli, A., Merakou, C., Tavarini, S., Bagnoli, F., Serruto, D., and Unnikrishnan, M. (2014). Staphylococcal Esx proteins modulate apoptosis and release of intracellular *Staphylococcus aureus* during infection in epithelial cells. Infect. Immun. *82*, 4144–4153. https://doi.org/10.1128/IAI.01576-14.

Koronakis, V., Sharff, A., Koronakis, E., Luisi, B., and Hughes, C. (2000). Crystal structure of the bacterial membrane protein TolC central to multidrug efflux and protein export. Nature *405*, 914–919. https://doi.org/10.1038/35016007.

Korotkov, K.V., Sandkvist, M., and Hol, W.G. (2012). The type II secretion system: biogenesis, molecular architecture and mechanism. Nat. Rev. Microbiol. *10*, 336–351. https://doi.org/10.1038/nrmicro2762.

Korotkov, K.V., Johnson, T.L., Jobling, M.G., Pruneda, J., Pardon, E., Héroux, A., Turley, S., Steyaert, J., Holmes, R.K., Sandkvist, M., *et al.* (2011). Structural and functional studies on the interaction of GspC and GspD in the type II secretion system. PLOS Pathog. 7, e1002228. https://doi.org/10.1371/journal.ppat.1002228.

Korotkova, N., Freire, D., Phan, T.H., Ummels, R., Creekmore, C.C., Evans, T.J., Wilmanns, M., Bitter, W., Parret, A.H., Houben, E.N., *et al.* (2014). Structure of the *Mycobacterium tuberculosis* type VII secretion system chaperone EspG5 in complex with PE25-PPE41 dimer. Mol. Microbiol. *94*, 367–382. https://doi.org/10.1111/mmi.12770.

Koster, M., Bitter, W., de Cock, H., Allaoui, A., Cornelis, G.R., and Tommassen, J. (1997). The outer membrane component, YscC, of the Yop secretion machinery of Yersinia enterocolitica forms a ring-shaped multimeric complex. Mol. Microbiol. *26*, 789–797. https://doi.org/10.1046/j.1365-2958.1997.6141981.x.

Krall, R., Schmidt, G., Aktories, K., and Barbieri, J.T. (2000). *Pseudomonas aeruginosa* ExoT is a Rho GTPase-activating protein. Infect. Immun. *68*, 6066–6068. https://doi.org/10.1128/IAI.68.10.6066-6068.2000.

Kube, S., Kapitein, N., Zimniak, T., Herzog, F., Mogk, A., and Wendler, P. (2014). Structure of the VipA/B type VI secretion complex suggests a contraction-state-specific recycling mechanism. Cell Rep. *8*, 20–30. https://doi.org/10.1016/j.celrep.2014.05.034.

Kubori, T., Matsushima, Y., Nakamura, D., Uralil, J., Lara-Tejero, M., Sukhan, A., Galán, J.E., and Aizawa, S.I. (1998). Supramolecular structure of the Salmonella typhimurium type III protein secretion system. Science *280*, 602–605. https://doi.org/10.1111/j.1365-2958.2008.06124.x.

Kubori, T., and Nagai, H. (2016). The Type IVB secretion system: an enigmatic chimera. Curr. Opin. Microbiol. *29*, 22–29. https://doi.org/10.1016/j.mib.2015.10.001.

Kubori, T., Sukhan, A., Aizawa, S.I., and Galán, J.E. (2000). Molecular characterization and assembly of the needle complex of the Salmonella typhimurium type III protein secretion system. Proc. Natl. Acad. Sci. U. S. A. *97*, 10225–10230. https://doi.org/10.1073/pnas.170128997.

Kubori, T., Hyakutake, A., and Nagai, H. (2008). Legionella translocates an E3 ubiquitin ligase that has multiple U-boxes with distinct functions. Mol. Microbiol. *67*, 1307–1319. https://doi.org/10.1111/j.1365-2958.2008.06124.x.

Kudryashev, M., Wang, R.Y., Brackmann, M., Scherer, S., Maier, T., Baker, D., DiMaio, F., Stahlberg, H., Egelman, E.H., and Basler, M. (2015). Structure of the type VI secretion system contractile sheath. Cell *160*, 952–962. https://doi.org/10.1016/j.cell.2015.01.037.

Kudryashov, D.S., Durer, Z.A., Ytterberg, A.J., Sawaya, M.R., Pashkov, I., Prochazkova, K., Yeates, T.O., Loo, R.R., Loo, J.A., Satchell, K.J., *et al.* (2008). Connecting actin monomers by iso-peptide bond is a toxicity mechanism of the *Vibrio cholerae* MARTX toxin. Proc. Natl. Acad. Sci. U.S.A. *105*, 18537–18542. https://doi.org/10.1073/pnas.0808082105.

Kumar, R.B., Xie, Y.H., and Das, A. (2000). Subcellular localization of the *Agrobacterium tumefaciens* T-DNA transport pore proteins: VirB8 is essential for the assembly of the transport pore. Mol. Microbiol. *36*, 608–617.

Kwok, T., Zabler, D., Urman, S., Rohde, M., Hartig, R., Wessler, S., Misselwitz, R., Berger, J., Sewald, N., König, W., *et al.* (2007). Helicobacter exploits integrin for type IV secretion and kinase activation. Nature *449*, 862–866.

Lally, E.T., Kieba, I.R., Sato, A., Green, C.L., Rosenbloom, J., Korostoff, J., Wang, J.F., Shenker, B.J., Ortlepp, S., Robinson, M.K., *et al.* (1997). RTX toxins recognize a beta2 integrin on the surface of human target cells. J. Biol. Chem. *272*, 30463–30469. https://doi.org/10.1074/jbc.272.48.30463.

Lara-Tejero, M., Kato, J., Wagner, S., Liu, X., and Galán, J.E. (2011). A sorting platform determines the order of protein secretion in bacterial type III systems. Science *331*, 1188–1191. https://doi.org/10.1126/science.1201476.

LaRock, C.N., and Cookson, B.T. (2012). The Yersinia virulence effector YopM binds caspase-1 to arrest inflammasome assembly and processing. Cell Host Microbe *12*, 799–805. https://doi.org/10.1016/j.chom.2012.10.020.

Larson, C.L., Beare, P. a, Howe, D., and Heinzen, R. a (2013). *Coxiella burnetii* effector protein subverts clathrin-mediated vesicular trafficking for pathogen vacuole biogenesis. Proc. Natl. Acad. Sci. U. S. A. *110*, E4770-9. https://doi.org/10.1073/pnas.1309195110.

Lawley, T.D., Klimke, W.A., Gubbins, M.J., and Frost, L.S. (2003). F factor conjugation is a true type IV secretion system. FEMS Microbiol. Lett. *224*, 1–15.

Lefebre, M.D., and Galán, J.E. (2014). The inner rod protein controls substrate switching and needle length in a Salmonella type III secretion system. Proc. Natl. Acad. Sci. U.S.A. *111*, 817–822. https://doi.org/10.1073/pnas.1319698111.

Leiman, P.G., Basler, M., Ramagopal, U.A., Bonanno, J.B., Sauder, J.M., Pukatzki, S., Burley, S.K., Almo, S.C., and Mekalanos, J.J. (2009). Type VI secretion apparatus and phage tail-associated protein complexes share a common evolutionary origin. Proc. Natl. Acad. Sci. *106*, 4154–4159. https://doi.org/10.1073/pnas.0813360106.

Leininger, E., Roberts, M., Kenimer, J.G., Charles, I.G., Fairweather, N., Novotny, P., and Brennan, M.J. (1991). Pertactin, an Arg-Gly-Asp-containing Bordetella pertussis surface protein that promotes adherence of mammalian cells. Proc. Natl. Acad. Sci. U. S. A. *88*, 345–349. https://doi.org/10.1073/pnas.88.2.345.

Leo, J.C., Grin, I., and Linke, D. (2012). Type V secretion: mechanism(s) of autotransport through the bacterial outer membrane. Phil. Trans. R. Soc. Lond. B. Biol. Sci. *367*, 1088–1101. https://doi.org/10.1098/rstb.2011.0208.

Létoffé, S., Ghigo, J.M., and Wandersman, C. (1994). Secretion of the Serratia marcescens HasA protein by an ABC transporter. J. Bacteriol. *176*, 5372–5377.

Lett, M.C., Sasakawa, C., Okada, N., Sakai, T., Makino, S., Yamada, M., Komatsu, K., and Yoshikawa, M. (1989). virG, a plasmid-coded virulence gene of *Shigella flexneri*: identification of the virG protein and determination of the complete coding sequence. J. Bacteriol. *171*, 353–359.
Leung, K.Y., Reisner, B.S., and Straley, S.C. (1990). YopM inhibits platelet aggregation and is necessary for virulence of *Yersinia pestis* in mice. Infect. Immun. *58*, 3262–3271.
Linhartová, I., Bumba, L., Mašín, J., Basler, M., Osička, R., Kamanová, J., Procházková, K., Adkins, I., Hejnová-Holubová, J., Sadílková, L., *et al.* (2010). RTX proteins: a highly diverse family secreted by a common mechanism. FEMS Microbiol. Rev. *34*, 1076–1112. https://doi.org/10.1111/j.1574-6976.2010.00231.x.
Litvak, Y., and Selinger, Z. (2007). Aeromonas salmonicida toxin AexT has a Rho family GTPase-activating protein domain. J. Bacteriol. *189*, 2558–2560. https://doi.org/10.1128/JB.01358-06.
Lomma, M., Dervins-Ravault, D., Rolando, M., Nora, T., Newton, H.J., Sansom, F.M., Sahr, T., Gomez-Valero, L., Jules, M., Hartland, E.L., et al. (2010). The *Legionella pneumophila* F-box protein Lpp2082 (AnkB) modulates ubiquitination of the host protein parvin B and promotes intracellular replication. Cell. Microbiol. *12*, 1272–1291. https://doi.org/10.1111/j.1462-5822.2010.01467.x.
Loquet, A., Sgourakis, N.G., Gupta, R., Giller, K., Riedel, D., Goosmann, C., Griesinger, C., Kolbe, M., Baker, D., Becker, S., *et al.* (2012). Atomic model of the type III secretion system needle. Nature *486*, 276–279. https://doi.org/10.1038/nature11079.
Low, H.H., Gubellini, F., Rivera-Calzada, A., Braun, N., Connery, S., Dujeancourt, A., Lu, F., Redzej, A., Fronzes, R., Orlova, E.V., *et al.* (2014). Structure of a type IV secretion system. Nature *508*, 550–553. https://doi.org/10.1038/nature13081.
Low, L.Y., Mukasa, T., Reed, J.C., and Pascual, J. (2007). Characterization of a TIR-like protein from Paracoccus denitrificans. Biochem. Biophys. Res. Commun. *356*, 481–486. https://doi.org/10.1016/j.bbrc.2007.03.003.
Luthra, A., Mahmood, A., Arora, A., and Ramachandran, R. (2008). Characterization of Rv3868, an essential hypothetical protein of the ESX-1 secretion system in *Mycobacterium tuberculosis*. J. Biol. Chem. *283*, 36532–36541. https://doi.org/10.1074/jbc.M807144200.
Lybarger, S.R., Johnson, T.L., Gray, M.D., Sikora, A.E., and Sandkvist, M. (2009). Docking and assembly of the type II secretion complex of *Vibrio cholerae*. J. Bacteriol. *191*, 3149–3161. https://doi.org/10.1128/JB.01701-08.
Madden, J.C., Ruiz, N., and Caparon, M. (2001). Cytolysin-mediated translocation (CMT): a functional equivalent of type III secretion in gram-positive bacteria. Cell *104*, 143–152. https://doi.org/10.1016/S0092-8674(01)00198-2.
Majlessi, L., Brodin, P., Brosch, R., Rojas, M.J., Khun, H., Huerre, M., Cole, S.T., and Leclerc, C. (2005). Influence of ESAT-6 secretion system 1 (RD1) of *Mycobacterium tuberculosis* on the interaction between mycobacteria and the host immune system. J. Immunol. *174*, 3570–3579.
Mallama, C.A., McCoy-Simandle, K., and Cianciotto, N.P. (2017). The Type II Secretion System of *Legionella pneumophila* Dampens the MyD88 and TLR2 Signaling Pathway in Infected Human Macrophages. Infect. Immun. https://doi.org/10.1128/IAI.00897-16.
Mansa, B., and Kilian, M. (1986). Retained antigen-binding activity of Fab(α) fragments of human monoclonal immunoglobulin A1 (IgA1) cleaved by IgA1 protease. Infect. Immun. *52*, 171–174.
Martinez, E., Allombert, J., Cantet, F., Lakhani, A., Yandrapalli, N., Neyret, A., Norville, I.H., Favard, C., Muriaux, D., and Bonazzi, M. (2016). *Coxiella burnetii* effector CvpB modulates phosphoinositide metabolism for optimal vacuole development. Proc. Natl. Acad. Sci. *113*, E3260–E3269. https://doi.org/10.1073/pnas.1522811113.
Matsuzawa, T., Kuwae, A., Yoshida, S., Sasakawa, C., and Abe, A. (2004). Enteropathogenic *Escherichia coli* activates the RhoA signaling pathway via the stimulation of GEF-H1. EMBO J. *23*, 3570–3582. https://doi.org/10.1038/sj.emboj.7600359.
McCoy-Simandle, K., Stewart, C.R., Dao, J., DebRoy, S., Rossier, O., Bryce, P.J., and Cianciotto, N.P. (2011). *Legionella pneumophila* type II secretion dampens the cytokine response of infected macrophages and epithelia. Infect. Immun. *79*, 1984–1997. https://doi.org/10.1128/IAI.01077-10.
McGhie, E.J., Hayward, R.D., and Koronakis, V. (2004). Control of actin turnover by a salmonella invasion protein. Mol. Cell *13*, 497–510. https://doi.org/10.1016/S1097-2765(04)00053-X.
McNamara, B.P., Koutsouris, A., O'Connell, C.B., Nougayréde, J.P., Donnenberg, M.S., and Hecht, G. (2001). Translocated EspF protein from enteropathogenic *Escherichia coli* disrupts host intestinal barrier function. J. Clin. Invest. *107*, 621–629. https://doi.org/10.1172/JCI11138.

McPhee, J.B., Mena, P., and Bliska, J.B. (2010). Delineation of regions of the Yersinia YopM protein required for interaction with the RSK1 and PRK2 host kinases and their requirement for interleukin-10 production and virulence. Infect. Immun. *78*, 3529–3539. https://doi.org/10.1128/IAI.00269-10.

Mehra, A., Zahra, A., Thompson, V., Sirisaengtaksin, N., Wells, A., Porto, M., Köster, S., Penberthy, K., Kubota, Y., Dricot, A., *et al.* (2013). *Mycobacterium tuberculosis* type VII secreted effector EsxH targets host ESCRT to impair trafficking. PLOS Pathog. *9*, e1003734. https://doi.org/10.1371/journal.ppat.1003734.

Ménard, R., Sansonetti, P., and Parsot, C. (1994). The secretion of the *Shigella flexneri* Ipa invasins is activated by epithelial cells and controlled by IpaB and IpaD. EMBO J. *13*, 5293–5302.

Michard, C., Sperandio, D., Baïlo, N., Pizarro-Cerdá, J., LeClaire, L., Chadeau-Argaud, E., Pombo-Grégoire, I., Hervet, E., Vianney, A., Gilbert, C., et al. (2015). The legionella kinase legk2 targets the arp2/3 complex to inhibit actin nucleation on phagosomes and allow bacterial evasion of the late endocytic pathway. MBio *6*, 1–14. https://doi.org/10.1128/mBio.00354-15.

Mishra, A.K., Del Campo, C.M., Collins, R.E., Roy, C.R., and Lambright, D.G. (2013). The *Legionella pneumophila* GTPase activating protein LepB accelerates Rab1 deactivation by a non-canonical hydrolytic mechanism. J. Biol. Chem. *288*, 24000–24011. https://doi.org/10.1074/jbc.M113.470625.

Mittal, R., Peak-Chew, S.Y., and McMahon, H.T. (2006). Acetylation of MEK2 and I kappa B kinase (IKK) activation loop residues by YopJ inhibits signaling. Proc. Natl. Acad. Sci. U.S.A. *103*, 18574–18579. https://doi.org/10.1073/pnas.0608995103.

Miyata, S.T., Kitaoka, M., Brooks, T.M., McAuley, S.B., and Pukatzki, S. (2011). *Vibrio cholerae* requires the type VI secretion system virulence factor vasx to kill dictyostelium discoideum. Infect. Immun. *79*, 2941–2949. https://doi.org/10.1128/IAI.01266-10.

Mougous, J.D., Cuff, M.E., Raunser, S., Shen, A., Zhou, M., Gifford, C.A., Goodman, A.L., Joachimiak, G., Ordoñez, C.L., Lory, S., et al. (2006). A virulence locus of *Pseudomonas aeruginosa* encodes a protein secretion apparatus. Science *312*, 1526–1530. https://doi.org/10.1126/science.1128393.

Murata, T., Delprato, A., Ingmundson, A., Toomre, D.K., Lambright, D.G., and Roy, C.R. (2006). The *Legionella pneumophila* effector protein DrrA is a Rab1 guanine nucleotide-exchange factor. Nat. Cell Biol. *8*, 971–977. https://doi.org/10.1038/ncb1463.

Myeni, S.K., and Zhou, D. (2010). The C terminus of SipC binds and bundles F-actin to promote Salmonella invasion. J. Biol. Chem. *285*, 13357–13363. https://doi.org/10.1074/jbc.M109.094045.

Myeni, S., Child, R., Ng, T.W., Kupko, J.J., Wehrly, T.D., Porcella, S.F., Knodler, L.A., and Celli, J. (2013). Brucella modulates secretory trafficking via multiple type IV secretion effector proteins. PLOS Pathog. *9*, e1003556. https://doi.org/10.1371/journal.ppat.1003556.

Nagamune, K., Yamamoto, K., Naka, A., Matsuyama, J., Miwatani, T., and Honda, T. (1996). In vitro proteolytic processing and activation of the recombinant precursor of El Tor cytolysin/hemolysin (pro-HlyA) of *Vibrio cholerae* by soluble hemagglutinin/protease of V. cholerae, trypsin, and other proteases. Infect. Immun. *64*, 4655–4658.

Nampiaparampil, D.E., and Shmerling, R.H. (2004). A review of fibromyalgia. Am. J. Manag. Care *10*, 794–800 2759 [pii].

Nans, A., Kudryashev, M., Saibil, H.R., and Hayward, R.D. (2015). Structure of a bacterial type III secretion system in contact with a host membrane in situ. Nat. Commun. *6*, 10114. https://doi.org/10.1038/ncomms10114.

Narita, Y., Sato, K., Yukitake, H., Shoji, M., Daisuke, N., Nagano, K., Yoshimura, F., Naito, M., and Nakayama, K. (2014). Lack of a surface layer in Tannerella forsythia mutants deficient in the type IX secretion system. Microbiology (United Kingdom) *160*, 2295–2303. https://doi.org/10.1099/mic.0.080192-0.

Newman, R.M., Salunkhe, P., Godzik, A., and Reed, J.C. (2006). Identification and characterization of a novel bacterial virulence factor that shares homology with mammalian Toll/interleukin-1 receptor family proteins. Infect. Immun. *74*, 594–601.

Newton, H.J., Pearson, J.S., Badea, L., Kelly, M., Lucas, M., Holloway, G., Wagstaff, K.M., Dunstone, M.A., Sloan, J., Whisstock, J.C., *et al.* (2010). The type III effectors NleE and NleB from enteropathogenic E. coli and OspZ from Shigella block nuclear translocation of NF-kappaB p65. PLOS Pathog. *6*, e1000898. https://doi.org/10.1371/journal.ppat.1000898.

Nicaud, J.M., Mackman, N., Gray, L., and Holland, I.B. (1986). The C-terminal, 23 kDa peptide of E. coli haemolysin 2001 contains all the information necessary for its secretion by the haemolysin (Hly) export machinery. FEBS Lett. *204*, 331–335. https://doi.org/10.1016/0014-5793(86)80838-9.

O'Callaghan, D. (1999). A homologue of the *Agrobacterium tumefaciens* VirB and Bordetella pertussis Ptl type IV secretion systems is essential for intracellular survival of *Brucella suis*. Mol. Microbiol. 33, 1210–1220.

Ochman, H., and Groisman, E.A. (1996). Distribution of pathogenicity islands in Salmonella spp. Infect. Immun. *64*, 5410–5412.

Odenbreit, S., Püls, J., Sedlmaier, B., Gerland, E., Fischer, W., and Haas, R. (2000). Translocation of *Helicobacter pylori* CagA into gastric epithelial cells by type IV secretion. Science *287*, 1497–1500. https://doi.org/10.1126/science.287.5457.1497.

Ogawa, M., Yoshimori, T., Suzuki, T., Sagara, H., Mizushima, N., and Sasakawa, C. (2005). Escape of intracellular Shigella from autophagy. Science *307*, 727–731. https://doi.org/10.1126/science.1106036.

Ohol, Y.M., Goetz, D.H., Chan, K., Shiloh, M.U., Craik, C.S., and Cox, J.S. (2010). *Mycobacterium tuberculosis* MycP1 protease plays a dual role in regulation of ESX-1 secretion and virulence. Cell Host Microbe 7, 210–220. https://doi.org/10.1016/j.chom.2010.02.006.

Okujava, R., Guye, P., Lu, Y.Y., Mistl, C., Polus, F., Vayssier-Taussat, M., Halin, C., Rolink, A.G., and Dehio, C. (2014). A Translocated Effector Required for Bartonella Dissemination from Derma to Blood Safeguards Migratory Host Cells from Damage by Co-translocated Effectors. PLOS Pathog. *10*. https://doi.org/10.1371/journal.ppat.1004187.

Olsén, A., Jonsson, A., and Normark, S. (1989). Fibronectin binding mediated by a novel class of surface organelles on *Escherichia coli*. Nature *338*, 652–655. https://doi.org/10.1038/338652a0.

Orchard, R.C., Kittisopikul, M., Altschuler, S.J., Wu, L.F., Süel, G.M., and Alto, N.M. (2012). Identification of F-actin as the dynamic hub in a microbial-induced GTPase polarity circuit. Cell *148*, 803–815. https://doi.org/10.1016/j.cell.2011.11.063.

Pallen, M.J. (2002). The ESAT-6/WXG100 superfamily – and a new Gram-positive secretion system? Trends Microbiol. *10*, 209–212.

Pallen, M.J., Beatson, S.A., and Bailey, C.M. (2005). Bioinformatics, genomics and evolution of non-flagellar type-III secretion systems: a Darwinian perspective. FEMS Microbiol. Rev. *29*, 201–229.

Pallen, M.J., Bailey, C.M., and Beatson, S.A. (2006). Evolutionary links between FliH/YscL-like proteins from bacterial type III secretion systems and second-stalk components of the FoF1 and vacuolar ATPases. Protein Sci. *15*, 935–941. https://doi.org/10.1126/science.273.5279.1231.

Paquette, N., Conlon, J., Sweet, C., Rus, F., Wilson, L., Pereira, a., Rosadini, C.V., Goutagny, N., Weber, a. N.R., Lane, W.S., et al. (2012). Serine/threonine acetylation of TGF-activated kinase (TAK1) by *Yersinia pestis* YopJ inhibits innate immune signaling. Proc. Natl. Acad. Sci. *109*, 12710–12715. https://doi.org/10.1073/pnas.1008203109.

Park, K.S., Ono, T., Rokuda, M., Jang, M.H., Okada, K., Iida, T., and Honda, T. (2004). Functional characterization of two type III secretion systems of Vibrio parahaemolyticus. Infect. Immun. *72*, 6659–6665. https://doi.org/10.1073/pnas.0706532104.

Parsot, C. (2009). Shigella type III secretion effectors: how, where, when, for what purposes? Curr. Opin. Microbiol. *12*, 110–116. https://doi.org/10.1016/j.mib.2008.12.002.

Pavloff, N., Potempa, J., Pike, R.N., Prochazka, V., Kiefer, M.C., Travis, J., and Barr, P.J. (1995). Molecular cloning and structural characterization of the Arg-gingipain proteinase of Porphyromonas gingivalis: Biosynthesis as a proteinase-adhesin polyprotein. J. Biol. Chem. *270*, 1007–1010.

Pawar, D.M., Rossman, M.L., and Chen, J. (2005). Role of curli fimbriae in mediating the cells of enterohaemorrhagic *Escherichia coli* to attach to abiotic surfaces. J. Appl. Microbiol. *99*, 418–425. https://doi.org/10.1073/pnas.94.3.802.

Pechmann, S., Chartron, J.W., and Frydman, J. (2014). Local slowdown of translation by nonoptimal codons promotes nascent-chain recognition by SRP in vivo. Nat. Struct. Mol. Biol. *21*, 1100–1105. https://doi.org/10.1038/nsmb.2919.

Persson, C., Carballeira, N., Wolf-Watz, H., and Fällman, M. (1997). The PTPase YopH inhibits uptake of Yersinia, tyrosine phosphorylation of p130(Cas) and FAK, and the associated accumulation of these proteins in peripheral focal adhesions. EMBO J. *16*, 2307–2318. https://doi.org/10.1093/emboj/16.9.2307.

Pettersson, J., Nordfelth, R., Dubinina, E., Bergman, T., Gustafsson, M., Magnusson, K.E., and Wolf-Watz, H. (1996). Modulation of virulence factor expression by pathogen target cell contact. Science *273*, 1231–1233. https://doi.org/10.1051/vetres/2009007.

Pizarro-Cerdá, J., Méresse, S., Parton, R.G., Van Der Goot, G., Sola-Landa, A., Lopez-Goñi, I., Moreno, E., and Gorvel, J.P. (1998). Brucella abortus transits through the autophagic pathway and replicates in the endoplasmic reticulum of nonprofessional phagocytes. Infect. Immun. *66*, 5711–5724.

Plaut, R.D., and Carbonetti, N.H. (2008). Retrograde transport of pertussis toxin in the mammalian cell. Cell. Microbiol. *10*, 1130–1139. https://doi.org/10.1111/j.1462-5822.2007.01115.x.

Plaut, A.G., Gilbert, J.V., Artenstein, M.S., and Capra, J.D. (1975). *Neisseria gonorrhoeae* and *Neisseria meningitidis*: extracellular enzyme cleaves human immunoglobulin A. Science *190*, 1103–1105. https://doi.org/10.1126/science.810892.

Pohlner, J., Halter, R., Beyreuther, K., and Meyer, T.F. (1987). Gene structure and extracellular secretion of *Neisseria gonorrhoeae* IgA protease. Nature *325*, 458–462. https://doi.org/10.1038/325458a0.

Pukatzki, S., Ma, A.T., Revel, A.T., Sturtevant, D., and Mekalanos, J.J. (2007). Type VI secretion system translocates a phage tail spike-like protein into target cells where it cross-links actin. Proc. Natl. Acad. Sci. U.S.A. *104*, 15508–15513. https://doi.org/10.1152/ajplung.00482.2004.

Pulliainen, A.T., Pieles, K., Brand, C.S., Hauert, B., Böhm, A., Quebatte, M., Wepf, A., Gstaiger, M., Aebersold, R., Dessauer, C.W., *et al.* (2012). Bacterial effector binds host cell adenylyl cyclase to potentiate Gαs-dependent cAMP production. Proc. Natl. Acad. Sci. U.S.A. *109*, 9581–9586. https://doi.org/10.1073/pnas.1117651109.

Qiu, J., Sheedlo, M.J., Yu, K., Tan, Y., Nakayasu, E.S., Das, C., Liu, X., and Luo, Z.Q. (2016). Ubiquitination independent of E1 and E2 enzymes by bacterial effectors. Nature *533*, 120–124. https://doi.org/10.1038/nature17657.

Radics, J., Königsmaier, L., and Marlovits, T.C. (2014). Structure of a pathogenic type 3 secretion system in action. Nat. Struct. Mol. Biol. *21*, 82–87. https://doi.org/10.1038/nsmb.2722.

Rana, R.R., Simpson, P., Zhang, M., Jennions, M., Ukegbu, C., Spear, A.M., Alguel, Y., Matthews, S.J., Atkins, H.S., and Byrne, B. (2011). *Yersinia pestis* TIR-domain protein forms dimers that interact with the human adaptor protein MyD88. Microb. Pathog. *51*, 89–95. https://doi.org/10.1016/j.micpath.2011.05.004.

Randall, L.L., Topping, T.B., Hardy, S.J., Pavlov, M.Y., Freistroffer, D.V., and Ehrenberg, M. (1997). Binding of SecB to ribosome-bound polypeptides has the same characteristics as binding to full-length, denatured proteins. Proc. Natl. Acad. Sci. U.S.A. *94*, 802–807.

Ratner, D., Orning, M.P.A., Starheim, K.K., Marty-Roix, R., Proulx, M.K., Goguen, J.D., and Lien, E. (2016). Manipulation of IL-1β and IL-18 production by *Yersinia pestis* effectors YopJ and YopM and redundant impact on virulence. J. Biol. Chem. jbc.M115.697698. https://doi.org/10.1074/jbc.M115.697698.

Reichow, S.L., Korotkov, K.V., Hol, W.G., and Gonen, T. (2010). Structure of the cholera toxin secretion channel in its closed state. Nat. Struct. Mol. Biol. *17*, 1226–1232. https://doi.org/10.1038/nsmb.1910.

Ren, C.P., Chaudhuri, R.R., Fivian, A., Bailey, C.M., Antonio, M., Barnes, W.M., and Pallen, M.J. (2004). The ETT2 gene cluster, encoding a second type III secretion system from *Escherichia coli,* is present in the majority of strains but has undergone widespread mutational attrition. J. Bacteriol. *186*, 3547–3560. https://doi.org/10.1128/JB.186.11.3547-3560.2004.

Rhomberg, T.A., Truttmann, M.C., Guye, P., Ellner, Y., and Dehio, C. (2009). A translocated protein of Bartonella henselae interferes with endocytic uptake of individual bacteria and triggers uptake of large bacterial aggregates via the invasome. Cell. Microbiol. *11*, 927–945. https://doi.org/10.1111/j.1462-5822.2009.01302.x.

Rolando, M., Sanulli, S., Rusniok, C., Gomez-Valero, L., Bertholet, C., Sahr, T., Margueron, R., and Buchrieser, C. (2013). *Legionella pneumophila* effector RomA uniquely modifies host chromatin to repress gene expression and promote intracellular bacterial replication. Cell Host Microbe *13*, 395–405. https://doi.org/10.1016/j.chom.2013.03.004.

Rolando, M., Escoll, P., Nora, T., Botti, J., Boitez, V., Bedia, C., Daniels, C., Abraham, G., Stogios, P.J., Skarina, T., *et al.* (2016). *Legionella pneumophila* S1P-lyase targets host sphingolipid metabolism and restrains autophagy. Proc. Natl. Acad. Sci. U.S.A. *113*, 1901–1906. https://doi.org/10.1073/pnas.1522067113.

Rosenberg, O.S., Dovala, D., Li, X., Connolly, L., Bendebury, A., Finer-Moore, J., Holton, J., Cheng, Y., Stroud, R.M., and Cox, J.S. (2015). Substrates Control Multimerization and Activation of the Multi-Domain ATPase Motor of Type VII Secretion. Cell *161*, 501–512. https://doi.org/10.1016/j.cell.2015.03.040.

Rosqvist, R., Bölin, I., and Wolf-Watz, H. (1988). Inhibition of phagocytosis in Yersinia pseudotuberculosis: a virulence plasmid-encoded ability involving the Yop2b protein. Infect. Immun. *56*, 2139–2143.

Rossiter, A.E., Leyton, D.L., Tveen-Jensen, K., Browning, D.F., Sevastsyanovich, Y., Knowles, T.J., Nichols, K.B., Cunningham, A.F., Overduin, M., Schembri, M.A., et al. (2011). The essential β-barrel assembly machinery complex components bamd and bama are required for autotransporter biogenesis. J. Bacteriol. *193*, 4250–4253. https://doi.org/10.1128/JB.00192-11.

Ruiz-Perez, F., Henderson, I.R., Leyton, D.L., Rossiter, A.E., Zhang, Y., and Nataro, J.P. (2009). Roles of periplasmic chaperone proteins in the biogenesis of serine protease autotransporters of Enterobacteriaceae. J. Bacteriol. *191*, 6571–6583. https://doi.org/10.1128/JB.00754-09.

Russo, T.A., Davidson, B.A., Genagon, S.A., Warholic, N.M., Macdonald, U., Pawlicki, P.D., Beanan, J.M., Olson, R., Holm, B.A., and Knight, P.R. (2005). E. coli virulence factor hemolysin induces neutrophil apoptosis and necrosis/lysis in vitro and necrosis/lysis and lung injury in a rat pneumonia model. Am. J. Physiol. Lung Cell. Mol. Physiol. *289*, L207–16.

Salacha, R., Kovacić, F., Brochier-Armanet, C., Wilhelm, S., Tommassen, J., Filloux, A., Voulhoux, R., and Bleves, S. (2010). The *Pseudomonas aeruginosa* patatin-like protein PlpD is the archetype of a novel Type V secretion system. Environ. Microbiol. *12*, 1498–1512. https://doi.org/10.1111/j.1462-2920.2010.02174.x.

Salcedo, S.P., Marchesini, M.I., Degos, C., Terwagne, M., Von Bargen, K., Lepidi, H., Herrmann, C.K., Santos Lacerda, T.L., Imbert, P.R.C., Pierre, P., et al. (2013). BtpB, a novel Brucella TIR-containing effector protein with immune modulatory functions. Front. Cell. Infect. Microbiol. *3*, 28. https://doi.org/10.3389/fcimb.2013.00028.

Sana, T.G., Baumann, C., Merdes, A., Soscia, C., Rattei, T., Hachani, A., Jones, C., Bennett, K.L., Filloux, A., Superti-Furga, G., *et al.* (2015). Internalization of *Pseudomonas aeruginosa* strain PAO1 into Epithelial cells is promoted by interaction of a T6SS effector with the microtubule network. MBio *6*, e00712. https://doi.org/10.1128/mBio.00712-15.

Sandkvist, M. (2001). Type II secretion and pathogenesis. Infect. Immun. *69*, 3523–3535. https://doi.org/10.1128/IAI.69.6.3523-3535.2001.

Sandkvist, M., Bagdasarian, M., Howard, S.P., and DiRita, V.J. (1995). Interaction between the autokinase EpsE and EpsL in the cytoplasmic membrane is required for extracellular secretion in *Vibrio cholerae*. EMBO J. *14*, 1664–1673.

Sandkvist, M., Michel, L.O., Hough, L.P., Morales, V.M., Bagdasarian, M., Koomey, M., Dirita, V.J., and Bagdasarian, M. (1997). General secretion pathway (eps) genes required for toxin secretion and outer membrane biogenesis in *Vibrio cholerae*. J. Bacteriol. *179*, 6994–7003. https://doi.org/10.1111/j.1745-6584.1984.tb01414.x.

Sato, K., Naito, M., Yukitake, H., Hirakawa, H., Shoji, M., McBride, M.J., Rhodes, R.G., and Nakayama, K. (2010a). A protein secretion system linked to bacteroidete gliding motility and pathogenesis. Proc. Natl. Acad. Sci. U.S.A. *107*, 276–281. https://doi.org/10.1073/pnas.0912010107.

Sato, K., Naito, M., Yukitake, H., Hirakawa, H., Shoji, M., McBride, M.J., Rhodes, R.G., and Nakayama, K. (2010b). A protein secretion system linked to bacteroidete gliding motility and pathogenesis. Proc. Natl. Acad. Sci. U.S.A. *107*, 276–281. https://doi.org/10.1073/pnas.0912010107.

Schmiederer, M., and Anderson, B. (2000). Cloning, sequencing, and expression of three *Bartonella henselae* genes homologous to the *Agrobacterium tumefaciens* VirB region. DNA Cell Biol. *19*, 141–147. https://doi.org/10.1089/104454900314528.

Schneewind, O., and Missiakas, D.M. (2012). Protein secretion and surface display in Gram-positive bacteria. Phil. Trans. R. Soc. Lond. B. Biol. Sci. *367*, 1123–1139. https://doi.org/10.1098/rstb.2011.0210.

Schraidt, O., and Marlovits, T.C. (2011). Three-dimensional model of Salmonella's needle complex at subnanometer resolution. Science *331*, 1192–1195. https://doi.org/10.1126/science.1199358.

Schulein, R., Guye, P., Rhomberg, T.A., Schmid, M.C., Schröder, G., Vergunst, A.C., Carena, I., and Dehio, C. (2005). A bipartite signal mediates the transfer of type IV secretion substrates of Bartonella henselae into human cells. Proc. Natl. Acad. Sci. U.S.A. *102*, 856–861. https://doi.org/10.1073/pnas.0406796102.

Schwarz, S., Singh, P., Robertson, J.D., LeRoux, M., Skerrett, S.J., Goodlett, D.R., West, T.E., and Mougous, J.D. (2014). VgrG-5 is a Burkholderia type VI secretion system-exported protein required for multinucleated giant cell formation and virulence. Infect. Immun. *82*, 1445–1452. https://doi.org/10.1128/IAI.01368-13.

Segal, G., and Shuman, H.A. (1997). Characterization of a new region required for macrophage killing by *Legionella pneumophila*. Infect. Immun. *65*, 5057–5066.

Segal, G., Feldman, M., and Zusman, T. (2005). The Icm/Dot type-IV secretion systems of *Legionella pneumophila* and *Coxiella burnetii*. FEMS Microbiol. Rev. *29*, 65–81. https://doi.org/10.1016/j.femsre.2004.07.001.

Selbach, M., Moese, S., Hauck, C.R., Meyer, T.F., and Backert, S. (2002). Src is the kinase of the *Helicobacter pylori* CagA protein in vitro and in vivo. J. Biol. Chem. *277*, 6775–6778. https://doi.org/10.1074/jbc.C100754200.

Selyunin, A.S., Sutton, S.E., Weigele, B.A., Reddick, L.E., Orchard, R.C., Bresson, S.M., Tomchick, D.R., and Alto, N.M. (2011). The assembly of a GTPase-kinase signalling complex by a bacterial catalytic scaffold. Nature *469*, 107–111. https://doi.org/10.1038/nature09593.

Sender, R., Fuchs, S., and Milo, R. (2016). Revised estimates for the number of human and bacteria cells in the body. PLOS Biol. *14*, e1002533. https://doi.org/10.1371/journal.pbio.1002533.

Serafini, A., Boldrin, F., Palù, G., and Manganelli, R. (2009). Characterization of a *Mycobacterium tuberculosis* ESX-3 conditional mutant: essentiality and rescue by iron and zinc. J. Bacteriol. *191*, 6340–6344. https://doi.org/10.1128/JB.00756-09.

Shamaei-Tousi, A., Cahill, R., and Frankel, G. (2004). Interaction between protein subunits of the type IV secretion system of *Bartonella henselae*. J. Bacteriol. *186*, 4796–4801. https://doi.org/10.1128/JB.186.14.4796-4801.2004.

Shaw, R.K., Smollett, K., Cleary, J., Garmendia, J., Straatman-Iwanowska, A., Frankel, G., and Knutton, S. (2005). Enteropathogenic *Escherichia coli* type III effectors EspG and EspG2 disrupt the microtubule network of intestinal epithelial cells. Infect. Immun. *73*, 4385–4390. https://doi.org/10.1128/IAI.73.7.4385-4390.2005.

Sheahan, K.L., and Satchell, K.J. (2007). Inactivation of small Rho GTPases by the multifunctional RTX toxin from *Vibrio cholerae*. Cell. Microbiol. *9*, 1324–1335.

Sheahan, K.L., Cordero, C.L., and Satchell, K.J. (2004). Identification of a domain within the multifunctional *Vibrio cholerae* RTX toxin that covalently cross-links actin. Proc. Natl. Acad. Sci. U.S.A. *101*, 9798–9803. https://doi.org/10.1073/pnas.0401104101.

Shere, K.D., Sallustio, S., Manessis, A., D'Aversa, T.G., and Goldberg, M.B. (1997). Disruption of IcsP, the major *Shigella* protease that cleaves IcsA, accelerates actin-based motility. Mol. Microbiol. *25*, 451–462.

Shneider, M.M., Buth, S.A., Ho, B.T., Basler, M., Mekalanos, J.J., and Leiman, P.G. (2013). PAAR-repeat proteins sharpen and diversify the type VI secretion system spike. Nature *500*, 350–353. https://doi.org/10.1038/nature12453.

Sijbrandi, R., Urbanus, M.L., Ten Hagen-Jongman, C.M., Bernstein, H.D., Oudega, B., Otto, B.R., and Luirink, J. (2003). Signal recognition particle (SRP)-mediated targeting and Sec-dependent translocation of an extracellular *Escherichia coli* protein. J. Biol. Chem. *278*, 4654–4659. https://doi.org/10.1074/jbc.M211630200.

Simeone, R., Bobard, A., Lippmann, J., Bitter, W., Majlessi, L., Brosch, R., and Enninga, J. (2012). Phagosomal rupture by *Mycobacterium tuberculosis* results in toxicity and host cell death. PLOS Pathog. *8*, e1002507. https://doi.org/10.1371/journal.ppat.1002507.

Smith, J., Manoranjan, J., Pan, M., Bohsali, A., Xu, J., Liu, J., McDonald, K.L., Szyk, A., LaRonde-LeBlanc, N., and Gao, L.Y. (2008). Evidence for pore formation in host cell membranes by ESX-1-secreted ESAT-6 and its role in Mycobacterium marinum escape from the vacuole. Infect. Immun. *76*, 5478–5487. https://doi.org/10.1128/IAI.00614-08.

Solomonson, M., Setiaputra, D., Makepeace, K.A., Lameignere, E., Petrotchenko, E.V., Conrady, D.G., Bergeron, J.R., Vuckovic, M., DiMaio, F., Borchers, C.H., *et al.* (2015). Structure of EspB from the ESX-1 type VII secretion system and insights into its export mechanism. Structure *23*, 571–583. https://doi.org/10.1016/j.str.2015.01.002.

Souza, D.P., Oka, G.U., Alvarez-Martinez, C.E., Bisson-Filho, A.W., Dunger, G., Hobeika, L., Cavalcante, N.S., Alegria, M.C., Barbosa, L.R., Salinas, R.K., *et al.* (2015). Bacterial killing via a type IV secretion system. Nat. Commun. *6*, 6453. https://doi.org/10.1038/ncomms7453.

Stanley, P., Packman, L.C., Koronakis, V., and Hughes, C. (1994). Fatty acylation of two internal lysine residues required for the toxic activity of *Escherichia coli* hemolysin. Science *266*, 1992–1996. https://doi.org/10.1126/science.7801126.

Starr, T., Child, R., Wehrly, T.D., Hansen, B., Hwang, S., López-Otin, C., Virgin, H.W., and Celli, J. (2012). Selective subversion of autophagy complexes facilitates completion of the *Brucella* intracellular cycle. Cell Host Microbe *11*, 33–45. https://doi.org/10.1016/j.chom.2011.12.002.

Stender, S., Friebel, A., Linder, S., Rohde, M., Mirold, S., and Hardt, W.D. (2000). Identification of SopE2 from Salmonella typhimurium, a conserved guanine nucleotide exchange factor for Cdc42 of the host cell. Mol. Microbiol. *36*, 1206–1221. https://doi.org/10.1046/j.1365-2958.2000.01933.x.

Strozen, T.G., Stanley, H., Gu, Y., Boyd, J., Bagdasarian, M., Sandkvist, M., and Howard, S.P. (2011). Involvement of the GspAB complex in assembly of the type II secretion system secretin of Aeromonas and Vibrio species. J. Bacteriol. *193*, 2322–2331. https://doi.org/10.1128/JB.01413-10.

Suarez, G., Sierra, J.C., Erova, T.E., Sha, J., Horneman, A.J., and Chopra, A.K. (2010). A type VI secretion system effector protein, VgrG1, from Aeromonas hydrophila that induces host cell toxicity by ADP ribosylation of actin. J. Bacteriol. *192*, 155–168. https://doi.org/10.1128/JB.01260-09.

Suzuki, T., Lett, M.C., and Sasakawa, C. (1995). Extracellular transport of VirG protein in *Shigella*. J. Biol. Chem. *270*, 30874–30880. https://doi.org/10.1074/jbc.270.52.30874.

Thomas, S., Bakkes, P.J., Smits, S.H., and Schmitt, L. (2014). Equilibrium folding of pro-HlyA from *Escherichia coli* reveals a stable calcium ion dependent folding intermediate. Biochim. Biophys. Acta. *1844*, 1500–1510. https://doi.org/10.1016/j.bbapap.2014.05.006.

Toesca, I.J., French, C.T., and Miller, J.F. (2014). The type VI secretion system spike protein VgrG5 mediates membrane fusion during intercellular spread by pseudomallei group Burkholderia species. Infect. Immun. *82*, 1436–1444. https://doi.org/10.1128/IAI.01367-13.

Toulabi, L., Wu, X., Cheng, Y., and Mao, Y. (2013). Identification and structural characterization of a Legionella phosphoinositide phosphatase. J. Biol. Chem. *288*, 24518–24527. https://doi.org/10.1074/jbc.M113.474239.

Truttmann, M.C., Guye, P., and Dehio, C. (2011). BID-F1 and BID-F2 domains of *Bartonella henselae* effector protein BepF trigger together with BepC the formation of invasome structures. PLOS ONE *6*, e25106. https://doi.org/10.1371/journal.pone.0025106.

Uhlén, P., Laestadius, A., Jahnukainen, T., Söderblom, T., Bäckhed, F., Celsi, G., Brismar, H., Normark, S., Aperia, A., and Richter-Dahlfors, A. (2000a). Alpha-haemolysin of uropathogenic *E. coli* induces Ca^{2+} oscillations in renal epithelial cells. Nature *405*, 694–697. https://doi.org/10.1038/35015091.

Uhlén, P., Laestadius, A., Jahnukainen, T., Söderblom, T., Bäckhed, F., Celsi, G., Brismar, H., Normark, S., Aperia, A., and Richter-Dahlfors, A. (2000b). Alpha-haemolysin of uropathogenic *E. coli* induces Ca2+ oscillations in renal epithelial cells. Nature *405*, 694–697. https://doi.org/10.1038/35015091.

van der Wel, N., Hava, D., Houben, D., Fluitsma, D., van Zon, M., Pierson, J., Brenner, M., and Peters, P.J. (2007). M. tuberculosis and M. leprae translocate from the phagolysosome to the cytosol in myeloid cells. Cell *129*, 1287–1298. https://doi.org/10.1016/j.cell.2007.05.059.

van Ulsen, P., van Alphen, L., ten Hove, J., Fransen, F., van der Ley, P., and Tommassen, J. (2003). A Neisserial autotransporter NalP modulating the processing of other autotransporters. Mol. Microbiol. *50*, 1017–1030. https://doi.org/10.1046/j.1365-2958.2003.03773.x.

van Ulsen, P., Rahman, S.u., Jong, W.S., Daleke-Schermerhorn, M.H., and Luirink, J. (2014). Type V secretion: from biogenesis to biotechnology. Biochim. Biophys. Acta. *1843*, 1592–1611. https://doi.org/10.1016/j.bbamcr.2013.11.006.

Vance, R.E., Isberg, R.R., and Portnoy, D.A. (2009). Patterns of pathogenesis: discrimination of pathogenic and nonpathogenic microbes by the innate immune system. Cell Host Microbe *6*, 10–21. https://doi.org/10.1016/j.chom.2009.06.007.

Veith, P.D., Nor Muhammad, N.A., Dashper, S.G., Likić, V.A., Gorasia, D.G., Chen, D., Byrne, S.J., Catmull, D.V., and Reynolds, E.C. (2013). Protein substrates of a novel secretion system are numerous in the Bacteroidetes phylum and have in common a cleavable C-terminal secretion signal, extensive post-translational modification, and cell-surface attachment. J. Proteome Res. *12*, 4449–4461. https://doi.org/10.1021/pr400487b.

Vetter, I.R., and Wittinghofer, A. (2001). The guanine nucleotide-binding switch in three dimensions. Science *294*, 1299–1304. https://doi.org/10.1126/science.1062023.

Vincent, C.D., Friedman, J.R., Jeong, K.C., Buford, E.C., Miller, J.L., and Vogel, J.P. (2006). Identification of the core transmembrane complex of the Legionella Dot/Icm type IV secretion system. Mol. Microbiol. *62*, 1278–1291. https://doi.org/10.1111/j.1365-2958.2006.05446.x.

Vincent, T.S., Fraylick, J.E., McGuffie, E.M., and Olson, J.C. (1999). ADP-ribosylation of oncogenic Ras proteins by *pseudomonas aeruginosa* exoenzyme S in vivo. Mol. Microbiol. 32, 1054–1064. https://doi.org/10.1046/j.1365-2958.1999.01420.x.

Vojtova, J., Kamanova, J., and Sebo, P. (2006). Bordetella adenylate cyclase toxin: a swift saboteur of host defense. Curr. Opin. Microbiol. *9*, 69–75. https://doi.org/10.1016/j.mib.2005.12.011.

Vonaesch, P., Sellin, M.E., Cardini, S., Singh, V., Barthel, M., and Hardt, W.D. (2014). The SalmonellaTyphimurium effector protein SopE transiently localizes to the early SCV and contributes to intracellular replication. Cell. Microbiol. *16*, 1723–1735. https://doi.org/10.1111/cmi.12333.

Voulhoux, R., Ball, G., Ize, B., Vasil, M.L., Lazdunski, A., Wu, L.F., and Filloux, A. (2001). Involvement of the twin-arginine translocation system in protein secretion via the type II pathway. EMBO J. *20*, 6735–6741. https://doi.org/10.1093/emboj/20.23.6735.

Wagner, J.M., Evans, T.J., and Korotkov, K.V. (2014). Crystal structure of the N-terminal domain of $EccA_1$ ATPase from the ESX-1 secretion system of *Mycobacterium tuberculosis*. Proteins *82*, 159–163. https://doi.org/10.1002/prot.24351.

Wagner, J.M., Chan, S., Evans, T.J., Kahng, S., Kim, J., Arbing, M.A., Eisenberg, D., and Korotkov, K.V. (2016). Structures of EccB1 and EccD1 from the core complex of the mycobacterial ESX-1 type VII secretion system. BMC Struct. Biol. *16*, 5. https://doi.org/10.1186/s12900-016-0056-6.

Wang, T., Si, M., Song, Y., Zhu, W., Gao, F., Wang, Y., Zhang, L., Zhang, W., Wei, G., Luo, Z.Q., *et al.* (2015). Type VI Secretion System Transports Zn2+ to Combat Multiple Stresses and Host Immunity. PLOS Pathog. *11*, e1005020. https://doi.org/10.1371/journal.ppat.1005020.

Ward, J.E., Akiyoshi, D.E., Regier, D., Datta, A., Gordon, M.P., and Nester, E.W. (1988). Characterization of the virB operon from an *Agrobacterium tumefaciens* Ti plasmid. J. Biol. Chem. *263*, 5804–5814.

Waters, C.M., and Bassler, B.L. (2005). Quorum sensing: cell-to-cell communication in bacteria. Annu. Rev. Cell Dev. Biol. *21*, 319–346. https://doi.org/10.1146/annurev.cellbio.21.012704.131001.

Weber, M.M., Faris, R., van Schaik, E.J., McLachlan, J., Wright, W.U., Tellez, A., Roman, V.A., Rowin, K., Di Russo Case, E., Luo, Z.-Q., et al. (2016). The type IV secreted effector protein CirA stimulates the GTPase activity of RhoA and is required for virulence in a mouse model of *Coxiella burnetii* infection. Infect. Immun. IAI.01554-15. https://doi.org/10.1128/IAI.01554-15.

Weiner, J.H., Bilous, P.T., Shaw, G.M., Lubitz, S.P., Frost, L., Thomas, G.H., Cole, J.A., and Turner, R.J. (1998). A novel and ubiquitous system for membrane targeting and secretion of cofactor-containing proteins. Cell 93, 93–101. https://doi.org/10.1016/S0092-8674(00)81149-6.

Welch, R.A., Forestier, C., Lobo, A., Pellett, S., Thomas, W., and Rowe, G. (1992). The synthesis and function of the *Escherichia coli* hemolysin and related RTX exotoxins. FEMS Microbiol. Immunol. *5*, 29–36. https://doi.org/10.1016/0378-1097(92)90072-V.

Westermann, M., Pop, O.I., Gerlach, R., Appel, T.R., Schlörmann, W., Schreiber, S., and Müller, J.P. (2006). The TatAd component of the Bacillus subtilis twin-arginine protein transport system forms homo-multimeric complexes in its cytosolic and membrane embedded localisation. Biochim. Biophys. Acta – Biomembr. *1758*, 443–451. https://doi.org/10.1016/j.bbamem.2006.03.018.

White, R.C., and Cianciotto, N.P. (2016). Type II secretion is necessary for the optimal association of the *Legionella*-containing Vacuole with macrophage Rab1B but mainly enhances intracellular replication by Rab1B-independent mechanisms. Infect. Immun. *84*, IAI.00750-16. https://doi.org/10.1128/IAI.00750-16.

Yadav, M., Zhang, J., Fischer, H., Huang, W., Lutay, N., Cirl, C., Lum, J., Miethke, T., and Svanborg, C. (2010). Inhibition of TIR domain signaling by TcpC: MyD88-dependent and independent effects on *Escherichia coli* virulence. PLOS Pathog. *6*, e1001120. https://doi.org/10.1371/journal.ppat.1001120.

Yahr, T.L., Goranson, J., and Frank, D.W. (1996). Exoenzyme S of *Pseudomonas aeruginosa* is secreted by a type III pathway. Mol. Microbiol. *22*, 991–1003. https://doi.org/10.1046/j.1365-2958.1996.01554.x.

Yang, J.G., and Rees, D.C. (2015). The allosteric regulatory mechanism of the *Escherichia coli* MetNI methionine ATP binding cassette (ABC) transporter. J. Biol. Chem. *290*, 9135–9140. https://doi.org/10.1074/jbc.M114.603365.

Yoshida, S., Katayama, E., Kuwae, A., Mimuro, H., Suzuki, T., and Sasakawa, C. (2002). *Shigella* deliver an effector protein to trigger host microtubule destabilization, which promotes Rac1 activity and efficient bacterial internalization. EMBO J. *21*, 2923–2935. https://doi.org/10.1093/emboj/cdf319.

Yu, X.J., McGourty, K., Liu, M., Unsworth, K.E., and Holden, D.W. (2010). pH sensing by intracellular *Salmonella* induces effector translocation. Science *328*, 1040–1043. https://doi.org/10.1126/science.1189000.

Zarivach, R., Deng, W., Vuckovic, M., Felise, H.B., Nguyen, H.V., Miller, S.I., Finlay, B.B., and Strynadka, N.C. (2008). Structural analysis of the essential self-cleaving type III secretion proteins EscU and SpaS. Nature 453, 124–127. https://doi.org/10.1038/nature06832.

Zhang, W., Rong, C., Chen, C., and Gao, G.F. (2012). Type-IVC secretion system: a novel subclass of type IV secretion system (T4SS) common existing in gram-positive genus Streptococcus. PLOS ONE 7, e46390. https://doi.org/10.1371/journal.pone.0046390.

Zhang, X.L., Li, D.F., Fleming, J., Wang, L.W., Zhou, Y., Wang, D.C., Zhang, X.E., and Bi, L.J. (2015). Core component EccB1 of the *Mycobacterium tuberculosis* type VII secretion system is a periplasmic ATPase. FASEB J. *29*, 4804–4814. https://doi.org/10.1096/fj.15-270843.

Zhou, D., Mooseker, M.S., and Galán, J.E. (1999). An invasion-associated *Salmonella* protein modulates the actin-bundling activity of plastin. Proc. Natl. Acad. Sci. U.S.A. *96*, 10176–10181. https://doi.org/10.1073/pnas.96.18.10176.

Zhou, H., Monack, D.M., Kayagaki, N., Wertz, I., Yin, J., Wolf, B., and Dixit, V.M. (2005). *Yersinia* virulence factor YopJ acts as a deubiquitinase to inhibit NF-kappa B activation. J. Exp. Med. *202*, 1327–1332. https://doi.org/10.1084/jem.20051194.

Zhou, L., Tan, A., and Hershenson, M.B. (2004). Yersinia YopJ inhibits pro-inflammatory molecule expression in human bronchial epithelial cells. Respir. Physiol. Neurobiol. *140*, 89–97. https://doi.org/10.1016/j.resp.2003.12.003.

Zoued, A., Durand, E., Bebeacua, C., Brunet, Y.R., Douzi, B., Cambillau, C., Cascales, E., and Journet, L. (2013). TssK is a trimeric cytoplasmic protein interacting with components of both phage-like and membrane anchoring complexes of the type VI secretion system. J. Biol. Chem. *288*, 27031–27041. https://doi.org/10.1074/jbc.M113.499772.

Zoued, A., Durand, E., Brunet, Y.R., Spinelli, S., Douzi, B., Guzzo, M., Flaugnatti, N., Legrand, P., Journet, L., Fronzes, R., *et al.* (2016). Priming and polymerization of a bacterial contractile tail structure. Nature *531*, 59–63. https://doi.org/10.1038/nature17182

Within-host Envelope Remodelling and its Impact in Bacterial Pathogen Recognition

2

M. Graciela Pucciarelli[1,2] and Francisco García-del Portillo[1*]

[1]Laboratory of Intracellular Bacterial Pathogens, Department of Microbial Biotechnology, Centro Nacional de Biotecnología-Consejo Superior de Investigaciones Científicas (CNB-CSIC), Madrid, Spain.
[2]Centro de Biología Molecular Severo Ochoa (CBMSO-CSIC), Department de Molecular Biology, Universidad Autónoma de Madrid, Madrid, Spain.

*Correspondence: fgportillo@cnb.csic.es

http://doi.org/10.21775/9781910190692.02

Abstract

Following colonization of host tissues, bacterial pathogens encounter new niches in which they must gain access to nutrients and cope with stresses and defence signals generated by the host. For some pathogens, the adaptation to a new 'within-host' lifestyle involves modifications of envelope components that bear molecular patterns normally recognized by the host innate immune system. These new modified patterns limit host recognition, therefore promoting immune evasion and pathogenicity. In this review, we describe how envelope components like the peptidoglycan or lipopolysaccharide can be altered within the host to impair responses triggered by pattern recognition receptors (PRRs). We also discuss the few cases reported to date of chemical modifications that occur in the envelope of some intracellular bacterial pathogens when they reside inside eukaryotic cells. These envelope alterations may have evolved due to the sentinel role performed by PRRs over pathogen-specific molecular patterns. The available data indicate that only selected pathogens seem to evade recognition due to 'within-host' envelope changes, with most of them displaying such patterns also in non-host environments. Given the importance of these alterations, future studies should focus in the responsible pathogen regulators – most yet unknown – that could be targeted to prevent immune evasion.

Introduction

Bacterial pathogens confront a plethora of adverse conditions and diverse defences imposed by the host after their initial encounter. Both partners continuously evolve based in this confrontation and this fact is illustrated by the multiple mechanisms of immune evasion characterized to date. Immune evasion encompasses strategies directed to target key elements of the host defence machinery ('direct attack' strategy) or to interfere with host recognition ('masking' strategy). In this review, we mostly discuss cases of impaired recognition, including those examples derived from changes in cell envelope components that occur 'within the host'. We emphasize changes in molecular patterns of these components that are normally detected by dedicated host receptors and, therefore, understood as changes that has consequences such as immune evasion.

Due to the bulk of data accumulated in the recent years, the review describes immune evasion exclusively in bacterial pathogens. We nevertheless recommend to the reader other recent reviews that address immune evasion in viral (Beachboard and Horner, 2016; Messaoudi *et al.*, 2015); parasitic (Flávia Nardy *et al.*, 2015; Nakada-Tsukui and Nozaki, 2016; Radolf *et al.*, 2016; Stijlemans *et al.*, 2016; Wright and Rayner, 2014); and fungal infections (Jiménez-López and Lorenz, 2013). Of hallmark importance is also the immune evasion associated to tumour development in various types of cancer (van der Burg *et al.*, 2016).

Immune evasion strategies based on a 'direct attack' to innate host defences

Mammalian hosts recognize intruder organisms using innate defences that act as primary barrier to control infections. These defences include the complement cascade and the phagocytosis by activated macrophages and dendritic cells. Many pathogen proteins alter in one or another way diverse activities in the complement and can interfere with the phagocytic process. These strategies are mostly exploited by some Gram-positive pathogens, which use numerous immune evasion proteins either located in the pathogen surface or actively secreted to the milieu (Koymans *et al.*, 2016). The literature provides many examples of blockage of immune receptors, proteases or antimicrobial peptides; active degradation of immune receptors, antimicrobial peptides or DNA (e.g. neutrophil extracellular traps, NET); host cell lysis triggered by a variety of toxins; interference in the coagulation process and overstimulation of host defences with proteins acting as super-antigens (Foster *et al.*, 2014; Koymans *et al.*, 2016). In the specific case of *Staphylococcus aureus*, it has been estimated that the pathogen could use up to 35 distinct secreted proteins to evade host immunity. A few examples include inhibitors of neutrophil serine proteases (Stapels *et al.*, 2016); surface proteins that impede opsono-phagocytosis and, in addition, act as super-antigens (Pauli *et al.*, 2014); proteins that induce tumour-necrosis factor 1 receptor (TNFR1) in neutrophils, which neutralizes circulating TNF-α (Hong *et al.*, 2016); proteins promoting apoptosis in B cells (Hong *et al.*, 2016); and, proteins that assemble into 'sequestering' complexes with factor H (FH) and the complement protein C3b (Amdahl *et al.*, 2013). A similar 'trapping' mechanism on complement proteins has been shown for the surface protein Fhb of *Streptococcus suis* (Li *et al.*, 2016). Group B streptococci have also been shown to use an ectonucleotidase, CdnP, to degrade cyclic nucleotides such as cyclic-di-AMP that stimulate the cGAS-STING cascade involved in production of interferon beta (INF-β) (Andrade *et al.*, 2016). For a more complete analysis of these evasion strategies falling in the category of 'direct attack' we recommend excellent recent reviews (Foster *et al.*, 2014; Geoghegan and Foster, 2015).

The masking strategy: immune evasion based on the interference with host receptors recognizing microbe-associated molecular patterns

The Toll-like receptor family

The innate immune system also relies in receptors that recognize microbe-associated molecular patterns (MAMP). These 'pattern recognition receptors' (PRR) differentiate based on the specific molecular pattern they recognize and their subcellular location (Caruso *et al.*, 2014; Strober *et al.*, 2006). An important class of PRR are the membrane-embedded Toll-like receptor (TLR) family, which bind pathogen's molecules either at the plasma membrane or in membranes of endocytic compartments (Strober *et al.*, 2006).

The most extensively characterized member of the TLR family is TLR4, which, together with its co-receptor MD2, recognizes the lipid A component of the lipopolysaccharide, LPS (Maeshima and Fernandez, 2013). It has been proposed that individual LPS molecules are those that are recognized by TLR4, which implies previous extraction from supramolecular lamellar membrane aggregates released from the outer membrane (Gioannini *et al.*, 2004). Other studies point to a contribution of the aggregate structure to TLR4 stimulation (Zughaier *et al.*, 2007). LPS is extracted from bacterial membranes by LPS-binding protein (LPB), a serum protein that transfers the cargo to the surface protein CD14. It is at this stage when CD14 subsequently deliver LPS to facilitate assembly of the TLR–LPS–MD2 tripartite complex (Maeshima and Fernandez, 2013). Following LPS binding at the host cells surface, the complex dimerizes to stimulate downstream signalling by interaction with the adaptor proteins TIRAP and MyD88. The TLR–LPS–MD2 complex is also know to signal once it has been internalized in endosomes, a process involving the adaptor proteins TRAM and TRIF (Maeshima and Fernandez, 2013). The stimulation derived from the plasma membrane leads to 'early' activation of the host pro-inflammatory regulator nuclear factor kappa B (NFκB) whereas the activation of the TRAM-TRIF pathway results in type I interferon production and a 'delayed' NFκB response. In both pathways, the cytosolic region of TLR4 binds to the distinct adaptor proteins once the TLR4–MD2-LPS complex is formed. Structural studies with distinct lipid A variants have demonstrated distinct signalling fluxes depending on the stability of this tripartite complex (Maeshima and Fernandez, 2013). Changes in lipid A structure are, in fact, a common strategy used by some Gram-negative pathogens to dampen first lines of host innate immune response (see Maeshima and Fernandez, 2013; Needham and Trent, 2013).

Besides TLR4, other important cell surface TLR involved in recognizing MAMP include those targeting bacterial lipoproteins (TLR1, TLR2 and TLR6) and flagellin (TLR5). Intracellular TLR located in the membrane of endosomal compartments include TLR3, TLR7, TLR8 and TLR9, which recognize nucleic acids from virus or bacteria (Kawasaki *et al.*, 2011).

The NLR receptor family

A second major line of innate immune defence involving PRR are the members of the nucleotide-binding oligomerization domain (NOD)-like receptor (NLR) family. NLR locate in intracellular locations where they recognize diverse molecules of invasive pathogens (Caruso *et al.*, 2014; Franchi *et al.*, 2009). NLR have three well defined regions consisting of a C-terminal leucine rich repeat (LRR) domain, a central nucleotide-binding domain (NACTH) and a variable N-terminal domain. This latter domain differentiates the distinct NLR subfamilies, named NLRA, NLRB, NLRC and NLRP (Feerick and McKernan, 2017). The respective N-terminal domains bear an acidic transactivation domain (NLRA subfamily); baculovirus inhibitor repeat (NLRB subfamily); caspase recruitment domain, CARD (NLRC subfamily); or a pyrin domain (NLRP subfamily).

This review will focus mainly in the NLRC receptors NOD1 and NOD2, which bear CARD domains and recognize peptidoglycan (PG)-derived fragments, triggering production of inflammatory cytokines and antimicrobial molecules (Caruso *et al.*, 2014; Kaparakis-Liaskos, 2015; Strober *et al.*, 2006). A large number of data have confirmed a discriminating ligand specificity for these two receptors (Chamaillard *et al.*, 2003; Girardin *et al.*, 2003; Hasegawa *et al.*, 2006). Thus, NOD1 recognizes PG fragments bearing the γ-D-glutamy l-meso-diaminopimelic (DAP) (iE-DAP) motif present in stem peptides of PG from many Gram-negative bacteria and some Gram-positive bacteria such as *Listeria monocytogenes* and *Bacillus subtilis*. NOD2 recognizes N-acetyl-muramyl-L-alanyl-γ–D-glutamic acid (muramyl-di-peptide, MDP), which is widely conserved in PG of many Gram-positive and Gram-negative bacteria. Whereas NOD1 is expressed ubiquitously in all cell types, NOD2 expression is restricted to immune cells, including macrophages, dendritic cells, neutrophils and T cells (Kaparakis-Liaskos, 2015; Moreira and Zamboni, 2012). NOD2 expression can increase in response to stimuli such as LPS, tumour necrosis factor (TNF) or interferon gamma (INF-γ).

NOD1 and NOD2 reside in an auto-inhibited monomeric state in the cytosol. Upon ligand binding, they multimerize in a state that exposes their CARD domains for interacting with the receptor-interacting serine/threonine kinase 2 (RIPK2) (Caruso *et al.*, 2014). Despite their cytosolic location, recent studies have shown that recognition of PG fragments mediated by NOD1/NOD2 and their interaction with RIPK2 occurs in association to endomembranes (Irving *et al.*, 2014; Nakamura *et al.*, 2014). Thus, NOD1 binds PG fragments internalized via bacterial outer membrane vesicles (OMV) and interacts with RIPK2 in early endosomes (Irving *et al.*, 2014). PG recognition mediated by NOD1 can also promote the formation of autophagosomes in a RIPK2-dependent manner (Irving *et al.*, 2014). Intriguingly, stimulation of autophagy was observed exclusively in non-immune cells, underscoring the importance of this innate defence mechanism in non-phagocytic cells invaded by intracellular bacterial pathogens. Unlike NOD1, NOD2 recognizes PG and interacts with RIPK2 in endosomal platforms enriched in the late endosome-lysosomal markers LAMP1 and LAMP2 (Nakamura *et al.*, 2014). Of interest, this latter study also demonstrated the presence in these endosomes of transporters such as SLC15A3 and SL15A4, capable of transporting the NOD2 ligand MDP from *Salmonella*-containing phagosomes to the cytosol (Nakamura *et al.*, 2014). Therefore, PG fragments can reach the host cell cytosol in infections caused by invasive intracellular pathogens directly, as it is the case of pathogens adapted to a cytosolic lifestyle; or, indirectly, following transport from the phagosomal compartment by host transporters such as SLC15A3 and SL15A4. These transporters were reported also to contribute to release of PG fragments to the cytosol following uptake of OMVs or endocytosis of soluble PG fragments (Lee *et al.*, 2009). An alternative route for delivery of PG fragments to the cytosol is supposed to exist due to the connection demonstrated between the activity of specialized type III and IV secretion systems and NOD1-dependent host responses. This association was reported in infections caused by extracellular pathogens such as *Citrobacter rodentium* (LeBlanc *et al.*, 2008) and *Helicobacter pylori* (Viala *et al.*, 2004).

A third line of innate defence involved in recognition of pathogen molecules: the inflammasome

In addition to the important role played by TLR and NLR receptors in the control of bacterial infections, mammals have developed additional innate host defences to monitor intracellular infections. This defence mechanism relies in supramacromolecular platforms, named inflammasomes, which upon recognition of specific pathogen molecules activate the pro-inflammatory caspases caspase-1 and caspase-11 (this latter, identified in mouse, with orthologue caspases-4 and -5 in humans). Activation of these caspases, which takes place following processing of the respective pro-caspase precursors, leads to production of the inflammatory cytokines IL-1β and IL-18 (Storek and Monack, 2015; Stowe *et al.*, 2015). Importantly, lysis of the infected cell in which inflammasome undergoes activation results in release of these pro-inflammatory molecules, an event known as pyroptosis. Two alternative pathways of inflammasome activation are known: (i) the canonical, which relies in the consecutive action of a PRR that binds the pathogen 'danger' molecule to further interact with one bridge adaptor protein that ultimately activates pro-caspase-1 (Crowley *et al.*, 2016; Storek and Monack, 2015); and, (ii) the non-canonical, involving in the last stage the contribution of caspase-11 or its human caspases-4/5 counterparts (Crowley *et al.*, 2016). Inflammasome activation has been extensively studied in *Salmonella* infections, involving receptors such as NAIP1 and NAIP2, which bind pathogen proteins of the type III secretion machinery (PrgI, PrgJ); and NAIP5/NAIP6, which recognize flagellin (FliC/FljB). These receptors converge in the PRR protein NLRC4, which via its CARD domain activate subsequently pro-caspase-1, which processes itself auto catalytically. Processed caspase-1 then contributes to the production of IL-18 and 1L-1β by proteolysis of their precursor forms. Pro-caspase-1 can also be activated by the adaptor protein ASC (apoptosis-associated speck-like protein) after stimulation of NLRP3, which responds to increased levels of mitochondrial reactive oxidative species (mROS), often a signal associated to intracellular bacterial infections (Crowley *et al.*, 2016). Unlike the canonical pathway of inflammasome activation, much less is known regarding the non-canonical activation route involving caspase-11 and its human counterpart caspase-4/5. Interestingly, a recent study provided evidence for direct binding of LPS by these 'non-canonical' caspases (Shi *et al.*, 2014), which makes apparently unnecessary the intervention of upstream receptors. Another matter for further studies involves the modes of inflammasome activation in different host cell types. Data obtained in mouse infection *in vivo* model using *Salmonella* support the requirement of both caspase-1 and caspase-11 to control pathogen proliferation in intestinal epithelial cells (IEC) (Knodler *et al.*, 2014; Sellin *et al.*, 2014). Of interest is also the fact that inflammasome activation in neutrophils is not accompanied by pyroptosis (Crowley *et al.*, 2016).

Immune evasion and the alteration of envelope components

Previous sections summarized mechanisms by which innate immunity controls bacterial infections based on recognition of distinct MAMPs. Receptors such as TLR4, NOD1/NOD2, and inflammasome-related receptors target distinct envelope molecules. This 'multifactorial' defence increases the probability of an effective control of the infection. Despite this arsenal of host receptors, successful pathogens have evolved mechanisms to evade them. A common strategy is the alteration of the envelope chemistry, rendering it invisible to host PRR. This review mostly focuses in this 'masking' strategy, differentiating modifications known to occur in LPS and peptidoglycan that different studies have shown to impair host recognition. We will emphasize those advances in the field attempting to characterize envelope changes that occur 'within the host'.

Evasion due to impaired LPS recognition by the TLR4–MD2 complex

The LPS has a complex chemistry in which three regions are clearly differentiated. One of these is the lipid-A component, recognized by the TLR4–MD2 complex. Due to this property, the lipid A is the LPS region with immunomodulatory activity (Maeshima and Fernandez, 2013; Maldonado *et al.*, 2016; Neyen and Lemaitre, 2016). The other LPS regions include the central core oligosaccharide and the O-antigen, this latter harbouring a variable number of repeating oligosaccharide units in contact with the external milieu (Whitfield and Trent, 2014).

The lipid A, which is embedded in the outer membrane, has a di-glucosamine backbone decorated with acyl chains (carbons 2, 3, 2′ and 3′) and phosphate groups (carbons 1 and 4′) (Whitfield and Trent, 2014). Lipid A chemistry has been extensively studied in *Escherichia coli*, in which it is normally hexa-acylated. Four of these acyl groups are bound to the 2, 3, 2′ and 3′ positions of the di-glucosamine whereas the other two acyl groups are secondary chains bound to the hydroxyl groups of the 2′-and 3′-linked chains. Importantly, the biosynthetic pathway of the lipid-A molecule, which involves acylation and phosphorylation reactions, is interconnected with the addition of the 3-deoxy-D-manno-oct-2-ulosonic acid (Kdo_2), the core oligosaccharide in the LPS of *E. coli*. Thus, addition of the secondary acyl chains to the tetra-acylated lipid-A occurs only after the two phosphate groups have been added to the 1 and 4′ positions and the Kdo_2 is incorporated (Whitfield and Trent, 2014). This hexa-acylated lipid A form of *E. coli* elicits a strong inflammatory response via the TLR4–MD2 complex (Maeshima and Fernandez, 2013). Variations in the chemistry of lipid A implying a reduced acylation state are normally associated to lesser degree of TLR4-MD2 activation (Maeshima and Fernandez, 2013; Needham and Trent, 2013). A recent study pointed to lower recognition of the lipid A by CD14 as a critical step to evade TLR4-MD2 signalling (Tan *et al.*, 2015). Reduction in the number of lipid A acyl chains is a strategy widely used to evade TLR-MD2 signalling by diverse bacterial pathogens including, among others *S. enterica* serovar Typhimurium, *H. pylori, Burkholderia cenocepacia, Pseudomonas aeruginosa, Bordetella bronchispetica,* and *Yersinia* spp. (Needham and Trent, 2013). The reserve, increased lipid-A acylation, can also result in decreased TLR-MD2 activation. Representative cases are those of *Klebsiella pneumoniae* and *S. enterica* serovar Typhimurium (Dalebroux and Miller, 2014; Llobet *et al.*, 2015). Other modifications in the lipid A chemistry are known to confer resistance to cationic antimicrobial peptides (CAMP) due to a reduction in the negative charge balance of the molecule. Changes altering the net charge of the molecule include the addition of phospho-ethanolamine, aminoarabinose, or galactosamine; dephosphorylation reactions in lipid A; the removal of outer Kdo region; and hydroxylation of lipid A acyl chains or the Kdo (reviewed by Needham and Trent, 2013).

Despite the extensive knowledge on lipid A chemical modifications, there is relatively scarce information on the exact changes in the chemistry of this important envelope component when the pathogen colonizes host tissues. Most of the modifications reported for lipid A modifications have been characterized in bacteria grown in laboratory conditions (Needham and Trent, 2013). This remark also applies for the corresponding experiments involving stimulation of host receptors, which are generally based on material purified grown bacteria grown in axenic cultures. It is also important to note that while some modifications -compared to the 'reference' lipid-A from *E. coli*- are constitutive in other bacteria (e.g. phosphorylation of Kdo in *Haemophilus influenzae, Vibrio cholerae,* or *Bortedella pertussis*), other modifications are tightly regulated. In most instances, these modifications are orchestrated by regulators involved in controlling expression of virulence genes. Among this class of regulators are two-component systems such as PhoPQ, PmrAB, ParRA, CprRS, which control distinct modification such as lipid-A acylation (addition of extra acyl chains), deacylation, or the incorporation of phosphoryl-ethanolamine (PEA) or 4-amino-4-deoxy-L-arabinose in the sugars of the di-glucosamine backbone (Dalebroux and Miller, 2014; Whitfield and Trent, 2014). Besides these type of regulators, there is also evidence that small non-coding RNA (sRNA) are regulatory pieces that contribute to LPS modification in some enteric bacteria (Klein and Raina, 2015). This regulation has been reported to occur at the level of the enzyme that introduces the modification into the lipid A; for example, the sRNA MgrR that silences EptB, an enzyme that incorporates PEA to the Kdo (Moon *et al.*, 2013). Other sRNA have been proposed to control indirectly LPS synthesis and modifications by modulating the availability of metabolic precursors (Klein and Raina, 2015). Recent studies have shown that these two-component regulatory systems could also be involved in controlling phospholipid content (Dalebroux *et al.*, 2014), which agrees with the view of global changes in the envelope architecture ensuring survival following exposure to diverse host defences, including CAMP.

Most of the known modifications in lipid-A subjected to regulation have been demonstrated in envelope material isolated from wild-type and mutants lacking the corresponding regulators grown in axenic cultures. Conversely, there is scarce information of the exact changes occurring in lipid A or other regions of the LPS following colonization of host cells or tissues. Although some authors addressed this important issue using *in vitro* cultures of macrophages infected with *S.* Typhimurium (Gibbons *et al.*, 2005), there is still no direct proof of the type of LPS produced by this pathogen *in vivo*. Nonetheless, many of the lipid-A modifications characterized in intracellular bacteria inside macrophages match those regulated by PhoPQ and PmrAB in bacteria grown in minimal media with acidic pH and low Mg^{2+} concentration (Dalebroux and Miller, 2014; Gibbons *et al.*, 2005). Thus, remodelling of *S.* Typhimurium lipid A within the host may follow similar trends to those characterized in axenic media mimicking some host environmental conditions. To recall that the PhoPQ-PmrAB-mediated modifications culminate in a less inflammatory LPS, which could be beneficial to avoid TLR4-MD2 signalling.

Other indirect evidences supporting modification of lipid-A chemistry *in vivo* induced by host signals come from studies showing inactivating mutations in genes encoding for lipid A acylases, as *lpxL*, in clinical isolates of *Neisseria meningitidis* (Fransen *et al.*, 2009). *H. pylori* is also known to express as sole LPS species a variant lacking both phosphate groups (one being replaced by PEA) and having four acyl chains. *H. pylori* LPS has low stimulatory capacity for TLR4-MD2 (Cullen *et al.*, 2011). For *Y. pestis* it was shown that LpxP, the acyl transferase that adds additional acyl chains to the tetra-acylated lipid A, is temperature sensitive not being functional at the host temperature, 37°C (Kawahara *et al.*, 2002). This tetra-acylated lipid A turned to be less stimulatory for TL4-MD2 (Kawahara *et al.*, 2002), explaining the ability of *Y. pestis* to evade this innate immune defence.

Direct analysis of lipid A produced by pathogens within the host

A recent study by Llobet *et al.* (2015) has established the first evidence of lipid A remodelling *in vivo* in response to host signals. These authors infected mice with the respiratory pathogen *K. pneumoniae* to subsequently recover bacteria from the colonized lungs. Using state-of-the-art chromatography techniques, a novel lipid A species bearing a 2-hydroxylated myristoyl chain not present in the bacteria used to infect the animals, was identified (Llobet *et al.*, 2015). The study also provided evidence for the enzyme responsible for this derivation, LpxO, and two important aspects that may explain the strong ability of *K. pneumoniae* to evade innate immune responses. First, the lipid A produced by this pathogen *in vivo* is less inflammatory. Second, it confers resistance to CAMP. Intriguingly, the authors reported that the lipid A produced by bacteria exposed to a CAMP named colistin resembles that found in bacteria obtained from lung homogenates. This observation prompts to reconsider the risk-benefit cost of using this antimicrobial compound in clinical settings giving the possibility of positive selection of bacteria producing a less inflammatory LPS and, as consequence, with higher probability of succeeding in the infection.

The peptidoglycan: a plastic molecule influencing host innate immune responses

The peptidoglycan (PG) is a giant macromolecule consisting in glycan chains of alternating N-acetyl-glucosamine (GlcNAc)-N-acetyl-muramic residues (MurNAc) cross-linked by stem peptides (Typas *et al.*, 2011). As abovementioned, innate defences directed by NOD1 and NOD2 rely on the recognition of specific PG fragments with defined motifs, such as iE-DAP and MDP, respectively. Experiments carried out *in vitro* with purified compounds showed that modification of these motifs has a strong effect in NOD1 and NOD2 sensing. Taking as reference the motif MurNAc-L-Ala-γ-D-Glu-*meso*-DAP-D-Ala (M-Tetra) and a large repertoire of derivate molecules, Girardin *et al.* (2003) showed that the preferred substrate for NOD2 is muramyl-dipeptide (M-Di) and that the presence of *meso*-DAP in third position interfers with signalling by this receptor. By contrast, preferred substrates for NOD1 included those having the γ-D-Glu-*meso*-DAP (iE-DAP) motif (Chamaillard *et al.*, 2003). Interestingly, while compounds as M-Tri (MurNAc-L-Ala-γ-D-Glu-*meso*-DAP), Tri (L-Ala-γ-D-Glu-*meso*-DAP) and Di (γ-D-Glu-*meso*-DAP) stimulated NOD1 signalling, that was not the case for M- Tetra (MurNAc-L-Ala-γ-D-Glu-*meso*-DAP-D-Ala) (Girardin *et al.*, 2003). This latter result underscores the importance of the *meso*-DAP residue for NOD1 signalling implying that residues linked to the L-chiral centre of this diamino acid may impose steric hindrance for NOD1 to properly bind the ligand. Although the most common strategy followed by pathogens to evade NOD1/NOD2 sensing relies in alterations of the PG structure (see below), PG recycling can also play an important role in evasion of innate immunity by limiting the amount of muropeptide shed from the pathogen to the extracellular environment. This was first shown with *Shigella flexneri* mutants partly defective in PG recycling, which triggered NOD1-dependent activation of NF-κB at higher extent than wild-type bacteria (Nigro *et al.*, 2008). Differences in PG recycling between *N. meningitidis* and *N. gonorrhoeae* are also associated to distinct NOD1-dependent NF-κB activities, with meningococci having more efficient PG recycling and, as consequence, shedding fewer immunostimulatory PG fragments (Woodhams *et al.*, 2013).

Peptidoglycan structural modifications leading to 'intrinsic' immune evasion

The PG structure of certain bacteria varies in defined sites respect to the PG of model bacteria as *E. coli*, which bears canonical NOD1 and NOD2 motifs. Some of these differences can impact host recognition notoriously. A striking example is that of pathogenic spirochaetes of the *Borrellia* and *Treponema* genera, which are Gram-negative bacteria displaying in their PG L-ornithine as third residue of the stem peptide instead of *meso*-DAP (Beck *et al.*, 1990; Yanagihara *et al.*, 1984). The presence of L-ornithine in this position impairs NOD1 signalling (Girardin *et al.*, 2003), which is consistent with the low immunogenicity associated to the PG of spirochaetes. Intriguingly, *in vitro* studies with purified muropeptide showed that L-ornithine in the third position of a muramyl-tripeptide can modulate positively NOD2 sensing (Girardin *et al.*, 2003). Future studies should clarify these observations assessing PG spirochaetal immunogenicity in the context of the infection.

Another important change in PG structure that can potentially affect host recognition involves amidation of the *meso*-DAP residue. This modification is prominent in Corynebacteriales (Levefaudes *et al.*, 2015). Structural analysis of the PG of *Corynebacterium glutamicum* revealed that 100% of the D-Glu residues (second position of stem peptide) were amidated whereas that chemical modification was also observed for ≈80% of the *meso*-DAP residues (Levefaudes *et al.*, 2015). A high percentage of *meso*-DAP residues present in the PG are also amidated in other Gram-positive bacteria such as the intracellular pathogen *L. monocytogenes* (Boneca *et al.*, 2007) and *B. subtilis* (Atrih *et al.*, 1999). This scenario of only a fraction of the respectively muropeptides harbouring chemical modifications in a defined position is recurrent for the PG of many bacteria (Cava and de Pedro, 2014). Of interest, a recent study showed that amidated *meso*-DAP is weakly recognized by NOD1 (Vijayrajratnam *et al.*, 2016), which could be compatible with the low immunogenicity shown for the PG of Corynebacteriales. As in the case of L-ornithine present in the PG of spricohetes, amidated *meso*-DAP increases slightly NOD2 sensing (Girardin *et al.*, 2003). Based on these data, it is tempting to postulate that evasion mechanisms directed to avoid NOD1 sensing might have opposite consequences on NOD2-based defences. Future studies should investigate this interesting scenario.

meso-DAP amidation has also been extensively analysed in pathogenic mycobacteria, especially in the case of *Mycobacterium tuberculosis*. The PG of this pathogen has unique structural modifications involving glycolylation (instead of acetylation) of a fraction of the muramic acid residues and amidation of some of the D-Glu and *meso*-DAP residues (Angala *et al.*, 2014). A recent study performed with synthetic muropeptides has shown that amidation of D-Glu and *meso*-DAP reduces NOD1 activation (Wang *et al.*, 2016). The same study provides evidence of partial impairment of NOD2 signalling associated to the glycolylation of the muramic acid residue (Wang *et al.*, 2016). This observation however contrasts with previous studies inferring a contribution of N-glycolylation of the muramyl residue to NOD2-dependent immunogenicity (Hansen *et al.*, 2014) and increased recognition of

N-glycolyl muramyl dipeptide by NOD2 (Coulombe *et al.*, 2009). Other recent studies with PG from *M. leprae* show that neither D-Glu amidation nor the presence of glycine instead of L-alanine as first residue of the stem peptide interfere with NOD2 signalling (Schenk *et al.*, 2016). Globally, it can be concluded that among those modifications reported for the mycobacterial PG, that involving *meso*-DAP amidation seems to affect at a higher degree recognition by innate immune defences.

Another widely characterized structural modification in the PG of some bacteria involves deacetylation and O-acetylation in specific sites of the muramic acid residue (Cava and de Pedro, 2014). These modifications confer lysozyme resistance (Rae *et al.*, 2011) and interfere with immune signalling. PgdA and OatA, the two enzymes involved in these modifications, play an important role in immune evasion by *L. monocytogenes* (Aubry *et al.*, 2011; Boneca *et al.*, 2007). This pathogen contains a fraction of the muramic acid residues either deaceylated or O-acetylated, and these modifications are tightly regulated by response regulators and dedicated sRNA that ensure up-regulation of these enzymes *in vivo* (Burke *et al.*, 2014).

The PG of *H. pylori* displays also features that could diminish its immunogenicity, including deacetylation of the N-acetyl-muramyl residue by PgdA (Wang *et al.*, 2010). An early study showed that when *H. pylori* remodels substantially the PG structure following entry into in stationary phase. In this condition, there is a marked increase in muropeptides bearing the L-Ala-γ-D-Glu dipeptide (Costa *et al.*, 1999). *H. pylori* has also other interesting features as a high percentage of muropeptides bearing a pentapeptide stem chain and the lack of L–D-type cross-links among stem peptides. *H. pylori* PG was postulated to have lower potential to stimulate NOD1 given the high percentage of muropeptides harbouring dipeptide stem peptides (up to 20% in growth cessation), which lack the essential *meso*-DAP residue. This conclusion agrees with the observation that a defect in the amidase AmiA, which results in decrease in the amount of muropeptides bearing dipeptide as stem peptide, correlated with increased immunogenicity (Chaput *et al.*, 2006). A further study in *H. pylori* supported the presence of additional peptidases acting in the stem peptide that could contribute to generate the characteristic helical rod shape of this pathogen and influence its capacity to colonize the stomach (Sycuro *et al.*, 2012). Noteworthy, this study showed that the absence of these peptidases has no effect in the pro-inflammatory response of the infected animal despite a 3-fold increase in tripeptides in the PG (Sycuro *et al.*, 2012). Further studies are needed to elucidate the exact contribution of the *H. pylori* PG to signalling and the relative contribution of the distinct fragments (bearing motifs with low- or high-immunogenicity) that are released during infection. Early studies also reported the capacity of *cag*$^{+}$ *H. pylori* to stimulate NOD1 via its specialized type IV secretion system (Viala *et al.*, 2004), which agrees with the higher susceptibility to *H. pylori* infection displayed by NOD1-deficient mice (Viala *et al.*, 2004).

H. pylori is also known to produce varied forms of co-existing lipid-A having distinct degree of acylation (Suda *et al.*, 2001) and an O-antigen that mimics Lewis (Le) blood group antigens (Lina *et al.*, 2014). Future studies, to be performed in bacteria extracted from animals, could provide insights into the relative enrichment of the molecular patterns having distinct immune-stimulatory features during the colonization of the gastric epithelium.

Modification of the peptidoglycan moiety 'within the host' can affect innate immune defences

To date, most studies focused in the elucidation of PG structure in bacterial pathogens have been carried out in laboratory conditions. Like the case of the lipid A, our knowledge of the exact chemistry of envelope components that are produced by pathogen within the host is still limited. This absence of data contrasts with the accumulating evidence supporting elaborated mechanisms involved in the regulation of enzymes that participate in the biosynthesis, hydrolysis or remodelling of the PG (Egan *et al.*, 2017). The PG structure can be remodelled in bacteria grown in axenic cultures and it is conceivable that similar alterations also occurs within the host (Fig. 2.1).

Some examples of responses affecting activity of PG enzymes are known for the case

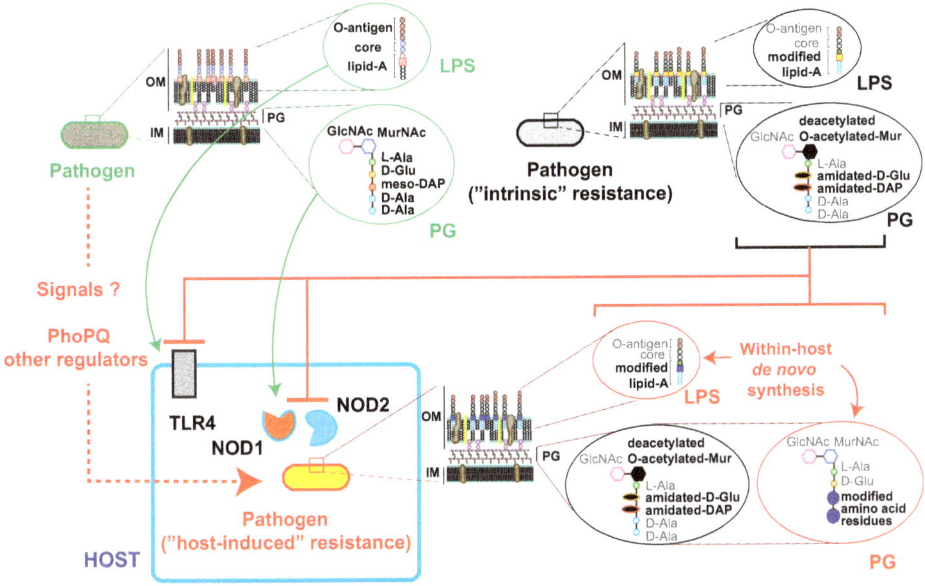

Figure 2.1 Evasion of TLR4 and NOD1/NOD2 signalling based on changes of lipid A and peptidoglycan (PG) structure. The scheme depicts a distinct chemistry of lipid A and PG present in some bacteria grown in non-host environments, which can affect recognition by TLR4 and/or NOD1/NOD2. This mode of evasion is referred as 'intrinsic resistance'. The scheme also denotes the synthesis *de novo* within the host of new molecular patterns in envelope components such as the lipid A component of lipopolysaccharide (LPS) and PG. This evasion mechanism, involving new molecular patterns can impair TLR4 and NOD1/NOD2 signalling, is referred as 'host-induced'. Note that the signals and regulators involved in the changes in these pathogen-associated molecular patterns occurring during the infection are mostly uncharacterized. One of the regulators known to be activated within the host and mediate changes in the LPS, membrane phospholipids and PG in several Gram-negative pathogens is the two-component system PhoP-PhoQ. Although the scheme refers to the envelope structure of Gram-negative, the concepts of 'intrinsic' and 'host-induced' capacity for evasion are equally applicable to Gram-positive bacterial pathogens. IM, inner membrane; OM, outer membrane; PG, peptidoglycan; LPS, lipopolysaccharide. See text for more details.

of the 'resuscitation-promoting factors' (Rpf) of mycobacteria (Kana and Mizrahi, 2010). These factors include glycosidases and endopeptidases that, in response to signals linked to nutrient availability, promote cultivability of dormant bacteria (Kana and Mizrahi, 2010).

In one of the few studies that accomplish the structural analysis of PG in intracellular bacteria, Quintela *et al.* showed that *S.* Typhimurium remodels substantially the PG when proliferates actively within cultured epithelial cells (Quintela *et al.*, 1997). These authors purified PG from intracellular bacteria and, when compared to that of bacteria used in the inoculum, they noticed a substantial increase in the relative proportion of cross-linked muropeptides lacking one of the GlcNAc-MurNAc disaccharide units. These muropeptides could be produced only by the action of an amidase acting in the MurNAc-L-Ala amide bond (Quintela *et al.*, 1997). Such an activity targeting un-cross-linked muropeptides could result in the release of stem peptides, with potential to target NOD1, ubiquitously expressed by all cell types. In this scenario, one could therefore assume a phenomenon favouring a pro-inflammatory response in the infected cell. Future investigations are also needed to clarify the source (either host or pathogen) of this yet uncharacterized amidase activity. Interestingly, a recent study has shown that intracellular *S.* Typhimurium up-regulates within the host cell an enzyme with D-L-endopeptidase activity against the γ-D-Glu-*meso*-DAP bond, which is regulated by the PhoP-PhoQ two component system (Rico-Pérez *et al.*, 2016). Such an activity could potentially attenuate the effect on NOD-1-mediated signalling by decreasing the number of peptides bearing the iE-DAP motif and shed by the pathogen inside the infected cell. These examples clearly manifest the ability of intracellular bacterial pathogens to remodel PG structure within the host.

Other recent studies have provided evidence of responses to 'host signals' that influence PG structure. When exposed to bile salts, *S.* Typhimurium alters PG structure by lowering the amount of lipoprotein bound covalently to the PG and decreasing the percentage of peptide bridges of the L-D-(*meso*-DAP-*meso*-DAP) type (Hernández *et al.*, 2015). Interestingly, mutants lacking the enzymes responsible for these modifications were shown to be more resistant to bile salts (Hernández *et al.*, 2015). These observations highlight the importance for the pathogen of negative regulation acting to remove enzymes probably unnecessary in the host environment.

Evasion of the inflammasome defence

The inflammasome has emerged as a highly elaborated mechanism to cope with invasive pathogen that colonize the intracellular niche of both phagocytic and non-phagocytic cells (Crowley *et al.*, 2016). Several forms of evading this important innate host defence have been postulated. One of them involves down-regulation of components recognized by the distinct receptors forming part of the canonical inflammasome. A relevant case is that of flagellin, which is strongly down-regulated by intracellular *S.* Typhimurium in persistent infections (Nuñez-Hernández *et al.*, 2013). This pathogen also switches the type III secretion system (T3SS) when colonizes the phagosomal compartment, changing from that encoded by the *Salmonella*-pathogenicity island 1 (SPI-1) to that encoded by SPI-2. Such reduction in the expression of structural proteins of the SPI-1 T3SS is proposed to lower the capability of the pathogen to stimulate the inflammasome at late infection times (Crowley *et al.*, 2016).

A recent report provided a new mechanism of inflammasome evasion based on maintaining envelope integrity and, in this way, avoiding release of potential stimulatory molecules recognized by the inflammasome receptors. In this work, Sampson *et al.* implicated Cas9, a component of the CRISPR-Cas system, in envelope homeostasis of *Francisella novicida* (Sampson *et al.*, 2014), an intracellular pathogen causing diseases in a variety of mammalian species, including humans. Infection with mutants defective in Cas9 resulted in overstimulation of the inflammasome system, a phenomenon that could be abrogated by a defect in ASC, a key adaptor required for activation of pro-caspase-1 or TLR2 (Sampson *et al.*, 2014). These data point to the importance that the control of envelope integrity has for the pathogen in intracellular infections. In this line, it is important to recall on the massive release of outer membrane vesicles (OMVs) that some intracellular pathogens experience, probably as result of confronting stresses imposed by the host cell in the cytosol or phagosomal locations. Pathogens such as *Legionella pneumophila* and *S.* Typhimurium release OMV within the infected cell and this process has been linked to processes such as pathogen intracellular replication or the delivery of protein effectors (García-del Portillo *et al.*, 1997; Jager *et al.*, 2015; Jung *et al.*, 2016). Of interest, production of OMV in *S.* Typhimurium is tightly linked to LPS remodelling that is controlled by the PhoP-PhoQ system (Elhenawy *et al.*, 2016). This correlation supports direct implication of virulence regulators in OMV formation. Indeed, OMV have been shown in a variety of pathogens to be important pieces in the infection process (Laughlin and Alaniz, 2016; O'Donoghue and Krachler, 2016; Pathirana and Kaparakis-Liaskos, 2016). Like in the case of lipid A and PG, the future challenge will be to characterize the content of OMV produced and released by the pathogen in extracellular or intracellular locations. This information is vital to understand the role played by OMV in immune evasion. Currently, there are on-going technical efforts in the proteomics field to reach this goal (Lee *et al.*, 2016).

Concluding remarks

Evolution has shaped host–pathogen interactions in many ways. A key step is the initial first encounter between pathogen and the host, in which molecular patterns recognized as 'foreign' by the host innate defences play a fundamental role. Despite the exhaustive analysis of envelope components in a myriad of bacterial pathogens, we are still far from knowing whether those components undergo significant changes inside the host and, as consequence, elicit one or another type of response in the host. We should be aware of this limitation since, to date, most of the studies rely in stress conditions artificially mounted in the laboratory and supposed to mimic the environment faced by the pathogen during host colonization. Another aspect to be evaluated in future studies is the known heterogeneity occurring in the multiple pathogen–host cell encounters as the infection proceeds. These can differ either stochastically or a result of specific host cell types invaded by the pathogen with unique molecular cross-talks (Bumann, 2015). Current technical advances should also allow to define at the molecular level those changes occurring in LPS, PG and other envelope components in extra- and intracellular infections. This should pave the way to identify the involved regulators and design new antimicrobial molecules. The work by Llobet *et al.*, which describes changes in the lipid A of *K. pneumoniae* LPS when the pathogen colonizes the lung and the low immunogenicity accompanying that modification (Llobet *et al.*, 2015), will certainly encourage investigators to focus on the characterization of pathogen envelope components within the host.

Acknowledgements

We apologize in advance for possible omission of citations due to the lack of space. Work in our laboratories is supported by grants BIO2014-55238-R (to M.G.P.) and, BIO2013-46281-P and BIO2016-77639-P-MINECO-FEDER (to F.G.-dP.) from the Spanish Ministry of Economy and Competitiveness and European Regional Development Funds (FEDER).

References

Amdahl, H., Jongerius, I., Meri, T., Pasanen, T., Hyvärinen, S., Haapasalo, K., van Strijp, J.A., Rooijakkers, S.H., and Jokiranta, T.S. (2013). Staphylococcal Ecb protein and host complement regulator factor H enhance functions of each other in bacterial immune evasion. J. Immunol. *191*, 1775–1784. https://doi.org/10.4049/jimmunol.1300638.

Andrade, W.A., Firon, A., Schmidt, T., Hornung, V., Fitzgerald, K.A., Kurt-Jones, E.A., Trieu-Cuot, P., Golenbock, D.T., and Kaminski, P.A. (2016). Group B Streptococcus Degrades Cyclic-di-AMP to Modulate STING-Dependent Type I Interferon Production. Cell Host Microbe *20*, 49–59. https://doi.org/10.1016/j.chom.2016.06.003.

Angala, S.K., Belardinelli, J.M., Huc-Claustre, E., Wheat, W.H., and Jackson, M. (2014). The cell envelope glycoconjugates of *Mycobacterium tuberculosis*. Crit. Rev. Biochem. Mol. Biol. *49*, 361–399. https://doi.org/10.3109/10409238.2014.925420.

Atrih, A., Bacher, G., Allmaier, G., Williamson, M.P., and Foster, S.J. (1999). Analysis of peptidoglycan structure from vegetative cells of Bacillus subtilis 168 and role of PBP 5 in peptidoglycan maturation. J. Bacteriol. *181*, 3956–3966.

Aubry, C., Goulard, C., Nahori, M.A., Cayet, N., Decalf, J., Sachse, M., Boneca, I.G., Cossart, P., and Dussurget, O. (2011). OatA, a peptidoglycan O-acetyltransferase involved in *Listeria monocytogenes* immune escape, is critical for virulence. J. Infect. Dis. *204*, 731–740. https://doi.org/10.1093/infdis/jir396.

Beachboard, D.C., and Horner, S.M. (2016). Innate immune evasion strategies of DNA and RNA viruses. Curr. Opin. Microbiol. *32*, 113–119. https://doi.org/10.1016/j.mib.2016.05.015.

Beck, G., Benach, J.L., and Habicht, G.S. (1990). Isolation, preliminary chemical characterization, and biological activity of Borrelia burgdorferi peptidoglycan. Biochem Biophys Res Commun *167*, 89-95.

Boneca, I.G., Dussurget, O., Cabanes, D., Nahori, M.A., Sousa, S., Lecuit, M., Psylinakis, E., Bouriotis, V., Hugot, J.P., Giovannini, M., *et al.* (2007). A critical role for peptidoglycan N-deacetylation in Listeria evasion from the host innate immune system. Proc. Natl. Acad. Sci. U.S.A. *104*, 997–1002.

Bumann, D. (2015). Heterogeneous host-pathogen encounters: act locally, think globally. Cell Host Microbe *17*, 13–19. https://doi.org/10.1016/j.chom.2014.12.006.

Burke, T.P., Loukitcheva, A., Zemansky, J., Wheeler, R., Boneca, I.G., and Portnoy, D.A. (2014). *Listeria monocytogenes* is resistant to lysozyme through the regulation, not the acquisition, of cell wall-modifying enzymes. J. Bacteriol. *196*, 3756–3767. https://doi.org/10.1128/JB.02053-14.

Caruso, R., Warner, N., Inohara, N., and Núñez, G. (2014). NOD1 and NOD2: signaling, host defense, and inflammatory disease. Immunity *41*, 898–908. https://doi.org/10.1016/j.immuni.2014.12.010.

Cava, F., and de Pedro, M.A. (2014). Peptidoglycan plasticity in bacteria: emerging variability of the murein sacculus and their associated biological functions. Curr Opin Microbiol *18*, 46-53.

Chamaillard, M., Hashimoto, M., Horie, Y., Masumoto, J., Qiu, S., Saab, L., Ogura, Y., Kawasaki, A., Fukase, K., Kusumoto, S., *et al.* (2003). An essential role for NOD1 in host recognition of bacterial peptidoglycan containing diaminopimelic acid. Nat. Immunol. *4*, 702–707. https://doi.org/10.1038/ni945.

Chaput, C., Ecobichon, C., Cayet, N., Girardin, S.E., Werts, C., Guadagnini, S., Prévost, M.C., Mengin-Lecreulx, D., Labigne, A., and Boneca, I.G. (2006). Role of AmiA in the morphological transition of *Helicobacter pylori* and in immune escape. PLOS Pathog. *2*, e97.

Costa, K., Bacher, G., Allmaier, G., Dominguez-Bello, M.G., Engstrand, L., Falk, P., de Pedro, M.A., and García-del Portillo, F. (1999). The morphological transition of *Helicobacter pylori* cells from spiral to coccoid is preceded by a substantial modification of the cell wall. J Bacteriol *181*, 3710-3715.

Coulombe, F., Divangahi, M., Veyrier, F., de Léséleuc, L., Gleason, J.L., Yang, Y., Kelliher, M.A., Pandey, A.K., Sassetti, C.M., Reed, M.B., *et al.* (2009). Increased NOD2-mediated recognition of N-glycolyl muramyl dipeptide. J. Exp. Med. *206*, 1709–1716. https://doi.org/10.1084/jem.20081779.

Crowley, S.M., Knodler, L.A., and Vallance, B.A. (2016). Salmonella and the inflammasome: battle for intracellular dominance. Curr. Top. Microbiol. Immunol. *397*, 43–67. https://doi.org/10.1007/978-3-319-41171-2_3.
Cullen, T.W., Giles, D.K., Wolf, L.N., Ecobichon, C., Boneca, I.G., and Trent, M.S. (2011). *Helicobacter pylori* versus the host: remodeling of the bacterial outer membrane is required for survival in the gastric mucosa. PLOS Pathog. *7*, e1002454. https://doi.org/10.1371/journal.ppat.1002454.
Dalebroux, Z.D., Matamouros, S., Whittington, D., Bishop, R.E., and Miller, S.I. (2014). PhoPQ regulates acidic glycerophospholipid content of the Salmonella Typhimurium outer membrane. Proc Natl Acad Sci U S A *111*, 1963-1968.
Dalebroux, Z.D., and Miller, S.I. (2014). Salmonellae PhoPQ regulation of the outer membrane to resist innate immunity. Curr. Opin. Microbiol. *17*, 106–113. https://doi.org/10.1016/j.mib.2013.12.005.
Egan, A.J., Cleverley, R.M., Peters, K., Lewis, R.J., and Vollmer, W. (2017). Regulation of bacterial cell wall growth. FEBS J. *284*, 851-867,
Elhenawy, W., Bording-Jorgensen, M., Valguarnera, E., Haurat, M.F., Wine, E., and Feldman, M.F. (2016). LPS Remodeling triggers formation of outer membrane vesicles in *Salmonella*. MBio *7*, e00940–16. https://doi.org/10.1128/mBio.00940-16.
Feerick, C.L., and McKernan, D.P. (2017). Understanding the regulation of pattern recognition receptors in inflammatory diseases – a 'Nod' in the right direction. Immunology.*150*, 237-247.
Flávia Nardy, A., Freire-de-Lima, C.G., and Morrot, A. (2015). Immune evasion strategies of *Trypanosoma cruzi*. J. Immunol. Res. *2015*, 178947. https://doi.org/10.1155/2015/178947.
Foster, T.J., Geoghegan, J.A., Ganesh, V.K., and Höök, M. (2014). Adhesion, invasion and evasion: the many functions of the surface proteins of *Staphylococcus aureus*. Nat. Rev. Microbiol. *12*, 49–62. https://doi.org/10.1038/nrmicro3161.
Franchi, L., Warner, N., Viani, K., and Nuñez, G. (2009). Function of Nod-like receptors in microbial recognition and host defense. Immunol. Rev. *227*, 106–128. https://doi.org/10.1111/j.1600-065X.2008.00734.x.
Fransen, F., Heckenberg, S.G., Hamstra, H.J., Feller, M., Boog, C.J., van Putten, J.P., van de Beek, D., van der Ende, A., and van der Ley, P. (2009). Naturally occurring lipid A mutants in *neisseria meningitidis* from patients with invasive meningococcal disease are associated with reduced coagulopathy. PLOS Pathog *5*, e1000396.
García-del Portillo, F., Stein, M.A., and Finlay, B.B. (1997). Release of lipopolysaccharide from intracellular compartments containing Salmonella typhimurium to vesicles of the host epithelial cell. Infect Immun *65*, 24-34.
Geoghegan, J.A., and Foster, T.J. (2015). Cell Wall-Anchored Surface Proteins of *Staphylococcus aureus*: many proteins, multiple functions. Curr. Top. Microbiol. Immunol. [Epub ahead of print]. https://doi.org/10.1007/82_2015_5002.
Gibbons, H.S., Kalb, S.R., Cotter, R.J., and Raetz, C.R. (2005). Role of Mg2+ and pH in the modification of Salmonella lipid A after endocytosis by macrophage tumour cells. Mol. Microbiol. *55*, 425–440.
Gioannini, T.L., Teghanemt, A., Zhang, D., Coussens, N.P., Dockstader, W., Ramaswamy, S., and Weiss, J.P. (2004). Isolation of an endotoxin-MD-2 complex that produces Toll-like receptor 4-dependent cell activation at picomolar concentrations. Proc. Natl. Acad. Sci. U.S.A. *101*, 4186–4191. https://doi.org/10.1073/pnas.0306906101.
Girardin, S.E., Travassos, L.H., Hervé, M., Blanot, D., Boneca, I.G., Philpott, D.J., Sansonetti, P.J., and Mengin-Lecreulx, D. (2003). Peptidoglycan molecular requirements allowing detection by Nod1 and Nod2. J. Biol. Chem. *278*, 41702–41708. https://doi.org/10.1074/jbc.M307198200.
Hansen, J.M., Golchin, S.A., Veyrier, F.J., Domenech, P., Boneca, I.G., Azad, A.K., Rajaram, M.V., Schlesinger, L.S., Divangahi, M., Reed, M.B., *et al.* (2014). N-glycolylated peptidoglycan contributes to the immunogenicity but not pathogenicity of *Mycobacterium tuberculosis*. J. Infect. Dis. *209*, 1045–1054. https://doi.org/10.1093/infdis/jit622.
Hasegawa, M., Yang, K., Hashimoto, M., Park, J.H., Kim, Y.G., Fujimoto, Y., Nuñez, G., Fukase, K., and Inohara, N. (2006). Differential release and distribution of Nod1 and Nod2 immunostimulatory molecules among bacterial species and environments. J. Biol. Chem. *281*, 29054–29063.
Hernández, S.B., Cava, F., Pucciarelli, M.G., García-Del Portillo, F., de Pedro, M.A., and Casadesús, J. (2015). Bile-induced peptidoglycan remodelling in *Salmonella enterica*. Environ. Microbiol. *17*, 1081–1089. https://doi.org/10.1111/1462-2920.12491.
Hong, X., Qin, J., Li, T., Dai, Y., Wang, Y., Liu, Q., He, L., Lu, H., Gao, Q., Lin, Y., *et al.* (2016). Staphylococcal protein A promotes colonization and immune evasion of the epidemic healthcare-associated MRSA ST239. Front. Microbiol. *7*, 951. https://doi.org/10.3389/fmicb.2016.00951.

Irving, A.T., Mimuro, H., Kufer, T.A., Lo, C., Wheeler, R., Turner, L.J., Thomas, B.J., Malosse, C., Gantier, M.P., Casillas, L.N., *et al.* (2014). The immune receptor NOD1 and kinase RIP2 interact with bacterial peptidoglycan on early endosomes to promote autophagy and inflammatory signaling. Cell Host Microbe *15*, 623–635. https://doi.org/10.1016/j.chom.2014.04.001.

Jager, J., Keese, S., Roessle, M., Steinert, M., and Schromm, A.B. (2015). Fusion of *Legionella pneumophila* outer membrane vesicles with eukaryotic membrane systems is a mechanism to deliver pathogen factors to host cell membranes. Cell Microbiol *17*, 607-620.

Jiménez-López, C., and Lorenz, M.C. (2013). Fungal immune evasion in a model host-pathogen interaction: *Candida albicans* versus macrophages. PLOS Pathog. *9*, e1003741. https://doi.org/10.1371/journal.ppat.1003741.

Jung, A.L., Stoiber, C., Herkt, C.E., Schulz, C., Bertrams, W., and Schmeck, B. (2016). *Legionella pneumophila*-Derived Outer Membrane Vesicles Promote Bacterial Replication in Macrophages. PLOS Pathog *12*, e1005592.

Kana, B.D., and Mizrahi, V. (2010). Resuscitation-promoting factors as lytic enzymes for bacterial growth and signaling. FEMS Immunol. Med. Microbiol. *58*, 39–50. https://doi.org/10.1111/j.1574-695X.2009.00606.x.

Kaparakis-Liaskos, M. (2015). The intracellular location, mechanisms and outcomes of NOD1 signaling. Cytokine *74*, 207–212. https://doi.org/10.1016/j.cyto.2015.02.018.

Kawahara, K., Tsukano, H., Watanabe, H., Lindner, B., and Matsuura, M. (2002). Modification of the structure and activity of lipid A in *Yersinia pestis* lipopolysaccharide by growth temperature. Infect. Immun. *70*, 4092–4098.

Kawasaki, T., Kawai, T., and Akira, S. (2011). Recognition of nucleic acids by pattern-recognition receptors and its relevance in autoimmunity. Immunol. Rev. *243*, 61–73. https://doi.org/10.1111/j.1600-065X.2011.01048.x.

Klein, G., and Raina, S. (2015). Regulated Control of the Assembly and Diversity of LPS by Noncoding sRNAs. Biomed. Res. Int. *2015*, 153561. https://doi.org/10.1155/2015/153561.

Knodler, L.A., Crowley, S.M., Sham, H.P., Yang, H., Wrande, M., Ma, C., Ernst, R.K., Steele-Mortimer, O., Celli, J., and Vallance, B.A. (2014). Noncanonical inflammasome activation of caspase-4/caspase-11 mediates epithelial defenses against enteric bacterial pathogens. Cell Host Microbe *16*, 249-256.

Koymans, K.J., Vrieling, M., Gorham, R.D., and van Strijp, J.A. (2016). Staphylococcal immune evasion proteins: structure, function, and host adaptation. Curr. Top. Microbiol. Immunol. [Epub ahead of print]. https://doi.org/10.1007/82_2015_5017.

Laughlin, R.C., and Alaniz, R.C. (2016). Outer membrane vesicles in service as protein shuttles, biotic defenders, and immunological doppelgängers. Gut Microbes 7, 450–454. https://doi.org/10.1080/19490976.2016.1222345.

LeBlanc, P.M., Yeretssian, G., Rutherford, N., Doiron, K., Nadiri, A., Zhu, L., Green, D.R., Gruenheid, S., and Saleh, M. (2008). Caspase-12 modulates NOD signaling and regulates antimicrobial peptide production and mucosal immunity. Cell Host Microbe 3, 146–157. https://doi.org/10.1016/j.chom.2008.02.004.

Lee, J., Kim, O.Y., and Gho, Y.S. (2016). Proteomic profiling of Gram-negative bacterial outer membrane vesicles: Current perspectives. Proteomics Clin Appl *10*, 897-909.

Lee, J., Tattoli, I., Wojtal, K.A., Vavricka, S.R., Philpott, D.J., and Girardin, S.E. (2009). pH-dependent internalization of muramyl peptides from early endosomes enables Nod1 and Nod2 signaling. J. Biol. Chem. *284*, 23818–23829. https://doi.org/10.1074/jbc.M109.033670.

Levefaudes, M., Patin, D., de Sousa-d'Auria, C., Chami, M., Blanot, D., Hervé, M., Arthur, M., Houssin, C., and Mengin-Lecreulx, D. (2015). Diaminopimelic acid amidation in Corynebacteriales: new insights into the role of LtsA in peptidoglycan modification. J. Biol. Chem. *290*, 13079–13094. https://doi.org/10.1074/jbc.M115.642843.

Li, X., Liu, P., Gan, S., Zhang, C., Zheng, Y., Jiang, Y., and Yuan, Y. (2016). Mechanisms of host-pathogen protein complex formation and bacterial immune evasion of Streptococcus suis protein Fhb. J. Biol. Chem. *291*, 17122–17132. https://doi.org/10.1074/jbc.M116.719443.

Lina, T.T., Alzahrani, S., Gonzalez, J., Pinchuk, I.V., Beswick, E.J., and Reyes, V.E. (2014). Immune evasion strategies used by *Helicobacter pylori*. World J. Gastroenterol. *20*, 12753–12766. https://doi.org/10.3748/wjg.v20.i36.12753.

Llobet, E., Martinez-Moliner, V., Moranta, D., Dahlstrom, K.M., Regueiro, V., Tomas, A., Cano, V., Perez-Gutierrez, C., Frank, C.G., Fernandez-Carrasco, H., *et al.* (2015). Deciphering tissue-induced Klebsiella pneumoniae lipid A structure. Proc Natl Acad Sci U S A *112*, E6369-6378.

Maeshima, N., and Fernandez, R.C. (2013). Recognition of lipid A variants by the TLR4-MD-2 receptor complex. Front. Cell. Infect. Microbiol. *3*, 3. https://doi.org/10.3389/fcimb.2013.00003.

Maldonado, R.F., Sá-Correia, I., and Valvano, M.A. (2016). Lipopolysaccharide modification in Gram-negative bacteria during chronic infection. FEMS Microbiol. Rev. *40*, 480–493. https://doi.org/10.1093/femsre/fuw007.

Messaoudi, I., Amarasinghe, G.K., and Basler, C.F. (2015). Filovirus pathogenesis and immune evasion: insights from Ebola virus and Marburg virus. Nat. Rev. Microbiol. *13*, 663–676. https://doi.org/10.1038/nrmicro3524.

Moon, K., Six, D.A., Lee, H.J., Raetz, C.R., and Gottesman, S. (2013). Complex transcriptional and post-transcriptional regulation of an enzyme for lipopolysaccharide modification. Mol. Microbiol. *89*, 52–64. https://doi.org/10.1111/mmi.12257.

Moreira, L.O., and Zamboni, D.S. (2012). NOD1 and NOD2 Signaling in Infection and Inflammation. Front. Immunol. *3*, 328. https://doi.org/10.3389/fimmu.2012.00328.

Nakada-Tsukui, K., and Nozaki, T. (2016). Immune Response of Amebiasis and Immune Evasion by *Entamoeba histolytica*. Front. Immunol. *7*, 175. https://doi.org/10.3389/fimmu.2016.00175.

Nakamura, N., Lill, J.R., Phung, Q., Jiang, Z., Bakalarski, C., de Mazière, A., Klumperman, J., Schlatter, M., Delamarre, L., and Mellman, I. (2014). Endosomes are specialized platforms for bacterial sensing and NOD2 signalling. Nature *509*, 240–244. https://doi.org/10.1038/nature13133.

Needham, B.D., and Trent, M.S. (2013). Fortifying the barrier: the impact of lipid A remodelling on bacterial pathogenesis. Nat. Rev. Microbiol. *11*, 467–481. https://doi.org/10.1038/nrmicro3047.

Neyen, C., and Lemaitre, B. (2016). Sensing Gram-negative bacteria: a phylogenetic perspective. Curr. Opin. Immunol. *38*, 8–17. https://doi.org/10.1016/j.coi.2015.10.007.

Nigro, G., Fazio, L.L., Martino, M.C., Rossi, G., Tattoli, I., Liparoti, V., De Castro, C., Molinaro, A., Philpott, D.J., and Bernardini, M.L. (2008). Muramylpeptide shedding modulates cell sensing of *Shigella flexneri*. Cell. Microbiol. *10*, 682–695.

Nuñez-Hernández, C., Tierrez, A., Ortega, A.D., Pucciarelli, M.G., Godoy, M., Eisman, B., Casadesús, J., and García-del Portillo, F. (2013). Genome expression analysis of nonproliferating intracellular *Salmonella enterica* serovar Typhimurium unravels an acid pH-dependent PhoP-PhoQ response essential for dormancy. Infect Immun *81*, 154-165.

O'Donoghue, E.J., and Krachler, A.M. (2016). Mechanisms of outer membrane vesicle entry into host cells. Cell Microbiol *18*, 1508-1517.

Pathirana, R.D., and Kaparakis-Liaskos, M. (2016). Bacterial membrane vesicles: Biogenesis, immune regulation and pathogenesis. Cell. Microbiol. *18*, 1518–1524. https://doi.org/10.1111/cmi.12658.

Pauli, N.T., Kim, H.K., Falugi, F., Huang, M., Dulac, J., Henry Dunand, C., Zheng, N.Y., Kaur, K., Andrews, S.F., Huang, Y., *et al.* (2014). *Staphylococcus aureus* infection induces protein A-mediated immune evasion in humans. J. Exp. Med. *211*, 2331–2339. https://doi.org/10.1084/jem.20141404.

Quintela, J.C., de Pedro, M.A., Zollner, P., Allmaier, G., and García-del Portillo, F. (1997). Peptidoglycan structure of Salmonella typhimurium growing within cultured mammalian cells. Mol Microbiol 23, 693-704.

Radolf, J.D., Deka, R.K., Anand, A., Šmajs, D., Norgard, M.V., and Yang, X.F. (2016). *Treponema pallidum*, the syphilis spirochete: making a living as a stealth pathogen. Nat. Rev. Microbiol. *14*, 744–759. https://doi.org/10.1038/nrmicro.2016.141.

Rae, C.S., Geissler, A., Adamson, P.C., and Portnoy, D.A. (2011). Mutations of the *Listeria monocytogenes* peptidoglycan N-deacetylase and O-acetylase result in enhanced lysozyme sensitivity, bacteriolysis, and hyperinduction of innate immune pathways. Infect Immun 79, 3596-3606.

Rico-Pérez, G., Pezza, A., Pucciarelli, M.G., de Pedro, M.A., Soncini, F.C., and García-del Portillo, F. (2016). A novel peptidoglycan D,L-endopeptidase induced by Salmonella inside eukaryotic cells contributes to virulence. Mol Microbiol 99, 546-556.

Sampson, T.R., Napier, B.A., Schroeder, M.R., Louwen, R., Zhao, J., Chin, C.Y., Ratner, H.K., Llewellyn, A.C., Jones, C.L., Laroui, H., *et al.* (2014). A CRISPR-Cas system enhances envelope integrity mediating antibiotic resistance and inflammasome evasion. Proc Natl Acad Sci U S A *111*, 11163-11168.

Schenk, M., Mahapatra, S., Le, P., Kim, H.J., Choi, A.W., Brennan, P.J., Belisle, J.T., and Modlin, R.L. (2016). Human NOD2 recognizes structurally unique muramyl dipeptides from *Mycobacterium leprae*. Infect. Immun. *84*, 2429–2438. https://doi.org/10.1128/IAI.00334-16.

Sellin, M.E., Muller, A.A., Felmy, B., Dolowschiak, T., Diard, M., Tardivel, A., Maslowski, K.M., and Hardt, W.D. (2014). Epithelium-intrinsic NAIP/NLRC4 inflammasome drives infected enterocyte expulsion to restrict Salmonella replication in the intestinal mucosa. Cell Host Microbe *16*, 237-248.

Shi, J., Zhao, Y., Wang, Y., Gao, W., Ding, J., Li, P., Hu, L., and Shao, F. (2014). Inflammatory caspases are innate immune receptors for intracellular LPS. Nature *514*, 187–192. https://doi.org/10.1038/nature13683.

Stapels, D.A., Kuipers, A., von Kockritz-Blickwede, M., Ruyken, M., Tromp, A.T., Horsburgh, M.J., de Haas, C.J., van Strijp, J.A., van Kessel, K.P., and Rooijakkers, S.H. (2016). *Staphylococcus aureus* protects its immune-evasion proteins against degradation by neutrophil serine proteases. Cell Microbiol *18*, 536-545.

Stijlemans, B., Caljon, G., Van Den Abbeele, J., Van Ginderachter, J.A., Magez, S., and De Trez, C. (2016). Immune evasion strategies of *Trypanosoma brucei* within the mammalian host: progression to pathogenicity. Front. Immunol. 7, 233. https://doi.org/10.3389/fimmu.2016.00233.

Storek, K.M., and Monack, D.M. (2015). Bacterial recognition pathways that lead to inflammasome activation. Immunol. Rev. *265*, 112–129. https://doi.org/10.1111/imr.12289.

Stowe, I., Lee, B., and Kayagaki, N. (2015). Caspase-11: arming the guards against bacterial infection. Immunol. Rev. *265*, 75–84. https://doi.org/10.1111/imr.12292.

Strober, W., Murray, P.J., Kitani, A., and Watanabe, T. (2006). Signalling pathways and molecular interactions of NOD1 and NOD2. Nat. Rev. Immunol. *6*, 9–20.

Suda, Y., Kim, Y.M., Ogawa, T., Yasui, N., Hasegawa, Y., Kashihara, W., Shimoyama, T., Aoyama, K., Nagata, K., Tamura, T., *et al.* (2001). Chemical structure and biological activity of a lipid A component from *Helicobacter pylori* strain 206. J. Endotoxin Res. *7*, 95–104.

Sycuro, L.K., Wyckoff, T.J., Biboy, J., Born, P., Pincus, Z., Vollmer, W., and Salama, N.R. (2012). Multiple peptidoglycan modification networks modulate *Helicobacter pylori*'s cell shape, motility, and colonization potential. PLOS Pathog *8*, e1002603.

Tan, Y., Zanoni, I., Cullen, T.W., Goodman, A.L., and Kagan, J.C. (2015). Mechanisms of Toll-like receptor 4 endocytosis reveal a common immune-evasion strategy used by pathogenic and commensal bacteria. Immunity *43*, 909–922. https://doi.org/10.1016/j.immuni.2015.10.008.

Typas, A., Banzhaf, M., Gross, C.A., and Vollmer, W. (2011). From the regulation of peptidoglycan synthesis to bacterial growth and morphology. Nat. Rev. Microbiol. *10*, 123–136. https://doi.org/10.1038/nrmicro2677.

van der Burg, S.H., Arens, R., Ossendorp, F., van Hall, T., and Melief, C.J. (2016). Vaccines for established cancer: overcoming the challenges posed by immune evasion. Nat. Rev. Cancer *16*, 219–233. https://doi.org/10.1038/nrc.2016.16.

Viala, J., Chaput, C., Boneca, I.G., Cardona, A., Girardin, S.E., Moran, A.P., Athman, R., Mémet, S., Huerre, M.R., Coyle, A.J., *et al.* (2004). Nod1 responds to peptidoglycan delivered by the *Helicobacter pylori* cag pathogenicity island. Nat. Immunol. *5*, 1166–1174.

Vijayrajratnam, S., Pushkaran, A.C., Balakrishnan, A., Vasudevan, A., Biswas, R., and Mohan, C.G. (2016). Bacterial peptidoglycan with amidated meso -diaminopimelic acid evades NOD1 recognition: an insight on NOD1 structure-recognition study. Biochem J. *473*, 4573-4592.

Wang, G., Maier, S.E., Lo, L.F., Maier, G., Dosi, S., and Maier, R.J. (2010). Peptidoglycan deacetylation in *Helicobacter pylori* contributes to bacterial survival by mitigating host immune responses. Infect Immun *78*, 4660-4666.

Wang, Q., Matsuo, Y., Pradipta, A.R., Inohara, N., Fujimoto, Y., and Fukase, K. (2016). Synthesis of characteristic Mycobacterium peptidoglycan (PGN) fragments utilizing with chemoenzymatic preparation of meso-diaminopimelic acid (DAP), and their modulation of innate immune responses. Org Biomol Chem *14*, 1013-1023.

Whitfield, C., and Trent, M.S. (2014). Biosynthesis and export of bacterial lipopolysaccharides. Annu. Rev. Biochem. *83*, 99–128. https://doi.org/10.1146/annurev-biochem-060713-035600.

Woodhams, K.L., Chan, J.M., Lenz, J.D., Hackett, K.T., and Dillard, J.P. (2013). Peptidoglycan fragment release from *Neisseria meningitidis*. Infect. Immun. *81*, 3490–3498. https://doi.org/10.1128/IAI.00279-13.

Wright, G.J., and Rayner, J.C. (2014). *Plasmodium falciparum* erythrocyte invasion: combining function with immune evasion. PLOS Pathog. *10*, e1003943. https://doi.org/10.1371/journal.ppat.1003943.

Yanagihara, Y., Kamisango, K., Yasuda, S., Kobayashi, S., Mifuchi, I., Azuma, I., Yamamura, Y., and Johnson, R.C. (1984). Chemical compositions of cell walls and polysaccharide fractions of spirochetes. Microbiol. Immunol. *28*, 535–544.

Zughaier, S.M., Lindner, B., Howe, J., Garidel, P., Koch, M.H., Brandenburg, K., and Stephens, D.S. (2007). Physicochemical characterization and biological activity of lipooligosaccharides and lipid A from *Neisseria meningitidis*. J. Endotoxin Res. *13*, 343–357. https://doi.org/10.1177/0968051907084435

Subversion of Macrophage Functions by Bacterial Protein Toxins and Effectors

3

Muyang Wan, Yan Zhou* and Yongqun Zhu*

Life Sciences Institute and Innovation Center for Cell Signaling Network, Zhejiang University, Hangzhou, China.

*Correspondence: zhuyongqun@zju.edu.cn and zhouyanlsi@zju.edu.cn

http://doi.org/10.21775/9781910190692.03

Abstract

Macrophages represent one of the first lines of host immune defences against the invasion of pathogenic bacteria. Many receptors, immune signalling pathways and cellular processes in macrophages, including Toll-like receptors, Nod-like receptors, phagocytosis, autophagy and programmed cell death, are involved in combating the infection of bacterial pathogens. For efficient colonization in the host, bacterial pathogens have evolved diverse mechanisms to interfere with macrophage functions to evade host defences. The major weapons utilized by bacterial pathogens are protein toxins and effectors secreted via specific bacterial secretion systems, including type I–VII secretion apparatuses. In recent years, great advances have been achieved in understanding how bacterial toxins and effectors subvert immune signalling and cellular processes of macrophages. In this review, we focus on the toxins and effectors that modulate the phagocytosis, intracellular immune signalling pathways, autophagy and programmed cell death processes of macrophages from the bacterium *Legionella pneumophila, Shigella flexneri, Listeria monocytogenes, Salmonella* spp., *Yersinia* spp., enteropathogenic *E. coli* and *Mycobacterium tuberculosis.*

Introduction

As professional phagocytes, macrophages engulf invading bacteria as particles for degradation and counteract the bacterial infection as the first line of host immune defences. Macrophages recognize the invading bacteria via pathogen-associated molecular patterns (PAMPs) by pattern recognition receptors (PRRs), or indirectly by Opsonins, including complement components and immunoglobulin G (Flannagan *et al.*, 2009; Mosser, 1994). Among PRRs, Toll-like receptor (TLR) 1, TLR2, TLR4, TLR-5, TLR6 and TLR10 embed in the cell membrane and are responsible for the recognition of microorganisms on the cell surface, while TLR-3, TLR-7, TLR-8 and TLR-9 are present in endosomes (Bryant *et al.*, 2015; Medzhitov and Janeway, 2002). The extracellular N-terminal leucine-rich repeat (LRR) domains of TLRs form homo- or heterodimers to directly recognize various PAMPs, such as lipopolysaccharide (LPS), flagellin and CpG DNAs. For instance, TLR-4 senses the LPS of Gram-negative bacteria, while TLR-5 recognizes bacterial flagellin (Bryant *et al.*, 2015). The C-terminal Toll-interleukin-1 receptor (TIR) domain of TLRs exists in the intracellular cytoplasm and recruits adaptor molecules to activate downstream immune signalling pathways, including mitogen-activated protein kinase (MAPK), NF-κB and type I interferon pathways. Different from TLRs, NOD-like receptors (NLRs) are cytoplasmic receptors, which sense intracellular bacterial PAMPs and subsequently form large protein complexes known as inflammasomes (Bryant and Fitzgerald, 2009; Zhao and Shao, 2016). The human gene encodes more than 20 NLRs, including NALPs, IPAF and NAIPs. Similarly to TLRs, different NLRs sense distinct stimuli from bacterial pathogens to form specific inflammasomes, which activate caspase-1 and induce downstream inflammatory responses, such as the secretion of the pro-inflammatory cytokines, including IL-1β and IL-18, and pyroptosis. For instance, Naip2 and Naip5 serve as specific inflammasome receptors for T3SS rod proteins and bacterial flagellin, respectively (Zhao *et al.*, 2011). Recognition of T3SS rod proteins or flagellin promotes the formation of Naip2/5-NLRC4 inflammasome complexes and activation of caspase-1 to produce IL-1β and induce cell pyrotosis. Caspase-4 and caspase-11 function as specific intracellular LPS receptors to sense invading bacteria in the cytoplasm and activate downstream Gasdermin D (GSDMD) to induce pyroptosis (Shi *et al.*, 2015; Shi *et al.*, 2014). In addition, STING is a specific receptor in the endoplasmic reticulum (ER) for bacterial cyclic di-GMP (Burdette *et al.*, 2011). cGAS is a newly identified receptor for bacterial or viral DNAs, which are released into the cytoplasm (Sun *et al.*, 2013).

Phagocytosis is a major defence mechanism of macrophages, which is initiated by opsonin receptors, such as Fcγ receptors (FcγRs) and complement receptor 3 (CR3) (Flannagan *et al.*, 2009). These receptors recognize Opsonins, including immunoglobulin G (IgG) and complement component are attached to the bacterial surface, and start an intracellular signalling cascade that activates remodelling of the actin cytoskeleton and initiates phagocytosis. For example, for FcγRs, phagocytosis is initiated by Src-family kinases, which catalyse tyrosine phosphorylation on the immunoreceptor tyrosine-based activation motif (ITAM) within the receptor's cytoplasmic domain (Flannagan *et al.*, 2009). Phosphorylated tyrosines in ITAM act as docking sites for the SH2 domain of the tyrosine kinase SyK. SyK in turn recruits and activates downstream signalling molecules, including phosphatidylinositol 3-kinase (PI3K), phospholipase C-gamma (PLCγ) and the RhoGEF Vav, which ultimately activates actin cytoskeleton signalling to remodel the cell membrane and engulf pathogens (Crowley *et al.*, 1997; Mócsai *et al.*, 2010). Phosphatidylinositols (PtdIns), which are abundant components of the cell membrane, play important roles in actin cytoskeleton remodelling during phagocytosis (Pizarro-Cerdá *et al.*, 2015). PtdIns-kinases and phospholipases cooperate to produce different types of phosphoinositides. Phosphoinositides act as second messengers or directly recruit downstream proteins to change membrane morphology and remodel the actin cytoskeleton for phagosome formation and maturation. For example, phosphatidylinositol-4,5-bisphosphate (PtdIns(4,5)P2) and phosphatidylinositol-3,4,5-trisphosphate (PtdIns(3,4,5)P3) accumulate at sites of particle engagement and are essential for phagosome formation (Flannagan *et al.*, 2009). After internalization of a pathogen and sealing of the membrane, phagosome maturation starts immediately and is accomplished via sequential fusion with early endosomes and late endosomes of the endocytic pathway, and finally fuses with lysosomes to form phagolysosomes. Proteins and protein complexes in the endocytic pathway including Rab GTPases, Arf GTPases and endosomal-sorting complex required for transport (ESCRT), play important roles in the phagosome maturation process. There is an accumulation of PtdIns3P on the cytosolic leaflet of the limiting membrane of the early phagosome (Flannagan *et al.*, 2012). The v-ATPases inserted into the phagosome membrane translocates H^+ across the bilayer and acidify the phagosomal lumen. Phagolysosomes finally become extremely acidic and oxidative organelles, which eliminate and degrade bacterial pathogens. In addition to phagocytosis, the cellular process of autophagy and programed cell death also plays key roles in clearing invading bacteria in macrophages. Many chemokines, cytokines, anti-microbial effectors are released by macrophages to initiate pro-inflammatory responses upon recognition of invading pathogens. Through presenting pathogen-derived antigens to T cells, macrophages also promote host adaptive immunity responses to counteract the bacterial invasion.

Despite the essential roles in host innate immune defences, macrophages are also specific host cells for many intracellular bacterial pathogens, such as *Mycobacterium tuberculosis*, *Legionella pneumophila* and *Coxiella burnetii* for survival and replication. Some intracellular bacterial pathogens, such as *Listeria monocytogenes*, *Shigella* spp., *Salmonella* spp. and *Yersinia* spp., invade both epithelial cells and macrophages. These intracellular bacteria have evolved distinct strategies to counteract the defence pathways of macrophages for survival. For extracellular bacteria, such as enteropathogenic *E. coli* (EPEC) and enterohaemorrhagic *E. coli* (EHEC), they do not directly invade macrophages. To avoid degradation by macrophages, extracellular bacterial pathogens mainly inhibit phagocytosis and cytokine secretion of macrophages. Both extracellular and intracellular bacterial pathogens escape macrophage defences predominantly through protein toxins and effectors secreted via specifically evolved bacterial secretion systems, including type I to type VII secretion systems. In this review, we discuss how bacterial toxins or effector proteins from *L. pneumophila*, *S. flexneri*, *L. monocytogenes*, *Salmonella* spp., *Yersinia* spp., EHEC/EPEC and *M. tuberculosis* subvert various macrophage functions during infection (Tables 3.1 and 3.2).

Evasion of phagocytosis by macrophages

Because phagocytosis is a major defence mechanism of macrophages, many bacterial pathogens inhibit the macrophage phagocytosis. The extracellular pathogens, EPEC and EHEC, inject the type III effector EspJ to target Src kinase (Fig. 3.1 and Table 3.1). EspJ adopts a similar structural fold as the ADP-ribosyltransferase domain of the *Pseudomonas syringae* effector AvrPphF-ORF2 (Young *et al.*, 2014). Via a unique ADP-ribosyltransferase activity, EspJ catalyse the ADP-ribosylation and amidation on the residue E310 of Src. The novel modification on the residue E310 by EspJ blocks the kinase activity of Src, which transinhibits macrophage opsono-phagocytosis. SeoC, a homologue of EspJ from *S. enterica* subspecies salamae and arizonae, has the same ADP-ribosylation and amidation activity to inactivate Src and inhibits IgG-mediated phagocytosis of macrophage (Pollard *et al.*, 2016). The effector YopH from *Yersinia* spp. dephosphorylates the Crk-associated substrate $p130^{Cas}$ and Fyn-binding protein Fyb via protein tyrosine phosphatase activity (Black and Bliska, 1997; Deleuil *et al.*, 2003). Dephosphorylation of $p130^{Cas}$ and Fyn-binding protein Fyb

Table 3.1 Summary of the host targets, enzymatic activities and mechanisms of bacterial toxins and effectors that subvert macrophage phagocytosis and cytoplasmic immune signalling pathways

Effector	Bacteria	Host target	Enzymatic activity
Evasion of phagocytosis by macrophages			
EspJ	EPEC/EHEC	Src kinase	ADP-ribosyltransferase
YopH	*Yersinia* spp.	p130Cas, Fyb	Tyrosine phosphatas
LLO	*L. monocytogenes*	phagosome membrane	Pore-forming
IpaC	*S. flexneri*	phagosomal membrane	Lyse the phagosomal membrane
SifA	*S.* Typhimurium	PLEKHM1	Scaffold protein
SopD2	*S.* Typhimurium	Rab32	GAP
GtgE	*S.* Typhimurium	Rab32	Cysteine protease
SapM	*M. tuberculosis*	PtdIns3P	PtdIns3P phosphatase
EsxA	*M. tuberculosis*	Phagosomal membrane	Membrane-permeabilizing
EsxH	*M. tuberculosis*	Hrs	PROTEIN binding
DrrA/SidM	*L. pneumophila*	Rab1	Nucleotidyl transferase, GDF, GEF
LepB	*L. pneumophila*	Rab1	GAP
SidD	*L. pneumophila*	Rab1	DeAMPylase
AnkX	*L. pneumophila*	Rab1	Phosphocholine transferase
Lem3	*L. pneumophila*	Rab1	Phosphocholine hydrolase
CvpB	*C. burnetii*	PtdIns3P, phosphatidylserine	Lipid binding
Modulation of cytoplasmic immune signalling pathways			
OspI	*S. flexneri*	Ubc13	Deamidase
OspG	*S. flexneri*	Ubiquitin-conjugated E2	Half-kinase
IpaH9.8	*S. flexneri*	NEMO	Ubiquitin E3 ligase
IpaH1.4/ IpaH2.5	*S. flexneri*	HOIP	Ubiquitin E3 ligase
OspF	*S. flexneri*	MAPK	Phosphothreonine lyase
NleE	EPEC/EHEC	TAB2/3	Cysteine methyltransferase
NleB	EPEC/EHEC	GAPDH	N-acetyl-D-gulcosamine transferase
NleC	EPEC/EHEC	p65	Zinc metalloprotease
NleD	EPEC/EHEC	JNK	Zinc metalloprotease
Tir	EPEC/EHEC	TRAF, SHP-2	Protein binding
SpvC	*S.* Typhimurium	MAPK	Phosphothreonine lyase
PipA family	*S.* Typhimurium	p65, RelB	Protease
GogB	*S.* Typhimurium	SCF E3 ligase	Protein binding
YopJ	*Yersinia* spp.	MKK6, IKKβ	Acetyltransferase
PtpA	*M. tuberculosis*	JNK, p38, TAB3	Tyrosine phosphatase
LegK1	*L. pneumophila*	IκBα, p100	Serine/threonine kinase

GEF, Guanine nucleotide exchange factor; GAP, GTPase-activating protein; GDF, GDI-displacement factor.

Table 3.2 Summary of the host targets, enzymatic activities and mechanisms of bacterial toxins and effectors that interfere with autophagy and cell death signalling

Effector	Bacteria	Host target	Enzymatic activity
Interference with autophagy			
VirA	*S. flexneri*	Rab1	TBC-like GAP
IcsB	*S. flexneri*	Atg5	Protein binding
RavZ	*L. pneumophila*	LC3	Cysteine protease
LpSpl	*L. pneumophila*	Sphingosine-1 phosphate	Sphingosine-1 phosphate lyase
SapM	*M. tuberculosis*	PtdIns3P	PtdIns3P phosphatase
PE_PGRS47	*M. tuberculosis*	Unknown	Unknown
SseL	*S.* Typhimurium	ALIS	Deubiquitinase
Interference with cell death signalling			
SidF	*L. pneumophila*	BNIP3, Bcl-rambo, PtdIns(3,4)P2, PtdIns(3,4,5)P3	Phosphatidylinositol phosphatase
SdhA	*L. pneumophila*	LCV	Protect the LCV integrity
IpaB	*S. flexneri*	Cell membrane	Pore-forming
IpaD	*S. flexneri*	Mitochondrial	Membrane-permeabilizing
IpaH7	*S. flexneri*	GLMN	Ubiquitin E3 ligase
OspC3	*S. flexneri*	Caspase-4	Protein binding
SipB	*S.* Typhimurium	Caspase-1	Protein binding
YopM	*Yersinia* spp.	PRK1, PRK2	Protein binding
EsxA	*M. tuberculosis*	Cell membrane	Membrane-permeabilizing
CpnT	*M. tuberculosis*	NAD^+	NAD^+ glycohydrolase

ALIS, aggresome-like induced structures; GLMN, the glomulin/flagellar-associated protein 68; LCV, *Legionella*-containing vacuole.

destabilizes focal adhesion and inhibits bacterial internalization by macrophages (Deleuil *et al.*, 2003). *L. monocytogenes*, a facultative intracellular pathogen, escapes the phagosomes of macrophages by releasing the cholesterol-dependent cytolysin listeriolysin O (LLO). LLO is an essential virulence factor of *L. monocytogenes* with multiple functions. LLO oligomerizes and produces pores in the phagosome membrane, which induce the loss of H^+ and Ca^{2+} in phagosomes and inhibits the fusion of phagosome with endosomes and lysosomes (Birmingham *et al.*, 2008; Shaughnessy *et al.*, 2006). Destruction of the phagosome membrane enables *L. monocytogenes* to cross phagosomes and enter the cytoplasm for proliferation (Birmingham *et al.*, 2008).

The intracellular pathogen *S. flexneri* escapes from the phagosomes into the cytoplasm of infected cells using its type III secretion apparatus (Du *et al.*, 2016). The *Shigella* type III secretion apparatus is sufficient to promote efficient vacuole lysis and evasion of phagocytosis. IpaC, a translocator component and effector of the *Shigella* type III secretion apparatus mediates phagosome escape efficiency (Du *et al.*, 2016) (Fig. 3.1). However, the homologue

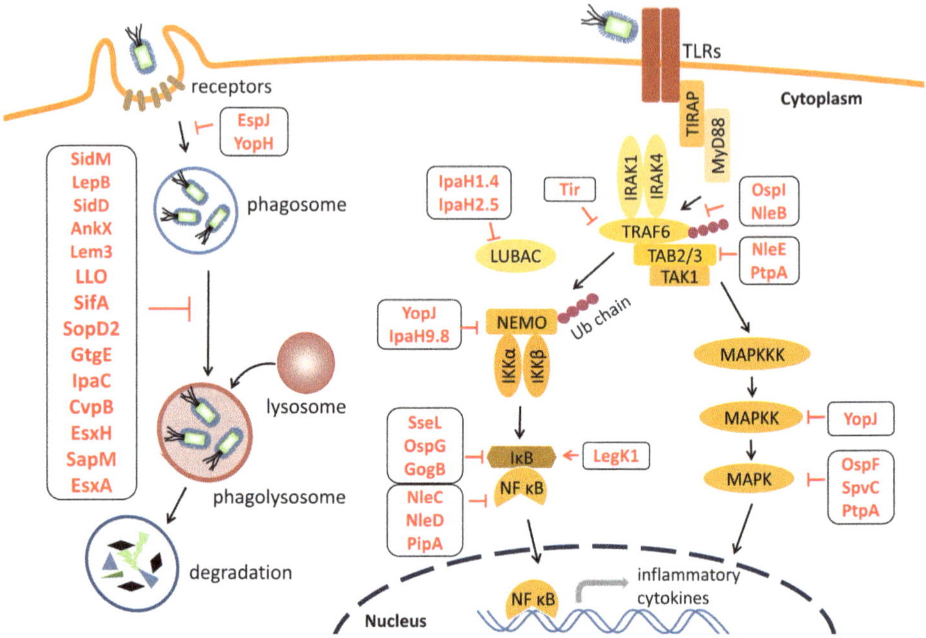

Figure 3.1 Bacterial evasion of phagocytosis by macrophages and modulation of cytoplasmic immune signalling pathways. The effectors SidM, LepB, SidD, AnkX, Lem3, IpaC, SifA, SopD2, GtgE, EsxH, EsxA, SapM, CvpB and the toxin LLO inhibit phagocytosis of macrophages. OspF, SpvC and PtpA inactivate MAPK. YopJ inactivates MAPKK and IKK. OspI, IpaH9.8, IpaH1.4, IpaH2.5, OspG, NleE, NleB, Tir and GogB inhibit NF-κB signalling, while LegK1 activates NF-κB signalling. LUBAC, linear ubiquitin chain assembly complex; MAPK, mitogen-activated protein kinase; TLR, Toll-like receptor; TIRAP, TIR domain containing adaptor protein.

of IpaC in *Salmonella*, SipC, cannot promote phagosome escape, suggesting a functional difference between IpaC and SipC (Osiecki *et al.*, 2001). In contrast to *Shigella*, *Salmonella* survive within both epithelial cells and macrophages by forming the *Salmonella*-containing vacuole (SCV) after entry into host cells. The formation of the SCV promotes the phagosome escape of *Salmonella*. Several type III effectors of *Salmonella* are involved in promoting the biogenesis of the SCV and inhibiting phagosome maturation. The *Salmonella* effector SifA directly interacts with Pleckstrin homology domain-containing protein family member 1 (PLEKHM1), a lysosome adaptor, to recruit Rab7 and the HOPS tethering complex, which transports phagolysosomal membranes to the SCV (McEwan *et al.*, 2015). The effectors, SopD2 and GtgE of *S.* Typhimurium, are a specific GAP and cysteine protease for Rab32, respectively (Spanò *et al.*, 2016). Rab32 mediates traffic to lysosome-related organelles in conjunction with components of the biogenesis of lysosome-related organelle complexes (BLOCs) (Spanò and Galán, 2012). SopD2 inactivates Rab32 by mimicking eukaryotic GAPs to catalyse GTP hydrolysis. Via the cysteine protease activity of the papain subfamily, GtgE cleaves Rab32 to disrupt the switch I region, which is required for interactions with downstream Rab effectors. SopD2 and GtgE redundantly inactivate Rab32 and prevent the recruitment of Rab32 to the SCV (Spano *et al.*, 2016).

M. tuberculosis secrets several effectors to arrest phagosomal maturation at different steps and to form the *Mycobacterium*-containing phagosomes (MCVs) in macrophages via type

VII secretion systems (also known as ESX systems). SapM, a secreted PtdIns3P phosphatase of *M. tuberculosis*, hydrolyses PtdIns3P to PtdIns, which arrests phagosomal maturation at an early stage (Puri *et al.*, 2013; Saleh and Belisle, 2000). EsxA (also known as ESAT-6), a key virulence factor of *M. tuberculosis*, is a membrane-permeabilizing protein and is essential for phagosome rupture and bacterial cytosolic translocation in macrophages (Ma *et al.*, 2015). EsxA lyses the membrane to arrest phagosomal maturation and to promote bacterial cytosolic escape and spread. An additional secreted protein EsxH of *M. tuberculosis* interacts with human hepatocyte growth factor-regulated tyrosine kinase substrate (Hgs/Hrs), a component of ESCRT (Mehra *et al.*, 2013). ESCRT is required to transport the MCVs to the lysosome and to restrict intracellular bacterial growth. The interaction between EsxH with Hrs disrupts ESCRT function and blocks the fusion of the MCV with the lysosome (Mehra *et al.*, 2013).

L. pneumophila reprogrammes the phagosomal maturation processes of macrophages to generate the *Legionella*-containing vacuoles (LCV) for intracellular survival and proliferation. *L. pneumophila* translocates nearly 300 effectors into macrophages via the Dot/Icm type IVb secretion system (T4BSS) during infection (Zhu *et al.*, 2011). Multiple effectors participate in the reprogramming of phagosome maturation and biogenesis of the LCV. The *Legionella* effector DrrA/SidM is a multifunctional protein possessing an N-terminal nucleotidyl transferase domain, a central dual-function domain with GDI-displacement factor (GDF) and RabGEF activities, and a C-terminal PtdIns4P-binding domain (Ingmundson *et al.*, 2007; Machner and Isberg, 2007). The C-terminal PtdIns4P-binding domain of SidM mediates the localization of the effector to the LCV. The central dual-function domain displaces GDI from the Rab1–GDI complex to release GDP-bound inactive Rab1. The dual-function domain subsequently catalyses the exchange of GDP in Rab1 to GTP for activation. The N-terminal nucleotidyl transferase domain of SidM AMPylates the GTP-bound active Rab1 for constitutive activation (Müller *et al.*, 2010). The effector SidD reverses the AMPylation modification of Rab1 via its deAMPylase activity (Neunuebel *et al.*, 2011; Tan and Luo, 2011). The deAMPylated Rab1 is further inactivated by the effector LepB, a GTPase-activating protein (GAP). An additional effector AnkX is a phosphocholine transferase that phosphocholinates Rab1 (Mukherjee *et al.*, 2011), while Lem3 removes the phosphocholination modification (Tan *et al.*, 2011). Via the collaboration of these effectors, *L. pneumophila* manipulates the membrane trafficking of ER-derived vesicles to the LCV and evades phagosome maturation. Similar to *L. pneumophila*, the Q fever bacterium *C. burnetii* exhibits a Dot/Icm type IVb secretion system to deliver effectors. *C. burnetii* renovates a lysosome into a mature *Coxiella*-containing vacuole (CCV), which makes it permissive to intracellular replication (van Schaik *et al.*, 2013). It has been reported that the *C. burnetii* effector Coxiella vacuolar protein B (CvpB) binds PtdIns3P and phosphatidylserine (PS) on the CCVs to perturb the activity of the phosphatidylinositol 5-kinase PIKfyve and manipulate PtdIns3P metabolism for optimal vacuole development (Martinez *et al.*, 2016). However, it is still largely unknown how the effectors of *C. burnetii* antagonize elimination by lysosomes.

Modulation of cytoplasmic immune signalling pathways

Cytoplasmic immune signalling pathways, including NF-κB, MAPK and type I interferon signalling pathways are activated upon bacterial recognition by immune receptors in macrophages. The NF-κB signalling pathway plays a pivotal role in the innate immune response by regulating the transcription of inflammation-related genes (Li and Verma, 2002). NF-κB signalling is initiated by cell surface receptors, including TLRs, and is transduced by cytosolic death-domain-containing adaptors to activate TRAF-mediated polyubiquitination, which sequentially activates the downstream IKK complex. The IKK complex phosphorylates IκB and induces its degradation by the proteasome. Upon IκB degradation, NF-κB translocates from the cytoplasm to the nucleus to initiate the gene expression of inflammation-related cytokines (Li and Verma, 2002). MAPK pathways consist of a three-tiered cascade of kinases. MAPK family kinases, including ERK1/2, c-Jun N-terminal kinase (JNK and p38, are phosphorylated and activated by MAPK kinase (MAPKK), which is activated by upstream MAPKK kinase kinase (MAPKKK) (Zheng and Guan, 1994). MAPK pathways regulate the expression of many inflammation-related cytokines. With the exception of the roles in immune defence, MAPK pathways regulate cell proliferation.

To inhibit the secretion of pro-inflammatory cytokines, bacterial effectors usually cooperate to target key signalling molecules of cytoplasmic immune signalling pathways in macrophages (Fig. 3.1 and Table 3.1). *S. flexneri* secretes several type III effectors to interfere with the NF-κB and MAPK signalling pathways. The *Shigella* effector OspI harbours a unique deamidase activity with a papain-like fold. OspI catalyses a specific deamidation reaction on the glutamine residue Q100 of Ubc13, a key ubiquitin-conjugating E2 enzyme in the NF-κB signalling pathway, which converts Q100 to a glutamic acid residue E100 (Fu *et al.*, 2013; Sanada *et al.*, 2012). The unique deamidation modification disrupts the interactions between Ubc13 and the ubiquitin E3 ligase TRAF6, and impairs the formation of Lys63-linked polyubiquitin chains, resulting in inhibition of the TRAF6-NF-κB signalling pathway. The *Shigella* effector OspG is a half-kinase and binds to ubiquitin-conjugated E2 enzymes (UbcH5 and UbcH7) to interfere with the ubiquitination-mediated degradation of IκB-α via the proteasome (Kim *et al.*, 2005). The ubiquitin E3 ligase effector IpaH9.8 of *Shigella* inhibits NF-κB signalling by inducing polyubiquitination and degradation of NEMO (Ashida *et al.*, 2010). In addition, the *Shigella* effector IpaH1.4 and IpaH2.5 are two E3 ligase effectors. Both effectors directly interact with the HOIL-1L subunit of the linear ubiquitin chain assembly complex (LUBAC) to ubiquitinate and degrade the subunit HOIP, which irreversibly inactivates linear ubiquitination and blocks NF-κB nuclear translocation (de Jong *et al.*, 2016). Moreover, MAPK pathways are inhibited by *Shigella* during infection. The *Shigella* effector OspF harbours a classic D-motif at the N-terminus to specifically target dual-phosphorylated active MAPKs. Via a novel β-elimination catalytic mechanism, OspF functions as a phosphothreonine lyase to dephosphorylate the phosphothreonine residue in the activation loop of active MAPKs to β-methyldehydroalanine. This 'dephosphorylation' irreversibly inactivates MAPKs and inhibits MAPK signalling pathways during *Shigella* infection (Li *et al.*, 2007; Zhu *et al.*, 2007).

EPEC/EHEC suppresses the NF-κB and MAPK signalling pathways via multiple effectors, including NleE, NleB, NleC, NleH, and Tir. NleE is a unique cysteine methyltransferase, which methylates the cysteine residue in the zinc finger domain of TAB2/3 (Zhang *et al.*, 2011). The methylation modification on the cysteine residue impairs the sensing and binding capability of TAB2/3 for the Lys63-linked polyubiquitin chains, which completely disable TRAF6-TAK1-NF-κB signalling. NleB is an N-acetyl-D-gulcosamine transferase and interacts with the host GAPDH. GAPDH regulates the NF-κB pathway by interacting with the ubiquitin E3 ligase TRAF2 (Gao *et al.*, 2013). NleB transfers N-acetylglucosamine from UDP-GlcNAc to modify GAPDH at multiple sites, which disrupts GAPDH–TRAF2 interactions, resulting in NF-κB inactivation. Both NleC and NleD are zinc metalloproteases, and specifically cleave p65 and JNK, respectively, to block NF-κB and AP-1 activation (Baruch *et al.*, 2011). The effector Tir plays an important role in regulating pedestal formation for EPEC/EHEC adhesion to epithelium cells by activating actin polymerization. In addition, Tir participates in the modulation of NF-κB signalling pathways. Tir inhibits NF-κB signalling by interacting with cytoplasm-located TRAF adaptor proteins to induce their proteasomal-independent degradation and the recruitment of SHP-1 to decrease TRAF6 ubiquitination (Ruchaud-Sparagano *et al.*, 2011; Yan *et al.*, 2012).

Salmonella spp., *Yersinia* spp., *M. tuberculosis* and *L. pneumophila* also interfere with NF-κB and MAPK signalling pathways. The *Salmonella* effector SpvC, a homologue of OspF, inhibits the p38 MAPK signalling pathway via phosphothreonine lyase activity during *Salmonella* infection (Haneda *et al.*, 2012; Li *et al.*, 2007). The PipA family effectors of *S.* Typhimurium, including PipA, GogA and GtgA, cleave the p65 and RelB subunits to inhibit NF-κB response (Sun *et al.*, 2016). SpvD specifically interacts with the exportin Xpo2 to block the nuclear translocation of p65, which subsequently blocks NF-κB activation (Rolhion *et al.*, 2016). The *Salmonella* effector GogB targets the SCF E3 ligase via interactions with Skp1 and FBXO22 to inhibit IκB degradation and NF-κB activation (Pilar *et al.*, 2012). The *Yersinia* effector YopJ is an acetyltransferase that blocks the activation of both MAPK and NF-κB pathways by acetylating serine and threonine residues on MAPKK6 and IKKβ (Mukherjee *et al.*, 2006). *M. tuberculosis* PtpA is a secreted tyrosine phosphatase and binds to the host ubiquitin. Binding to ubiquitin activates PtpA phosphatase activity, which dephosphorylates active JNK and p38 (Wang *et al.*, 2015). PtpA also inhibits NF-κB signalling in macrophages by competitively binding to the ubiquitin-interacting domain of TAB3 (Wang *et al.*, 2015). Unlike most bacterial secreted effectors inhibiting NF-κB signalling, the *L. pneumophila* effectors LnaB and LegK1 function as activators of the NF-κB pathway (Ge *et al.*, 2009; Losick *et al.*, 2010). LegK1 mimics the host IKK to phosphorylate IκBα and other inhibitors of the IκB family, including p100 via the serine/threonine kinase activity to activate NF-κB signalling (Ge *et al.*, 2009). However, how LnaB activates NF-κB signalling remains completely unknown.

Interference with autophagy

Autophagy is an intracellular degradation process that targets unfolded proteins, lipids and damaged organelles for cellular homeostasis. It also combats the invasion of various bacterial pathogens, thereby playing an important role in host immune defences (Huang and Brumell, 2014). Autophagy undergoes the following steps: signal induction, membrane nucleation, cargo targeting, autophagosome formation, lysosomal fusion, and cargo degradation. Induction of autophagy is triggered by a series of signals, including nutrient limitation, damaged organelles or invading bacteria (Parzych and Klionsky, 2014). Formation of the double-membrane autophagosome is a hallmark of autophagy. The cytoplasmic sequestering membrane (also known as a phagophore) at the early stage of autophagy is initiated in the endoplasmic reticulum (ER) and mitochondria. An autophagosome is formed by the subsequent expansion and closure of the phagophore. The autophagosome finally fuses with the lysosome to form an autolysosome, which degrades autophagic cargo (Parzych and Klionsky, 2014). Autophagy-related (Atg) genes mainly regulate the formation of the autophagosome. Several proteins involved in vesicle trafficking and protein ubiquitination, such as Rab GTPases, SNARE complexes, ubiquitin E3 ligases and deubiquitinases, are also involved in autophagosome formation and maturation. Two ubiquitination-like conjugation systems of Atg proteins are essential for autophagosome maturation (Fujita *et al.*, 2008; Mizushima *et al.*, 1998; Tanida *et al.*, 2004). Atg12 and Atg8 (also known as LC3) are two ubiquitin-like proteins in autophagy. Atg12 is conjugated to Atg5 in the presence of the E1-like enzyme Atg7 and the E2-like enzyme Atg10, which promotes phagophore expansion. Atg12–Atg5 forms a complex with Atg16L1 and associates with the expanding membrane of the phagophore. The Atg12–Atg5–Atg16 complex is then released from the autophagosome when the phagophore is closed. The Atg12–Atg5–Atg16L1 complex directs Atg8 to the phagosome membrane. In the presence of the E1-like enzyme Atg7 and the E2-like enzyme Atg3, Atg12-Atg5 functions as an E3-like enzyme and conjugates phosphatidylethanolamine to Atg8, the second ubiquitin-like conjugate (Atg8-PE). The cysteine protease Atg4 cleaves the carboxyl terminus of Atg8 and exposes a glycine residue for the covalent attachment of PE, and is thus essential for LC3-PE formation. Atg4 also deconjugates and releases a portion of the LC3–PE conjugates from the autophagosome membrane to recycle LC3 for the formation of new autophagosomes. Cargo receptors and adaptor proteins play important roles in cargo targeting of autophagy. Cargo receptors usually harbour an LC3-interacting region (LIR) that recruits LC3-decorated autophagosomes. There are several adaptor proteins, including p62, NDP52, NBR1 and optineurin, which have been identified (Fujita *et al.*, 2008; Kirkin *et al.*, 2009; Wild *et al.*, 2011). They consist of both a ubiquitin-binding domain and LIR to recognize ubiquitin-associated specific cargo for autophagosomes.

To counteract autophagic degradation, bacteria have developed diverse strategies to avoid or exploit autophagy by interfering with autophagy signalling via effector proteins (Fig. 3.2 and Table 3.2). The *Shigella* effector VirA is a RabGAP protein that harbours a TBC-like dual finger motif with a novel structural fold. Via a dual-figure catalytic mechanism, VirA specifically inactivates the host Rab1 GTPase. Inactivation of Rab1 by VirA blocks the autophagic defence of host cells to facilitate bacterial cytoplasmic survival (Dong *et al.*, 2012). IcsB, an effector of *S. flexneri*, also promotes the bacteria to evade host cell autophagy. It has been proposed that IcsB blocks the autophagic protein Atg5 binding to VirG, a surface protein of *S. flexneri* important for bacterial actin-based motility, to abolish autophagic recognition (Ogawa *et al.*, 2005). The *Legionella* effector RavZ is a cysteine protease and consists of an N-terminal catalytic domain with an Ulp family deubiquitinase-like fold and a C-terminal PtdIns3P-binding module (Choy *et al.*, 2012; Horenkamp *et al.*, 2015). RavZ binds to the autophagosome via the PtdIns3P-binding module. Through its cysteine protease activity, RavZ cleaves the amide bond between the carboxyl-terminal glycine residue and an adjacent aromatic residue in Atg8 proteins. The cleavage deconjugates Atg8 from PE and releases Atg8 from the phospholipid membranes (Choy *et al.*, 2012). The irreversible inactivation of Atg8 by RavZ inhibits host autophagy during *Legionella* infection. The *Legionella* effector sphingosine-1 phosphate lyase (LpSpl) also modulates autophagy (Rolando *et al.*, 2016). LpSp1 exhibits a similar structure as human sphingosine-1 phosphate lyase. LpSp1 degrades sphingosine-1 phosphate and modulates the host sphingosine biosynthesis in infected macrophages to inhibit autophagy during *Legionella* infection. The described lipid phosphatase SapM of *M. tuberculosis* hydrolyses PtdIns3P to maintain the MCVs PtdIns3P-free, which

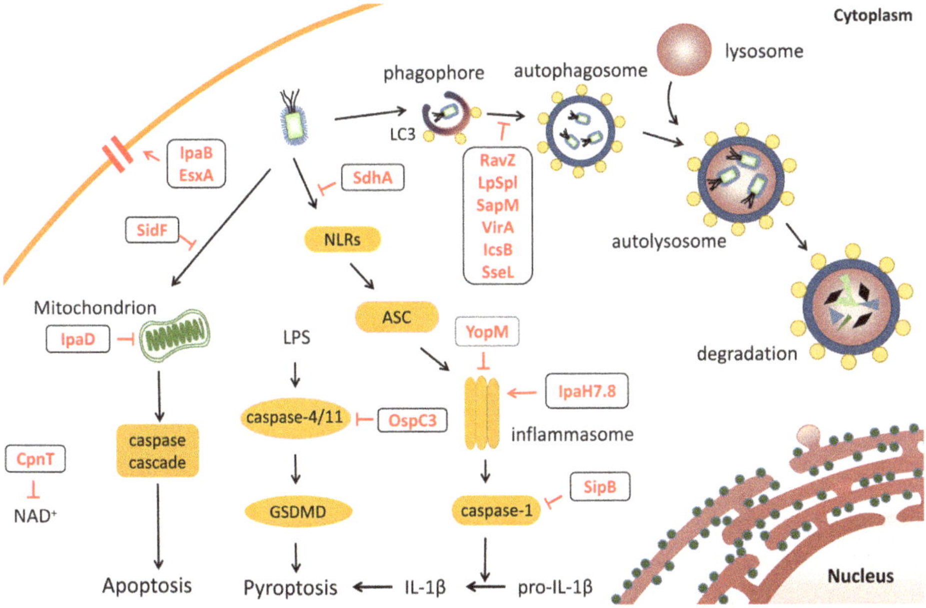

Figure 3.2 Interference with autophagy and cell death signalling by bacterial protein toxins and effectors. VirA, IcsB, RavZ, LpSp1, SapM and SseL inhibit host autophagy. SdhA, IpaH7.8, OspC3 and YopM inhibit pyroptosis, while IpaB induces pyroptosis. *Legionella* SidF antagonizes apoptosis, while *Shigella* IpaD activates apoptosis. EsxA and CpnT of *M. tuberculosis* induce macrophage necrosis. NLRs, NOD-like receptors; GSDMD, Gasdermin D.

inhibits phagocytosis and autophagosome formation in macrophages (Puri *et al.*, 2013; Saleh and Belisle, 2000). It was also reported that SapM blocks the autophagosome–lysosome fusion via binding Rab7 (Hu *et al.*, 2015). PE_PGRS47 (Rv2741), the other secreted effector of *M. tuberculosis*, has also been reported to inhibit macrophage autophagy, although its molecular mechanism remains unclear (Saini *et al.*, 2016). The *Salmonella* effector SseL functions as a deubiquitinase to clear the ubiquitinated structures, including SPI-2 T3SS-dependent aggregates and aggresome-like induced structures (ALIS) around the SCVs. Deubiquitination of ALIS prevents the recognition by the autophagy receptor p62 to recruit LC3 for autophagic degradation. The deubiquitinase activity of SseL lowers autophagic flux and favours intracellular *Salmonella* replication in macrophages (Mesquita *et al.*, 2012).

Interference with cell death signalling

The induction of host cell death is a conserved immune defence strategy throughout the animal and plant kingdoms. Apoptosis, necrosis and pyroptosis are three key cell death pathways in macrophages. Apoptosis is initiated by either intrinsic mitochondria/ER signalling or extrinsic death receptors. Both signalling pathways activate the downstream caspase cascade and induce cell shrinkage, membrane blebbing and the formation of apoptotic bodies. Necrosis is tightly regulated by RIP3, which phosphorylates the downstream pseudokinase MLKL. The phosphorylation modification relieves the autoinhibitory state of MLKL. The released N-terminal domain of MLKL oligomerizes in the cell membrane and disrupts the cellular ionic homeostasis via its pore-forming activity, which induces necrotic cell death (Sun and Wang, 2014). Pyroptosis is induced upon pathogen recognition by NLR receptors and the formation of corresponding inflammasomes (Zhao and Shao, 2016).

Bacterial pathogens modulate cell death pathways for intracellular survival and replication (Fig. 3.2 and Table 3.2). As previously described, the *L. pneumophila* effector LegK1 functions as a Ser/Thr kinase that phosphorylates IκBα and other inhibitors of the IκB family, which activates the NF-κB pathway (Ge *et al.*, 2009). Activation of NF-κB signalling prevents *Legionella*-infected macrophages from apoptotic cell death. The *Legionella* effector SidF, a phosphatidylinositol phosphatase, also blocks the apoptotic pathway by interacting with the Bcl2 family proapoptotic proteins, BNIP3 and Bcl-rambo (Banga *et al.*, 2007). The *Legionella* effector SdhA forms a unique ring-shaped membrane structure to protect the integrity of the LCV, which blocks DNA release from the LCV to the cytoplasm and prevents the DNA sensing by the AIM2 inflammasome (Creasey and Isberg, 2012; Ge *et al.*, 2012). Deletion of SdhA of *L. pneumophila* induces cell death of macrophages. In contrast to *L. pneumophila, S. flexneri* multiplies in the cytoplasm after entry into macrophages and induces rapid pyroptotic cell death. The bacteria released from dying macrophages then enter the intestinal epithelium via the basolateral surface of polarized enterocytes. The T3SS effector IpaB is a component of T3SS of *S. flexneri.* IpaB undergoes oligomerization on the cell membrane to form ion channels, which permeabilizes the host cell membrane and induces potassium influx. The potassium influx in turn activates the NLRC4/ASC/caspase-1 inflammasome and induces pyroptosis of macrophages (Senerovic *et al.*, 2012). In contrast to IpaB, *Salmonella* SipB, a homologous effector of IpaB, was reported to trigger macrophage cell death by directly binding to caspase-1 (Hernandez *et al.*, 2003). However, the real form of SipB-induced cell death of macrophages requires further validation and investigation. The translocator protein IpaD is an additional tip component of the T3SS apparatus needle of *Shigella.* IpaD activates a classical apoptotic pathway via the activation of host caspases and disruption of mitochondrial morphology to trigger apoptosis (Arizmendi *et al.*, 2016). The N-terminal domain of IpaD has been proposed to be required for apoptosis activation. Similarly, SipD, a structural homologue of IpaD from *Salmonella,* also triggers cell death in macrophages and exhibits the same cytotoxicity. However, the *Yersinia* counterpart LcrV, which lacks the N-terminal structures present in IpaD and SipD, cannot promote macrophage cell death (Arizmendi *et al.*, 2016). In addition, the effector IpaH7.8 of *Shigella* target the glomulin/flagellar-associated protein 68 (GLMN) for degradation via its ubiquitin E3 ligase activity. GLMN is a negative regulator of NLRP3 and NLRC4 inflammasomes. The IpaH7.8-mediated GLMN degradation activates NLRP3 and NLRC4 inflammasomes and caspase-1, resulting in macrophage pyroptotic cell death (Suzuki *et al.*, 2014). The *Shigella* effector OspC3 specifically interacts with the p19 subunit of caspase-4. The interaction inhibits the heterodimerization of the p19 subunit with the p10 subunit, which disrupts the activation of caspase-4 by LPS and antagonizes the detection of invading bacteria in the cytoplasm (Kobayashi *et al.*, 2013). *Y. pestis* and *Y. pseudotuberculosis* secrete the type III effector YopM to recruit and activate the host kinases PRK1 and PRK2, which inhibit activation of the pyrin inflammasome that is triggered by the RhoA-inactivating enzymatic activities of the *Yersinia* effectors, YopE and YopT (Chung *et al.*, 2016). *M. tuberculosis* activates macrophage necrosis during infection. EsxA (ESAT-6) secreted by the ESX-1 secretion system of *M. tuberculosis* produces pores in the MCV membranes via its pore-forming activity, facilitating bacterial escape from the vacuole. EsxA also induces necrosis of the macrophage and promotes bacterial cell-to-cell spread (Aguilo *et al.*, 2013; Simeone *et al.*, 2012). *M. tuberculosis* also secretes another endotoxin CpnT to induce necrotic cell death. The N-terminal domain of CpnT forms a channel to take up nutrients across the outer membrane of the bacteria. The C-terminal domain of CpnT harbours NAD^+ glycohydrolase activity to hydrolyse NAD^+, which causes NAD^+ depletion and necrosis of macrophages (Danilchanka *et al.*, 2014; Sun *et al.*, 2015).

Conclusion and future perspective

Infection of bacterial pathogens causes many diseases and elicits severe threats to global human health. As an important component of human innate immune system, macrophages are professional phagocytic cells that are present in diverse tissues to clear the invading bacteria. Following the long-term co-evolution between bacterial pathogens and host, bacterial pathogens have evolved diverse strategies to modulate macrophage functions to evade host defence. In this review, we discussed how bacterial pathogens, including *Legionella, Shigella, Salmonella, Listeria,* EPEC, *Yersinia* and *M. tuberculosis* subvert macrophage functions via their secreted toxins and effector proteins (Tables 3.1 and 3.2). On the basis of the distinct infection processes of these bacterial pathogens, the secreted toxins and effector proteins specifically target key signalling molecules of defence pathways to subtly interfere with macrophage functions, thereby promoting the establishment of safe niches for bacterial survival and proliferation. These toxins and effector proteins harbour unique biochemical activities or mimic eukaryotic proteins to target host substrates. The uncovered functions and mechanisms of these effectors enhance our understanding of bacterial pathogenesis and related eukaryotic signalling pathways.

The characterized effector proteins or toxins are also important for clinical therapies and biotechnological application. For example, the *Shigella* effector OspF harbours the phosphothreonine lyase activity and contains a classic 'D motif' to irreversibly inactivate the host MAPK (Li *et al.*, 2007; Zhu *et al.*, 2007). The unique biochemical activity and the specific targeting motif enable OspF to be applied in rewire kinase pathways in immune cells and the precise control of T-cell activation (Wei *et al.*, 2012). In addition, the *Shigella* effector VirA is a TBC-like RabGAP with a new structural fold (Dong *et al.*, 2012). VirA is essential for *Shigella* invasion and intracellular survival (Uchiya *et al.*, 1995). The gene deletion of VirA severely inhibits the *Shigella* infection. The novel structural fold of VirA and its essential role in infection indicate that VirA is a good candidate for designing new antibiotics to counteract the drug-resistant *Shigella*. As many effectors and toxins modulate the host immune pathways via novel biochemical activities, it will be very interesting to develop these bacterial virulence proteins as tools for clinical therapies and biomedical research.

Despite emerging progress, many questions still remain and require further investigation. For example, it has been revealed that *L. pneumophila* secrete LegK1 to activate NF-κB signalling and inhibit macrophage apoptosis (Ge *et al.*, 2009). Given that NF-κB signalling regulates the expression of pro-inflammation cytokines, it would be interesting to determine whether *Legionella* effectors inhibit the NF-κB pathway during infection. In addition, whether MAPK signalling pathways are involved in *Legionella* pathogenesis remains an open question. Intriguingly, *M. tuberculosis* interferes with macrophage cell death by EsxA-mediated and CpnT-induced necrosis (Danilchanka *et al.*, 2014; Ma *et al.*, 2015; Sun *et al.*, 2015). However, the roles of other types of cell death in *M. tuberculosis* modulation of macrophages are unknown. The bacterium *C. burnetii* invade and survival in macrophages, and contains a similar type IVb secretion system as *L. pneumophila*. However, the life cycle of *C. burnetii* is completely different from that of *L. pneumophila*. How *C. burnetii* modulates intracellular immune signalling pathways, autophagy and cell death of macrophages is only beginning to be illustrated. In addition, there are many functionally unknown toxins and effectors in *L. pneumophila, C. burnetii* and *M. tuberculosis*. Further efforts are required to investigate the mechanisms of these effectors in counteracting macrophage defence.

Acknowledgements

We apologize to colleagues whose work could not be cited due to space limitation. This work was supported by grants from National Natural Science Foundation of China (81561130162, 81322024, 81530068 and 31370722 to Y. Zhu, 81501717 to Y. Zhou), the Natural Science Foundation of Zhejiang province (LR13C050001 to Y. Zhu) and the Royal Society (NA140239 to Y. Zhu). M.W prepared the figures. M. W., Y. Zhou and Y. Zhu wrote the manuscript.

References

Aguilo, J.I., Alonso, H., Uranga, S., Marinova, D., Arbués, A., de Martino, A., Anel, A., Monzon, M., Badiola, J., Pardo, J., *et al.* (2013). ESX-1-induced apoptosis is involved in cell-to-cell spread of *Mycobacterium tuberculosis*. Cell. Microbiol. *15*, 1994–2005. https://doi.org/10.1111/cmi.12169.

Arizmendi, O., Picking, W.D., and Picking, W.L. (2016). Macrophage apoptosis triggered by IpaD from *Shigella flexneri*. Infect. Immun. *84*, 1857–1865. https://doi.org/10.1128/IAI.01483-15.

Ashida, H., Kim, M., Schmidt-Supprian, M., Ma, A., Ogawa, M., and Sasakawa, C. (2010). A bacterial E3 ubiquitin ligase IpaH9.8 targets NEMO/IKKgamma to dampen the host NF-kappaB-mediated inflammatory response. Nat. Cell Biol. *12*, 66-73. https://doi.org/10.1038/ncb2006.

Banga, S., Gao, P., Shen, X., Fiscus, V., Zong, W.X., Chen, L., and Luo, Z.Q. (2007). *Legionella pneumophila* inhibits macrophage apoptosis by targeting pro-death members of the Bcl2 protein family. Proc Natl Acad Sci U S A *104*, 5121-5126. https://doi.org/10.1073/pnas.0611030104.

Baruch, K., Gur-Arie, L., Nadler, C., Koby, S., Yerushalmi, G., Ben-Neriah, Y., Yogev, O., Shaulian, E., Guttman, C., Zarivach, R., *et al.* (2011). Metalloprotease type III effectors that specifically cleave JNK and NF-κB. EMBO J. *30*, 221–231. https://doi.org/10.1038/emboj.2010.297.

Birmingham, C.L., Canadien, V., Kaniuk, N.A., Steinberg, B.E., Higgins, D.E., and Brumell, J.H. (2008). Listeriolysin O allows *Listeria monocytogenes* replication in macrophage vacuoles. Nature *451*, 350–354. https://doi.org/10.1038/nature06479.

Black, D.S., and Bliska, J.B. (1997). Identification of p130Cas as a substrate of Yersinia YopH (Yop51), a bacterial protein tyrosine phosphatase that translocates into mammalian cells and targets focal adhesions. EMBO J. *16*, 2730–2744. https://doi.org/10.1093/emboj/16.10.2730.

Bryant, C., and Fitzgerald, K.A. (2009). Molecular mechanisms involved in inflammasome activation. Trends Cell Biol. *19*, 455–464. https://doi.org/10.1016/j.tcb.2009.06.002.

Bryant, C.E., Gay, N.J., Heymans, S., Sacre, S., Schaefer, L., and Midwood, K.S. (2015). Advances in Toll-like receptor biology: Modes of activation by diverse stimuli. Crit. Rev. Biochem. Mol. Biol. *50*, 359–379. https://doi.org/10.3109/10409238.2015.1033511.

Burdette, D.L., Monroe, K.M., Sotelo-Troha, K., Iwig, J.S., Eckert, B., Hyodo, M., Hayakawa, Y., and Vance, R.E. (2011). STING is a direct innate immune sensor of cyclic di-GMP. Nature *478*, 515–518. https://doi.org/10.1038/nature10429.

Choy, A., Dancourt, J., Mugo, B., O'Connor, T.J., Isberg, R.R., Melia, T.J., and Roy, C.R. (2012). The Legionella effector RavZ inhibits host autophagy through irreversible Atg8 deconjugation. Science *338*, 1072–1076. https://doi.org/10.1126/science.1227026.

Chung, L.K., Park, Y.H., Zheng, Y., Brodsky, I.E., Hearing, P., Kastner, D.L., Chae, J.J., and Bliska, J.B. (2016). The *Yersinia* virulence factor YopM hijacks host kinases to inhibit Type III effector-triggered activation of the pyrin inflammasome. Cell Host Microbe *20*, 296–306. https://doi.org/10.1016/j.chom.2016.07.018.

Creasey, E.A., and Isberg, R.R. (2012). The protein SdhA maintains the integrity of the Legionella-containing vacuole. Proc. Natl. Acad. Sci. U.S.A. *109*, 3481–3486. https://doi.org/10.1073/pnas.1121286109.

Crowley, M.T., Costello, P.S., Fitzer-Attas, C.J., Turner, M., Meng, F., Lowell, C., Tybulewicz, V.L., and DeFranco, A.L. (1997). A critical role for Syk in signal transduction and phagocytosis mediated by Fcgamma receptors on macrophages. J. Exp. Med. *186*, 1027–1039. https://doi.org/10.1084/jem.186.7.1027.

Danilchanka, O., Sun, J., Pavlenok, M., Maueröder, C., Speer, A., Siroy, A., Marrero, J., Trujillo, C., Mayhew, D.L., Doornbos, K.S., *et al.* (2014). An outer membrane channel protein of *Mycobacterium tuberculosis* with exotoxin activity. Proc. Natl. Acad. Sci. U.S.A. *111*, 6750–6755. https://doi.org/10.1073/pnas.1400136111.

de Jong, M.F., Liu, Z., Chen, D., and Alto, N.M. (2016). *Shigella flexneri* suppresses NF-κB activation by inhibiting linear ubiquitin chain ligation. Nat. Microbiol. *1*, 16084. https://doi.org/10.1038/nmicrobiol.2016.84.

Deleuil, F., Mogemark, L., Francis, M.S., Wolf-Watz, H., and Fällman, M. (2003). Interaction between the Yersinia protein tyrosine phosphatase YopH and eukaryotic Cas/Fyb is an important virulence mechanism. Cell. Microbiol. *5*, 53–64. https://doi.org/10.1046/j.1462-5822.2003.00236.x.

Dong, N., Zhu, Y., Lu, Q., Hu, L., Zheng, Y., and Shao, F. (2012). Structurally distinct bacterial TBC-like GAPs link Arf GTPase to Rab1 inactivation to counteract host defenses. Cell *150*, 1029–1041. https://doi.org/10.1016/j.cell.2012.06.050.

Du, J., Reeves, A.Z., Klein, J.A., Twedt, D.J., Knodler, L.A., and Lesser, C.F. (2016). The type III secretion system apparatus determines the intracellular niche of bacterial pathogens. Proc. Natl. Acad. Sci. U.S.A. *113*, 4794–4799. https://doi.org/10.1073/pnas.1520699113.

Flannagan, R.S., Cosío, G., and Grinstein, S. (2009). Antimicrobial mechanisms of phagocytes and bacterial evasion strategies. Nat. Rev. Microbiol. *7*, 355–366. https://doi.org/10.1038/nrmicro2128.

Flannagan, R.S., Jaumouillé, V., and Grinstein, S. (2012). The cell biology of phagocytosis. Annu. Rev. Pathol. *7*, 61–98. https://doi.org/10.1146/annurev-pathol-011811-132445.

Fu, P., Zhang, X., Jin, M., Xu, L., Wang, C., Xia, Z., and Zhu, Y. (2013). Complex structure of OspI and Ubc13: the molecular basis of Ubc13 deamidation and convergence of bacterial and host E2 recognition. PLOS Pathog. *9*, e1003322. https://doi.org/10.1371/journal.ppat.1003322.

Fujita, N., Itoh, T., Omori, H., Fukuda, M., Noda, T., and Yoshimori, T. (2008). The Atg16L complex specifies the site of LC3 lipidation for membrane biogenesis in autophagy. Mol. Biol. Cell *19*, 2092–2100. https://doi.org/10.1091/mbc.E07-12-1257.

Gao, X., Wang, X., Pham, T.H., Feuerbacher, L.A., Lubos, M.L., Huang, M., Olsen, R., Mushegian, A., Slawson, C., and Hardwidge, P.R. (2013). NleB, a bacterial effector with glycosyltransferase activity, targets GAPDH function to inhibit NF-κB activation. Cell Host Microbe *13*, 87–99. https://doi.org/10.1016/j.chom.2012.11.010.

Ge, J., Gong, Y.N., Xu, Y., and Shao, F. (2012). Preventing bacterial DNA release and absent in melanoma 2 inflammasome activation by a Legionella effector functioning in membrane trafficking. Proc. Natl. Acad. Sci. U.S.A. *109*, 6193–6198. https://doi.org/10.1073/pnas.1117490109.

Ge, J., Xu, H., Li, T., Zhou, Y., Zhang, Z., Li, S., Liu, L., and Shao, F. (2009). A Legionella type IV effector activates the NF-kappaB pathway by phosphorylating the IkappaB family of inhibitors. Proc. Natl. Acad. Sci. U.S.A. *106*, 13725–13730. https://doi.org/10.1073/pnas.0907200106.

Haneda, T., Ishii, Y., Shimizu, H., Ohshima, K., Iida, N., Danbara, H., and Okada, N. (2012). Salmonella type III effector SpvC, a phosphothreonine lyase, contributes to reduction in inflammatory response during intestinal phase of infection. Cell. Microbiol. *14*, 485–499. https://doi.org/10.1111/j.1462-5822.2011.01733.x.

Hernandez, L.D., Pypaert, M., Flavell, R.A., and Galán, J.E. (2003). A Salmonella protein causes macrophage cell death by inducing autophagy. J. Cell Biol. *163*, 1123–1131. https://doi.org/10.1083/jcb.200309161.

Horenkamp, F.A., Kauffman, K.J., Kohler, L.J., Sherwood, R.K., Krueger, K.P., Shteyn, V., Roy, C.R., Melia, T.J., and Reinisch, K.M. (2015). The Legionella anti-autophagy effector RavZ targets the autophagosome via PI3P- and curvature-sensing motifs. Dev. Cell *34*, 569–576. https://doi.org/10.1016/j.devcel.2015.08.010.

Hu, D., Wu, J., Wang, W., Mu, M., Zhao, R., Xu, X., Chen, Z., Xiao, J., Hu, F., Yang, Y., *et al.* (2015). Autophagy regulation revealed by SapM-induced block of autophagosome-lysosome fusion via binding RAB7. Biochem. Biophys. Res. Commun. *461*, 401–407. https://doi.org/10.1016/j.bbrc.2015.04.051.

Huang, J., and Brumell, J.H. (2014). Bacteria-autophagy interplay: a battle for survival. Nat. Rev. Microbiol. *12*, 101–114. https://doi.org/10.1038/nrmicro3160.

Ingmundson, A., Delprato, A., Lambright, D.G., and Roy, C.R. (2007). *Legionella pneumophila* proteins that regulate Rab1 membrane cycling. Nature *450*, 365–369. https://doi.org/10.1038/nature06336.

Kim, D.W., Lenzen, G., Page, A.L., Legrain, P., Sansonetti, P.J., and Parsot, C. (2005). The *Shigella flexneri* effector OspG interferes with innate immune responses by targeting ubiquitin-conjugating enzymes. Proc. Natl. Acad. Sci. U.S.A. *102*, 14046–14051. https://doi.org/10.1073/pnas.0504466102.

Kirkin, V., Lamark, T., Sou, Y.S., Bjørkøy, G., Nunn, J.L., Bruun, J.A., Shvets, E., McEwan, D.G., Clausen, T.H., Wild, P., *et al.* (2009). A role for NBR1 in autophagosomal degradation of ubiquitinated substrates. Mol. Cell 33, 505–516. https://doi.org/10.1016/j.molcel.2009.01.020.

Kobayashi, T., Ogawa, M., Sanada, T., Mimuro, H., Kim, M., Ashida, H., Akakura, R., Yoshida, M., Kawalec, M., Reichhart, J.M., *et al.* (2013). The Shigella OspC3 effector inhibits caspase-4, antagonizes inflammatory cell death, and promotes epithelial infection. Cell Host Microbe *13*, 570-583. https://doi.org/10.1016/j.chom.2013.04.012.

Li, H., Xu, H., Zhou, Y., Zhang, J., Long, C., Li, S., Chen, S., Zhou, J.M., and Shao, F. (2007). The phosphothreonine lyase activity of a bacterial type III effector family. Science *315*, 1000–1003.

Li, Q., and Verma, I.M. (2002). NF-kappaB regulation in the immune system. Nat. Rev. Immunol. *2*, 725–734. https://doi.org/10.1038/nri910.

Losick, V.P., Haenssler, E., Moy, M.Y., and Isberg, R.R. (2010). LnaB: a *Legionella pneumophila* activator of NF-kappaB. Cell. Microbiol. *12*, 1083–1097. https://doi.org/10.1111/j.1462-5822.2010.01452.x.

Ma, Y., Keil, V., and Sun, J. (2015). Characterization of *Mycobacterium tuberculosis* EsxA membrane insertion: roles of N- and C-terminal flexible arms and central helix-turn-helix motif. J Biol Chem *290*, 7314-7322. https://doi.org/10.1074/jbc.M114.622076.

Machner, M.P., and Isberg, R.R. (2007). A bifunctional bacterial protein links GDI displacement to Rab1 activation. Science *318*, 974–977. https://doi.org/10.1126/science.1149121.

Martinez, E., Allombert, J., Cantet, F., Lakhani, A., Yandrapalli, N., Neyret, A., Norville, I.H., Favard, C., Muriaux, D., and Bonazzi, M. (2016). *Coxiella burnetii* effector CvpB modulates phosphoinositide metabolism for optimal vacuole development. Proc Natl Acad Sci U S A *113*, E3260-3269. https://doi.org/10.1073/pnas.1522811113.

McEwan, D.G., Richter, B., Claudi, B., Wigge, C., Wild, P., Farhan, H., McGourty, K., Coxon, F.P., Franz-Wachtel, M., Perdu, B., *et al.* (2015). PLEKHM1 regulates Salmonella-containing vacuole biogenesis and infection. Cell Host Microbe *17*, 58–71. https://doi.org/10.1016/j.chom.2014.11.011.

Medzhitov, R., and Janeway, C.A., Jr. (2002). Decoding the patterns of self and nonself by the innate immune system. Science *296*, 298-300. https://doi.org/10.1126/science.1068883.

Mehra, A., Zahra, A., Thompson, V., Sirisaengtaksin, N., Wells, A., Porto, M., Köster, S., Penberthy, K., Kubota, Y., Dricot, A., *et al.* (2013). *Mycobacterium tuberculosis* type VII secreted effector EsxH targets host ESCRT to impair trafficking. PLOS Pathog. *9*, e1003734. https://doi.org/10.1371/journal.ppat.1003734.

Mesquita, F.S., Thomas, M., Sachse, M., Santos, A.J., Figueira, R., and Holden, D.W. (2012). The Salmonella deubiquitinase SseL inhibits selective autophagy of cytosolic aggregates. PLOS Pathog. *8*, e1002743. https://doi.org/10.1371/journal.ppat.1002743.

Mizushima, N., Noda, T., Yoshimori, T., Tanaka, Y., Ishii, T., George, M.D., Klionsky, D.J., Ohsumi, M., and Ohsumi, Y. (1998). A protein conjugation system essential for autophagy. Nature *395*, 395-398. https://doi.org/10.1038/26506.

Mócsai, A., Ruland, J., and Tybulewicz, V.L. (2010). The SYK tyrosine kinase: a crucial player in diverse biological functions. Nat. Rev. Immunol. *10*, 387–402. https://doi.org/10.1038/nri2765.

Mosser, D.M. (1994). Receptors on phagocytic cells involved in microbial recognition. Immunol. Ser. *60*, 99–114.

Mukherjee, S., Keitany, G., Li, Y., Wang, Y., Ball, H.L., Goldsmith, E.J., and Orth, K. (2006). *Yersinia* YopJ acetylates and inhibits kinase activation by blocking phosphorylation. Science *312*, 1211–1214. https://doi.org/10.1126/science.1126867.

Mukherjee, S., Liu, X., Arasaki, K., McDonough, J., Galán, J.E., and Roy, C.R. (2011). Modulation of Rab GTPase function by a protein phosphocholine transferase. Nature *477*, 103–106. https://doi.org/10.1038/nature10335.

Müller, M.P., Peters, H., Blümer, J., Blankenfeldt, W., Goody, R.S., and Itzen, A. (2010). The Legionella effector protein DrrA AMPylates the membrane traffic regulator Rab1b. Science *329*, 946–949. https://doi.org/10.1126/science.1192276.

Neunuebel, M.R., Chen, Y., Gaspar, A.H., Backlund, P.S., Yergey, A., and Machner, M.P. (2011). De-AMPylation of the small GTPase Rab1 by the pathogen *Legionella pneumophila*. Science 333, 453–456. https://doi.org/10.1126/science.1207193.

Ogawa, M., Yoshimori, T., Suzuki, T., Sagara, H., Mizushima, N., and Sasakawa, C. (2005). Escape of intracellular Shigella from autophagy. Science *307*, 727–731. https://doi.org/10.1126/science.1106036.

Osiecki, J.C., Barker, J., Picking, W.L., Serfis, A.B., Berring, E., Shah, S., Harrington, A., and Picking, W.D. (2001). IpaC from Shigella and SipC from Salmonella possess similar biochemical properties but are functionally distinct. Mol. Microbiol. *42*, 469–481. https://doi.org/10.1046/j.1365-2958.2001.02654.x.

Parzych, K.R., and Klionsky, D.J. (2014). An overview of autophagy: morphology, mechanism, and regulation. Antioxid. Redox Signal. *20*, 460–473. https://doi.org/10.1089/ars.2013.5371.

Pilar, A.V., Reid-Yu, S.A., Cooper, C.A., Mulder, D.T., and Coombes, B.K. (2012). GogB is an anti-inflammatory effector that limits tissue damage during Salmonella infection through interaction with human FBXO22 and Skp1. PLOS Pathog. *8*, e1002773. https://doi.org/10.1371/journal.ppat.1002773.

Pizarro-Cerdá, J., Kühbacher, A., and Cossart, P. (2015). Phosphoinositides and host-pathogen interactions. Biochim. Biophys. Acta *1851*, 911–918. https://doi.org/10.1016/j.bbalip.2014.09.011.

Pollard, D.J., Young, J.C., Covarelli, V., Herrera-Leon, S., Connor, T.R., Fookes, M., Walker, D., Echeita, A., Thomson, N.R., Berger, C.N., *et al.* (2016). The type III secretion system effector SeoC of *Salmonella enterica* subspecies salamae and arizonae ADP-ribosylates Src and inhibits opsono-phagocytosis. Infect Immun. https://doi.org/10.1128/IAI.00704-16.

Puri, R.V., Reddy, P.V., and Tyagi, A.K. (2013). Secreted acid phosphatase (SapM) of *Mycobacterium tuberculosis* is indispensable for arresting phagosomal maturation and growth of the pathogen in guinea pig tissues. PLOS ONE *8*, e70514. https://doi.org/10.1371/journal.pone.0070514.

Rolando, M., Escoll, P., Nora, T., Botti, J., Boitez, V., Bedia, C., Daniels, C., Abraham, G., Stogios, P.J., Skarina, T., *et al.* (2016). *Legionella pneumophila* S1P-lyase targets host sphingolipid metabolism and restrains autophagy. Proc. Natl. Acad. Sci. U.S.A. *113*, 1901–1906. https://doi.org/10.1073/pnas.1522067113.

Rolhion, N., Furniss, R.C., Grabe, G., Ryan, A., Liu, M., Matthews, S.A., and Holden, D.W. (2016). Inhibition of Nuclear Transport of NF-κB p65 by the Salmonella Type III Secretion System Effector SpvD. PLOS Pathog. *12*, e1005653. https://doi.org/10.1371/journal.ppat.1005653.

Ruchaud-Sparagano, M.H., Mühlen, S., Dean, P., and Kenny, B. (2011). The enteropathogenic E. coli (EPEC) Tir effector inhibits NF-κB activity by targeting TNFα receptor-associated factors. PLOS Pathog. *7*, e1002414. https://doi.org/10.1371/journal.ppat.1002414.

Saini, N.K., Baena, A., Ng, T.W., Venkataswamy, M.M., Kennedy, S.C., Kunnath-Velayudhan, S., Carreño, L.J., Xu, J., Chan, J., Larsen, M.H., *et al.* (2016). Suppression of autophagy and antigen presentation by *Mycobacterium tuberculosis* PE_PGRS47. Nat. Microbiol. *1*, 16133. https://doi.org/10.1038/nmicrobiol.2016.133.

Saleh, M.T., and Belisle, J.T. (2000). Secretion of an acid phosphatase (SapM) by *Mycobacterium tuberculosis* that is similar to eukaryotic acid phosphatases. J. Bacteriol. *182*, 6850–6853. https://doi.org/10.1128/JB.182.23.6850-6853.2000.

Sanada, T., Kim, M., Mimuro, H., Suzuki, M., Ogawa, M., Oyama, A., Ashida, H., Kobayashi, T., Koyama, T., Nagai, S., *et al.* (2012). The *Shigella flexneri* effector OspI deamidates UBC13 to dampen the inflammatory response. Nature *483*, 623–626. https://doi.org/10.1038/nature10894.

Senerovic, L., Tsunoda, S.P., Goosmann, C., Brinkmann, V., Zychlinsky, A., Meissner, F., and Kolbe, M. (2012). Spontaneous formation of IpaB ion channels in host cell membranes reveals how Shigella induces pyroptosis in macrophages. Cell Death Dis. *3*, e384. https://doi.org/10.1038/cddis.2012.124.

Shaughnessy, L.M., Hoppe, A.D., Christensen, K.A., and Swanson, J.A. (2006). Membrane perforations inhibit lysosome fusion by altering pH and calcium in *Listeria monocytogenes* vacuoles. Cell. Microbiol. *8*, 781–792.

Shi, J., Zhao, Y., Wang, K., Shi, X., Wang, Y., Huang, H., Zhuang, Y., Cai, T., Wang, F., and Shao, F. (2015). Cleavage of GSDMD by inflammatory caspases determines pyroptotic cell death. Nature *526*, 660–665. https://doi.org/10.1038/nature15514.

Shi, J., Zhao, Y., Wang, Y., Gao, W., Ding, J., Li, P., Hu, L., and Shao, F. (2014). Inflammatory caspases are innate immune receptors for intracellular LPS. Nature *514*, 187–192. https://doi.org/10.1038/nature13683.

Simeone, R., Bobard, A., Lippmann, J., Bitter, W., Majlessi, L., Brosch, R., and Enninga, J. (2012). Phagosomal rupture by *Mycobacterium tuberculosis* results in toxicity and host cell death. PLOS Pathog. *8*, e1002507. https://doi.org/10.1371/journal.ppat.1002507.

Spanò, S., and Galán, J.E. (2012). A Rab32-dependent pathway contributes to Salmonella typhi host restriction. Science 338, 960–963. https://doi.org/10.1126/science.1229224.

Spanò, S., Gao, X., Hannemann, S., Lara-Tejero, M., and Galán, J.E. (2016). A bacterial pathogen targets a host Rab-Family GTPase defense pathway with a GAP. Cell Host Microbe *19*, 216–226. https://doi.org/10.1016/j.chom.2016.01.004.

Sun, H., Kamanova, J., Lara-Tejero, M., and Galán, J.E. (2016). A Family of Salmonella Type III Secretion Effector Proteins Selectively Targets the NF-κB Signaling Pathway to Preserve Host Homeostasis. PLOS Pathog. *12*, e1005484. https://doi.org/10.1371/journal.ppat.1005484.

Sun, J., Siroy, A., Lokareddy, R.K., Speer, A., Doornbos, K.S., Cingolani, G., and Niederweis, M. (2015). The tuberculosis necrotizing toxin kills macrophages by hydrolyzing NAD. Nat. Struct. Mol. Biol. *22*, 672–678. https://doi.org/10.1038/nsmb.3064.

Sun, L., and Wang, X. (2014). A new kind of cell suicide: mechanisms and functions of programmed necrosis. Trends Biochem. Sci. *39*, 587–593. https://doi.org/10.1016/j.tibs.2014.10.003.

Sun, L., Wu, J., Du, F., Chen, X., and Chen, Z.J. (2013). Cyclic GMP-AMP synthase is a cytosolic DNA sensor that activates the type I interferon pathway. Science *339*, 786–791. https://doi.org/10.1126/science.1232458.

Suzuki, S., Mimuro, H., Kim, M., Ogawa, M., Ashida, H., Toyotome, T., Franchi, L., Suzuki, M., Sanada, T., Suzuki, T., *et al.* (2014). Shigella IpaH7.8 E3 ubiquitin ligase targets glomulin and activates inflammasomes to demolish macrophages. Proc. Natl. Acad. Sci. U.S.A. *111*, E4254–63. https://doi.org/10.1073/pnas.1324021111.

Tan, Y., Arnold, R.J., and Luo, Z.Q. (2011). *Legionella pneumophila* regulates the small GTPase Rab1 activity by reversible phosphorylcholination. Proc. Natl. Acad. Sci. U.S.A. *108*, 21212–21217. https://doi.org/10.1073/pnas.1114023109.

Tan, Y., and Luo, Z.Q. (2011). *Legionella pneumophila* SidD is a deAMPylase that modifies Rab1. Nature *475*, 506–509. https://doi.org/10.1038/nature10307.

Tanida, I., Ueno, T., and Kominami, E. (2004). LC3 conjugation system in mammalian autophagy. Int. J. Biochem. Cell Biol. *36*, 2503–2518. https://doi.org/10.1016/j.biocel.2004.05.009.

Uchiya, K., Tobe, T., Komatsu, K., Suzuki, T., Watarai, M., Fukuda, I., Yoshikawa, M., and Sasakawa, C. (1995). Identification of a novel virulence gene, virA, on the large plasmid of Shigella, involved in invasion and intercellular spreading. Mol. Microbiol. *17*, 241–250. https://doi.org/10.1111/j.1365-2958.1995.mmi_17020241.x.

van Schaik, E.J., Chen, C., Mertens, K., Weber, M.M., and Samuel, J.E. (2013). Molecular pathogenesis of the obligate intracellular bacterium *Coxiella burnetii*. Nat. Rev. Microbiol. *11*, 561–573. https://doi.org/10.1038/nrmicro3049.

Wang, J., Li, B.X., Ge, P.P., Li, J., Wang, Q., Gao, G.F., Qiu, X.B., and Liu, C.H. (2015). *Mycobacterium tuberculosis* suppresses innate immunity by coopting the host ubiquitin system. Nat. Immunol. *16*, 237–245. https://doi.org/10.1038/ni.3096.

Wei, P., Wong, W.W., Park, J.S., Corcoran, E.E., Peisajovich, S.G., Onuffer, J.J., Weiss, A., and Lim, W.A. (2012). Bacterial virulence proteins as tools to rewire kinase pathways in yeast and immune cells. Nature *488*, 384–388. https://doi.org/10.1038/nature11259.

Wild, P., Farhan, H., McEwan, D.G., Wagner, S., Rogov, V.V., Brady, N.R., Richter, B., Korac, J., Waidmann, O., Choudhary, C., *et al.* (2011). Phosphorylation of the autophagy receptor optineurin restricts Salmonella growth. Science *333*, 228–233. https://doi.org/10.1126/science.1205405.

Yan, D., Wang, X., Luo, L., Cao, X., and Ge, B. (2012). Inhibition of TLR signaling by a bacterial protein containing immunoreceptor tyrosine-based inhibitory motifs. Nat Immunol *13*, 1063-1071. https://doi.org/10.1038/ni.2417.

Young, J.C., Clements, A., Lang, A.E., Garnett, J.A., Munera, D., Arbeloa, A., Pearson, J., Hartland, E.L., Matthews, S.J., Mousnier, A., *et al.* (2014). The *Escherichia coli* effector EspJ blocks Src kinase activity via amidation and ADP ribosylation. Nat. Commun. *5*, 5887. https://doi.org/10.1038/ncomms6887.

Zhang, L., Ding, X., Cui, J., Xu, H., Chen, J., Gong, Y.N., Hu, L., Zhou, Y., Ge, J., Lu, Q., *et al.* (2011). Cysteine methylation disrupts ubiquitin-chain sensing in NF-κB activation. Nature *481*, 204–208. https://doi.org/10.1038/nature10690.

Zhao, Y., and Shao, F. (2016). Diverse mechanisms for inflammasome sensing of cytosolic bacteria and bacterial virulence. Curr. Opin. Microbiol. *29*, 37–42. https://doi.org/10.1016/j.mib.2015.10.003.

Zhao, Y., Yang, J., Shi, J., Gong, Y.N., Lu, Q., Xu, H., Liu, L., and Shao, F. (2011). The NLRC4 inflammasome receptors for bacterial flagellin and type III secretion apparatus. Nature *477*, 596–600. https://doi.org/10.1038/nature10510.

Zheng, C.F., and Guan, K.L. (1994). Cytoplasmic localization of the mitogen-activated protein kinase activator MEK. J. Biol. Chem. *269*, 19947–19952.

Zhu, W., Banga, S., Tan, Y., Zheng, C., Stephenson, R., Gately, J., and Luo, Z.Q. (2011). Comprehensive identification of protein substrates of the Dot/Icm type IV transporter of *Legionella pneumophila*. PLOS ONE *6*, e17638. https://doi.org/10.1371/journal.pone.0017638.

Zhu, Y., Li, H., Long, C., Hu, L., Xu, H., Liu, L., Chen, S., Wang, D.C., and Shao, F. (2007). Structural insights into the enzymatic mechanism of the pathogenic MAPK phosphothreonine lyase. Mol. Cell *28*, 899–913. https://doi.org/10.1016/j.molcel.2007.11.011

Manipulation of Autophagy by Bacterial Pathogens Impacts Host Immunity

4

Tobias C. Kunz[1,2], Flávia Viana[1,2], Carmen Buchrieser[1,2*] and Pedro Escoll[1,2*]

[1]Institut Pasteur, Biologie des Bactéries Intracellulaires, Paris, France.
[2]Centre National de la Recherche Scientifique (CNRS) UMR 3525, Paris, France.

*Correspondence: cbuch@pasteur.fr and pescoll@pasteur.fr

http://doi.org/10.21775/9781910190692.04

Abstract

Autophagy is a highly conserved catabolic process, degrading unnecessary or damaged components in the eukaryotic cell to maintain cellular homeostasis, but it is also an intrinsic cellular defence mechanism to remove invading pathogens. A cross-talk between autophagy and innate or adaptive immune responses has been recently reported, whereby autophagy influences both innate and adaptive immunity, like the production and secretion of pro-inflammatory cytokines or MHC class II antigen presentation to T cells. Pathogenic bacteria have evolved diverse strategies to manipulate autophagy, mechanisms that also impact host immune responses at different levels. Here we discuss the influence of autophagy on self-autonomous, innate and adaptive immunity and then focus on how bacterial mechanisms that shape autophagy may impact the host immune system.

Introduction

Maintaining cellular homeostasis requires a coordinated control of metabolic pathways. Thus the cell needs anabolic processes to build new components, and catabolic processes to discard long-lived or damaged components. Autophagy is one of these catabolic processes that leads to lysosomal degradation of unnecessary or damaged proteins, lipids and organelles (Ohsumi, 2014; Rubinsztein *et al.*, 2012). The term 'autophagy' describes mainly three different processes: macroautophagy, microautophagy and chaperone-mediated autophagy. In this review we will focus on macroautophagy (hereafter called autophagy), which is characterized by the sequestration of cargo into double-membrane vesicles called autophagosomes (Miller and Celli, 2016).

Autophagy can be divided into selective and non-selective autophagy: in non-selective autophagy a random portion of the cytoplasm is engulfed to degrade long-lived components or to provide nutrients during starvation, while selective autophagy targets selected components for degradation. This includes organelle-specific autophagy, such as removal of mitochondria called mitophagy, and xenophagy which is the removal of invading bacteria (Deretic *et al.*, 2013; Huang and Brumell, 2014).

Intracellular bacteria grow and replicate inside host cells. These bacteria can be divided into two groups: those that replicate in the cytosol, such as *Listeria* or *Shigella*, and those that replicate in membrane-derived pseudo-organelles called pathogen-containing vacuoles (PCVs), such as *Legionella*, *Mycobacterium* or *Salmonella*. PCVs facilitate the replication, survival and dormancy of intracellular bacteria. For their formation bacteria exploit the host's membrane system, in particular the dynamic machineries involved in exo/endocytic traffic and autophagy. To enhance survival and proliferation in PCVs, bacteria secrete effector proteins that inhibit their degradation in lysosomes and facilitate the acquisition of membrane sources and nutrients (Ray *et al.*, 2009). In order to ensure survival, intracellular bacteria also need to counteract self-defence mechanisms of the host cell. In animals, cellular self-defence (also known as cell-autonomous immunity) synergizes with the whole-body protection provided by 'conventional' immunity (innate and adaptative immunity) to grant resistance to pathogens. While professional immune cells patrol the body in search of pathogens, cell-autonomous immunity guards both individual immune and non-immune cells against the immediate threat of infection. The current paradigm shows that cell-autonomous immunity against intracellular pathogens is based on the tight and continuous control of the composition and behaviour of distinct cellular compartments, such as the cytosol, organelles, endosomes, autophagosomes and lysosomes (Randow *et al.*, 2013). As the autophagy machinery removes intracellular pathogens by directing them to lysosomal degradation, xenophagy has emerged as an important cellular self-defence process to protect host cells against intracellular bacteria. In this case, engulfment of bacteria by autophagosomes triggers an integrated response within the infected cell, leading to pathogen elimination while professional immune cells are warned about the threat.

As autophagy is an important cellular self-defence mechanism, bacteria evolved various mechanisms to modulate autophagy such as inhibiting autophagosome formation, self-masking with host proteins to avoid recognition, escaping targeting to autophagosomes or blocking fusion of autophagosomes with lysosomes (Escoll *et al.*, 2016; Huang and Brumell, 2014). As autophagy is linked to immune responses, bacteria-induced modulation of autophagy impacts also innate and adaptive immune responses (Kuballa *et al.*, 2012).

Here, we will first discuss the general knowledge about autophagosome formation, elongation and maturation, the cross-talk between autophagy and immunity and then focus on the mechanisms used by bacterial pathogens to manipulate autophagy and how these autophagy subversions by bacteria impact the host immune response.

Autophagosome formation, elongation and maturation is a complex process involving conserved host factors

The pathway leading to autophagy includes signal induction, nucleation and elongation of the phagophore to form the autophagosome, closure of the autophagosome and fusion with lysosomes (Parzych and Klionsky, 2014). More than 35 Autophagy-related proteins (ATG-proteins) that have metazoan orthologues have been described in yeast to coordinate autophagosome formation and maturation (Ohsumi, 2014) (Fig. 4.1).

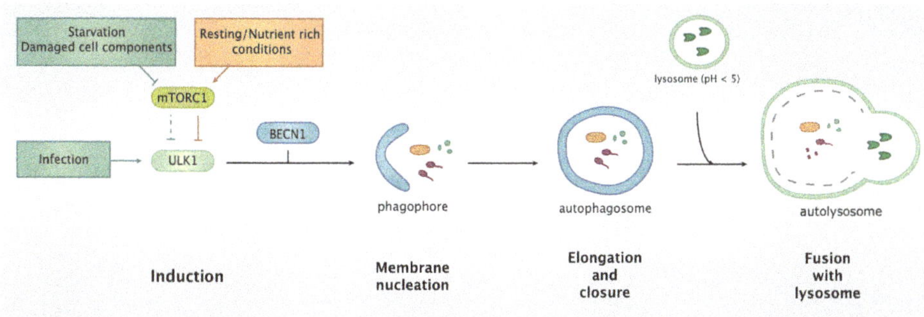

Figure 4.1 Schematic description of the autophagic pathway. Autophagy is triggered by starvation, damaged cellular components or cellular recognition of bacteria. These events activate the ULK1 complex. Under nutrient-rich conditions, mTORC1 inhibits autophagosome formation by repressing the ULK1 complex, however participation of mTORC1 in bacterial-induced autophagy is unclear. Once the ULK1 complex is activated, autophagosome formation begins with the recruitment of BECN1 and the nucleation of a small membrane called the phagophore, which starts to engulf the undesired material. The phagophore expands to form the autophagosome, a double-membrane compartment engulfing cyoplasmic targets (proteins, organelles or pathogens). This mature autophagosome then fuses with a lysosome for cargo degradation.

Autophagy induction can be triggered by different signals ranging from nutrient limitation (non-specific autophagy) to the recognition of pathogens or damaged cellular components (Deretic *et al.*, 2013). Most signals for the initiation of autophagy converge through the functions of the mammalian target of rapamycin complex 1 (mTORC1). Under nutrient-rich conditions the mTORC1 complex phosphorylates Unc-51-like kinase 1 (ULK1) and ATG13, which together with the FAK family kinase-interacting protein of 200kDa (FIP200) form the ULK1 complex. When mTORC1 phosphorylates ULK1 and ATG13, the complex cannot be recruited to the phagophore assembly site (PAS), therefore inhibiting autophagosome formation (Ganley *et al.*, 2009; Mizushima, 2010).

Under starvation or stress, mTORC1 becomes inactive and thus its inhibitory action on the ULK1 complex ceases which in turn phosphorylates FIP200 rendering the complex active (Ganley *et al.*, 2009). The activated ULK1 complex then translocates to the PAS and recruits other proteins such as beclin1 (BECN1, named in yeast Atg6) , ATG14-like protein (ATG14L) and phosphoinositide 3-kinase regulatory subunit 4 (PIK3R4; vacuolar protein sorting associated 15 Vps15 in yeast). Initiation of phagophore nucleation also results in engagement of PI3-kinase class III (PI3KC3; Vps34 in yeast), which produces PI3P at the phagophore and leads to the recruitment of WD-repeat domain phosphoinositide-interacting 1 (WIPI1; Atg18 in yeast) and WIPI2 (Atg21 in yeast). Like a cascade, these proteins recruit further downstream ATG proteins that aid phagophore assembly (Nair *et al.*, 2010; Polson *et al.*, 2014). The PAS needs a source of membranes to form the phagophore-expanding autophagosome. It has been shown that these are mainly mitochondrial-associated ER-membranes (MAMs) (Hamasaki *et al.*, 2013) although several other membranes have also been proposed as sites for nucleation of the phagophore, such as plasma membrane-derived vesicles, mitochondria and the Golgi apparatus (Tooze and Yoshimori, 2010).

Phagophore elongation is then mediated by two ubiquitination-like conjugation systems of ATG proteins. First, ATG7 (E1-like protein) and ATG10 (E2-like protein) conjugate

ATG12 and ATG5, promoting phagophore elongation (Mizushima *et al.*, 1998; Shintani *et al.*, 1999). ATG12–ATG5 forms a complex with ATG16L1, which associates with the expanding phagophore membrane and gets released after the autophagosome is formed (Mizushima *et al.*, 2001; 2003). The ATG12–ATG5–ATG16L1 complex is essential for the formation of the second ubiquitin-like conjugation system, where the protein light chain 3 (LC3; Atg8 in yeast) is conjugated to phosphatidylethanolamine (PE) in the phagophore membrane. This conjugation system relies on ATG7 (E1-like protein) and ATG3 (E2-like protein) that activate LC3. In addition, ATG4B cleaves the carboxy-terminus of LC3 and exposes a glycine residue. Activated LC3 is directed by ATG12–ATG5–ATG16L1 to the expanding phagophore, where the E3-like activity of ATG12–ATG5 conjugates PE to the previously exposed glycine residue of LC3 in the phagophore membrane. Once the phagophore is closed and the autophagosome is formed, the ATG12–ATG5–ATG16L1 complex is released from the autophagosome. ATG4B has also been shown to deconjugate a proportion of the LC3–PE complexes from mature autophagosomes, facilitating LC3 recycling for the formation of new autophagosomes (Fujita *et al.*, 2008; Tanida *et al.*, 2004). It is thought that recycling of LC3 is an important step in the maturation of fusion-capable autophagosomes. It has also been shown that LC3 mediates the hemifusion of vesicles and controls the size of autophagosomes in yeast (Nakatogawa *et al.*, 2007; Xie *et al.*, 2008).

Finally, a series of fusion events of autophagosomes with endosomes and lysosomes leads to their maturation into degrading autolysosomes. In mammalian cells the fusion of autophagosomes with lysosomes requires the small GTPase RAB7 (Ypt7 in yeast), the autophagosomal SNARE protein syntaxin 17, the lysosomal SNARE protein VAMP8 (vesicle-associated membrane protein 8) and several lysosomal membrane proteins, such as the lysosomal-associated membrane glycoprotein 2 (LAMP2). After autophagosome–lysosome fusion the cargo is degraded by the lysosomal hydrolases now present in the autolysosome (Eskelinen *et al.*, 2002; Itakura *et al.*, 2012; Jager *et al.*, 2004; Tanaka *et al.*, 2000).

For selective autophagy, such as xenophagy, an additional step of cargo selection is required, a process that is mediated by cargo receptors and adaptor proteins. Mammalian cargo specific receptors usually contain a LC3-interacting region (LIR) motif allowing the recruitment LC3-containing autophagosomes to the cargo (Liu *et al.*, 2012; Novak *et al.*, 2009; Polson *et al.*, 2014). In addition, many cargoes are ubiquitinated and then they are recognized by ubiquitin-binding protein adaptors that also contain LIRs. One of those proteins is p62 that, among other functions, directs bacteria to autophagosomes. Together with the adaptor proteins NDP52 and optineurin (OPTN), p62 is involved in *S.* Typhimurium recognition and its direction to autophagosomes (Boyle and Randow, 2013). NDP52 and p62 also target *Shigella flexneri* and *Listeria monocytogenes* to autophagosomes. Whereas p62 and NDP52 are recruited together during *S. flexneri* infection, p62 and NDP52 are recruited independently during *L. monocytogenes* infection, suggesting that *Shigella* and *Listeria* induce different pathways for selective autophagy (Mostowy *et al.*, 2011). Another example is the neighbour of BRCA1 gene 1 protein (NBR1) that is known to target *Francisella tularensis* to autophagy (Chong *et al.*, 2014) and also participates in the targeting of *S. flexneri* (Mostowy *et al.*, 2011), but it is not required for targeting of *S.* Typhimurium to autophagy (Zheng *et al.*, 2009). These studies highlight that selective autophagy uses distinct protein adaptors to target different types of bacteria.

Cross-talk between autophagy and the innate immune response

The first step of innate immune responses is the recognition of a pathogen that is mediated by a variety of pattern-recognition receptors (PRRs) like Toll-like receptors (TLRs), RIG-I-like receptors (RLRs) and NOD-like receptors (NLRs) that bind pathogen-associated molecular patterns (PAMPs) such as bacterial lipopolysaccharide (LPS), peptidoglycan, lipoproteins, flagellin or nucleid acids. As autophagic degradation of pathogens aids PAMP recognition by PRRs, autophagy is believed to have an important role in innate immunity. Additionally, binding of PAMPs to PRRs stimulates autophagy (Delgado *et al.*, 2009).

Toll-like receptors

TLRs are membrane-bound receptors found at the surface of the cell or in endosomal compartments. TLRs respond to LPS, lipotechoic acid, flagellin and bacterial nucleic acids (Deretic, 2012). After binding their cognate ligand, TLRs activate proinflammatory responses by triggering the production of cytokines such as tumour necrosis factor-α (TNF-α), interleukin 6 (IL-6) and IL-1β (Fig. 4.2).

Phagocytosed and cytosolic PAMPs are sequestered by autophagosomes and are delivered to the endosomally located and luminally oriented TLRs for PAMP recognition. Therefore autophagy takes part in the first steps of TLR activation (Desai *et al.*, 2015). On the other side, autophagy can also be induced upon TLR activation, through

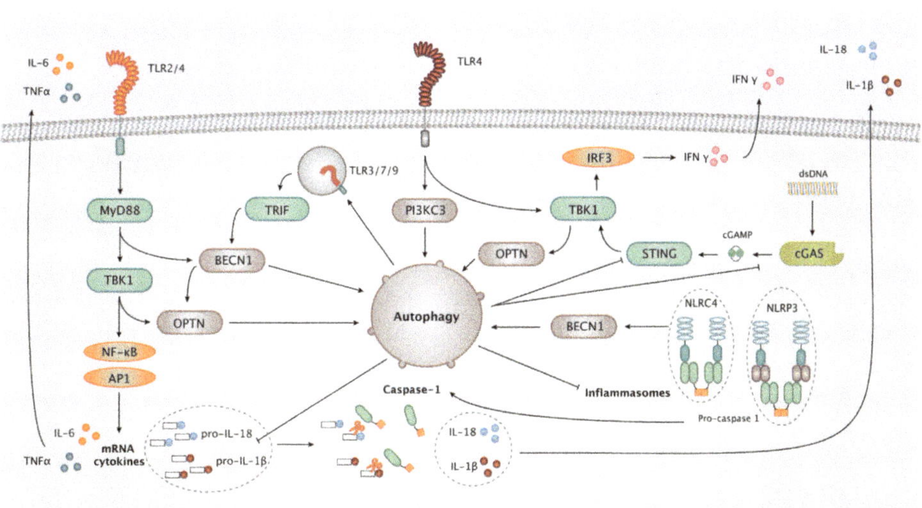

Figure 4.2 Cross-talk of autophagy and innate immunity. Activation of PRRs, such as TLRs, inflammasomes and cGAS, triggers signalling pathways (green components) that lead to activation of inflammatory transcription factors (orange components) and autophagy proteins that activate autophagy (brown components). Activation of inflammatory transcription factors leads to transcription and translation of cytokines such as TNF-α, IL-6 and IFN-γ, which are secreted, but also pro-IL-1β and pro-IL-18, which are then processed by inflammasome-activated caspase-1 into secreted IL-1β and IL-18. In turn, autophagy aids TLRs in meeting their cognate ligands, while it negatively regulates cytokine production by repressing cGAS synthesis of cGAMP and degrading precursor and mature forms of IL-1β and IL-18, and inflammasome components (dashed lines).

a MyD88/TRIF adaptor dependent process and their interaction with BECN1 (Shi *et al.*, 2008). TLR–autophagy cross-talk is supported by several other findings: TLR4 stimulates PI3KC3-dependent formation of cytosolic LC3 aggregates, enhancing the elimination of mycobacteria from macrophages (Xu *et al.*, 2007). Downstream of TLR4 activation, the TANK binding kinase 1 (TBK1) also links the TLR signalling pathway and autophagy as TBK1 phosphorylates the autophagic receptor OPTN, enhancing LC3 binding affinity and consequently the autophagic clearance of cytosolic *Salmonella enterica* (Wild *et al.*, 2011). TLR7 activation has also been shown to promote autophagic degradation of *Mycobacterium bovis* BCG (bacillus Calmette–Guérin) (Delgado *et al.*, 2008), and TLR2 induces autophagy in an ERK-dependent mechanism during *Listeria monocytogenes* infection (Anand *et al.*, 2011). Altogether these studies highlight that TLR activation elicits autophagy, thus activating cell-autonomous immunity after PAMP detection.

However, autophagy may also down-regulate TLR-induced responses. Inhibition of autophagy in macrophages and dendritic cells (DCs) through 3MA (3-methyladenine) or through siRNA knockdown of autophagy genes, leads to an increase in cytokine release in response to TLR3 or TLR4 agonists (Harris, 2011; Saitoh *et al.*, 2008). ATG16L1-deficient mice were shown to excessively activate caspase-1 in response to TLR4 stimulation by LPS, which led to increased IL-1β and IL-18 production. This suggests that ATG16L1 might have a regulatory function on the TLR4 signalling pathway and its depletion/malfunction may lead to increased inflammation (Saitoh *et al.*, 2008). Mature IL-1β protein can also be engulfed by autophagosomes in macrophages stimulated with TLR ligands, showing another way by which autophagy can down-regulate the production and secretion of cytokines (Harris *et al.*, 2011) (Fig. 4.2). Thus, autophagy influences TLR signalling and consequently impacts TLR-mediated cytokine production and secretion.

RIG-I-like receptors

RLRs sense cytosolic dsRNA or DNA, and thus recognize also nucleic acids from pathogens. PAMP recognition by RLRs as RIG-I and MDA5 triggers the production of type I interferon (IFN) by infected cells, as has been shown for cells infected with the intracellular pathogen *Legionella pneumophila* (Monroe *et al.*, 2009). This type I IFN activation pathway can be directly suppressed by several autophagy factors. For example, autophagy-defective *Atg5*$^{-/-}$ cells exhibit enhanced RLR signalling and increased IFN secretion, mostly due to the accumulation of dysfunctional mitochondria and increased generation of reactive oxygen species (ROS) which were largely responsible for the enhanced RLR signalling in *Atg5*$^{-/-}$ cells (Tal *et al.*, 2009).

cGAMP synthase

Bacterial or aberrant cytosolic DNA are also recognized by direct binding to cGAMP synthase (cGAS), a cytosolic protein that generates cyclic dinucleotides (CDNs) such as cGAMP within the host cytosol (Tao *et al.*, 2016). Host cell-generated CDNs activate the downstream stimulator of interferon genes (STING), a receptor that can be also activated by CDNs of intracellular bacteria (Burdette *et al.*, 2011). STING controls the activation of above-mentioned TBK1, which upon nucleic acid sensing and cGAMP synthesis triggers both IFN regulatory factor 3 (IRF3) phosporylation and type I IFN production. STING induced type I IFN production occurs during infection by intracellular bacteria such as *Mycobacterium tuberculosis* (Watson *et al.*, 2015). On the other hand, the autophagy activator protein ULK1 phosphorylates STING, inhibiting sustained type I IFN activation in response to dsDNA (Konno *et al.*, 2013) (Fig. 4.2). Furthermore, cGAS binding to BECN1 facilitates autophagic removal of cytosolic dsDNA and reduces excessive type I IFN responses (Liang *et al.*, 2014).

NOD-like receptors

NLRs are a class of receptors initiating a quick and potent inflammatory cytokine response to PAMPs. Upon sensing PAMPs, cytosolic NLRs form a signalling complex called the inflammasome. The inflammasome consists of several oligomerized NLRs that bind caspase-1 directly or through the adaptor protein called apoptosis-associated speck-like protein (ASC), which contains a caspase recruitment domain (CARD). These complexes cleave the protein precursor pro-caspase-1 into p10 and p20 subunits, activating caspase-1. Active caspase-1 then cleaves the presynthesized pro-IL-1β into the active form of the cytokine IL-1β, which is secreted (Rodgers *et al.*, 2014). IL-1β secretion relies on two different PRR activating signals: the NFκB-dependant expression of pro-IL-1β through TLR activation and a signal activating caspase-1 through inflammasomes to cleave pro-IL-1β (Eder, 2009) (Fig. 4.2). The inflammasome also cleaves pro-IL-18 into active IL-18 and orchestrates the programmed cell death known as pyroptosis (Miao *et al.*, 2011). Therefore, the inflammasome promotes inflammation by controlling the secretion of the strong proinflammatory cytokines IL-1β and IL-18 and avoiding that cells become a niche for the pathogens by activating pyroptotic cell death. These two cytokines are responsible for the recruitment of myeloid cells, including neutrophils, to sites of inflammation (Rider *et al.*, 2011).

NLRs that respond to PAMPs in macrophages and activate caspase-1 by inflammasome assembly are NLRP1, NLRP3, NLRC4, NLRP7, NLRP12 and AIM2 (absent in melanoma 2) (Latz *et al.*, 2013). Interestingly, a vast number of data strongly suggests that both machineries, autophagy and inflammasome, are highly interconnected and influence each other. Remnants of autophagosomes that were degraded during *S. flexneri* infection of epithelial cells are decorated with components of both the autophagy and the inflammasome machineries. Moreover, macrophages of caspase-$1^{-/-}$ mice display an excessive accumulation of autophagosomes during *S. flexneri* infection, suggesting that caspase-1 affects either autophagosome formation or maturation (Dupont *et al.*, 2009; Suzuki *et al.*, 2007). Activation of AIM2 with poly(dA:dT) or NLRP3 activation with uric acid crystals or nigericin leads to an increase in autophagosome formation in macrophages (Shi *et al.*, 2012). Autophagosomes have also been shown to sequester and degrade inflammasome components and cytokine precursors such as pro-IL-1β (Harris *et al.*, 2011; Shi *et al.*, 2012). Association between autophagy and inflammasome proteins was also shown in resting cells, where NLRP4 and NLRC4 formed a complex with BECN1, suggesting that NLRP4 sensing of bacteria leads to the initiation of BECN1-mediated autophagic responses (Jounai *et al.*, 2011). Autophagy and inflammation are also regulated by NOD1 and NOD2, a caspase-1 independent class of NLR proteins. NOD1 and NOD2 are intracellular sensors of peptidoglycan that induce autophagy by interacting with ATG16L1 and regulating IL-1β and IL-18 production through NFκB (Philpott *et al.*, 2013).

Key evidence for NLR signalling in anti-bacterial responses was gained from the analysis of mouse susceptibility to *L. pneumophila*. Indeed, only A/J mice are susceptible to infection due to a partial loss-of-function mutation in the Naip5 inflammasome, a component of the flagellin-NLCR4 pathway (Molofsky *et al.*, 2006; Zamboni *et al.*, 2006). Autophagosomes of macrophages from *L. pneumophila*-resistant C57BL/6J mice matured quickly and prevented efficient *L. pneumophila* replication when compared to autophagosomes of *L. pneumophila*-permissive Naip5 mutant A/J macrophages. This observation is reinforcing the idea that inflammasome activation and autophagy are intertwined processes during bacterial infection (Amer and Swanson, 2005). Additional results obtained in primary mouse macrophages support a model in which both caspase-1 and NLR components of inflammasomes are co-ordinately responding to *L. pneumophila* infection depending on the bacterial burden (Byrne *et al.*, 2013). According to this model, NLRC4 is complexed with autophagy component BECN1 in resting macrophages, inhibiting autophagy. Low levels of bacteria lead flagellin-bound NAIP5 to recruit NLRC4 to a complex containing the pro-caspase-1 protein, derepressing autophagy. When the capacity of autophagy to eliminate intracellular bacteria is exceeded, caspase-1 triggers pyroptosis to eliminate the pathogen's niche while initiating a potent inflammatory response (Byrne *et al.*, 2013).

Autophagy has recently been implied in maturation of natural killer cells (NKs). NK cells play an important role in early immune response through cytokine secretion and granule-mediated cytotoxicity of tumours and infected cells (Sun and Lanier, 2011). NK cells have been classified as members of one of the subsets (type 1) of innate lymphoid cells (ILCs), a recently identified group of cells from the lymphoid lineage that have emerging roles in mediating immune responses and in regulating tissue homeostasis (Walker *et al.*, 2013). Mice with a NK cell-specific deletion of ATG5 exhibited a reduced number of circulating NK cells compared to wild-type (WT) mice, whereas the amounts of peripheral T and B cells remained normal. Isolated NK progenitors from ATG5 deficient mice were unable to differentiate into immature NK or mature NK cells upon IL-15 exposure. The same effect was also observed by silencing ATG3 or ATG7 (O'Sullivan *et al.*, 2016; Wang *et al.*, 2016). Moreover lymphoid progenitors from *Atg5*$^{-/-}$ mice failed to generate mature type 2 and type 3 ILCs in the small intestine of recipient WT mice, showing the key role of autophagy in the development of ILCs *in vivo* (O'Sullivan *et al.*, 2016).

Autophagy cross-talks with adaptive immune responses

Considering the role of autophagy in the restriction and destruction of intracellular pathogens and its interplay with innate immunity, it is expected that autophagy is also a crucial process in adaptive immunity where it acts on the modulation of antigen processing and presentation to elicit the correct development and homeostasis of lymphocytes.

T cells are the main effectors of the adaptive immune system. They scan through the output of the proteolytic machineries of the cells to detect pathogen-derived peptides. CD4$^+$ helper T cells and cytotoxic CD8$^+$ T cells display a diverse receptor repertoire that allows them to recognize these peptides but this recognition is not direct as antigens need to be loaded to major histocompatibility class (MHC) molecules before being presented to T cells. MHC class I ligands are commonly generated by the proteasome and are presented to CD8$^+$ T cells while MHC class II loading peptides are produced by lysosomes and presented to CD4$^+$ T cells (Münz, 2010).

The classical concept suggests that intracellular antigens get processed and loaded onto MHC class I cells while extracellular ones go on MHC class II cells. However, as autophagy delivers cytoplasmic constituents for lysosomal degradation and MHC class II molecules are loaded with lysosomal products, autophagy also supports the processing of endogenous antigens for presentation by MHC class II (Dengjel *et al.*, 2005; Nimmerjahn *et al.*, 2003; Schmid *et al.*, 2007; Zhou *et al.*, 2005). Therefore, antigen-presenting cells such as macrophages and DCs can use the autophagy machinery to fuse autophagosomes containing bacteria-derived antigens with autolysosomes, which are afterwards loaded into MHC class II molecules for antigen presentation to CD4$^+$ T cells (Crotzer and Blum, 2009; Dengjel *et al.*, 2005). This antigen-presentation enhancing effect of autophagy has been used to significantly upgrade the efficiency of the BCG vaccine (Jagannath *et al.*, 2009). Autophagy also regulates exogenous antigen processing for presentation by MHC class II via the modification of the content and fate of phagosomes after LC3 recruitment to the phagosomal membranes (Shibutani *et al.*, 2015).

T cells may up-regulate autophagy upon T-cell receptor (TCR) stimulation, a process that seems to be essential not only for T-cell proliferation but also for their survival as autophagy-defective CD4$^+$ cells are more susceptible to apoptosis (Kovacs *et al.*, 2012; Pua *et al.*, 2007). One important physiological process in which autophagy-dependent endogenous antigen presentation by MHC class II is essential is the education of naive CD4+ T cells in the thymus, where thymic epithelial cells (TECs) present self-antigens on MHC molecules for the induction of T-cell tolerance. TECs display constitutive starvation-independent high levels of autophagy which appear to be crucial for correct negative selection of T cells, elimination of autoreactive T cells and the correct development of self-tolerance (Aichinger *et al.*, 2013; Nedjic *et al.*, 2008).

However, autophagy pathways engage in far more aspects of adaptive immunity than antigen presentation, as they also affect lymphocyte selection, maturation, proliferation and survival. Defects in autophagy lead to serious damage in the lymphoid lineage: mice lacking ATG5 displayed a reduced number of B and T cells, suggesting that autophagy regulates the activity of lymphoid precursors. Even though lethally irradiated mice get repopulated with haematopoietic cells of *Atg5*$^{-/-}$ mice, the CD4$^+$ and CD8$^+$ cells failed to undergo proliferation upon T-receptor stimulation. *Atg5*$^{-/-}$ T cells managed to repopulate the thymus but experienced high levels of cell death that prevented the repopulation of the periphery (Pua *et al.*, 2007). ATG7-deficient cells failed to reconstitute the haematopoietic system of lethally irradiated mice and the production of lymphoid progenitors was also impaired in the absence of ATG7 (Mortensen *et al.*, 2011). Selective autophagy of mitochondria (mitophagy), also seems to be essential for the development of T cells as the number of naive T cells is significantly lower if mitophagy is impaired (Farfariello *et al.*, 2012; Pua *et al.*, 2009). Furthermore, the absence of autophagy during B cell differentiation appears to negatively impact the numbers of B1 cells while the overall number of B cells remains unchanged, a process that seems to involve BECN1 and ATG5 (Arsov *et al.*, 2011; Miller *et al.*, 2008). Finally, autophagy is capital for the maintenance of plasma cells, which require autophagy for sustainable immunoglobulin production (Pengo *et al.*, 2013).

Autophagy regulates cytokine production during bacterial infection

Considering the cross-talk between autophagy and PRR signalling, it can be hypothesized that autophagy influences cytokine production during bacterial infection. Indeed, several groups investigated the effect of autophagy inhibition on cytokine production, either during *in vitro* infection of isolated cells or *in vivo* infection of mice with bacterial pathogens

Table 4.1 Effects of autophagy on cytokine production upon bacterial infection

Bacteria	Infection	Host model	Mechanism of autophagy-inhibition	Effect on autophagy compared with WT mice or non-treated cells	Cytokines	Reference
Mycobacterium tuberculosis	*In vitro*	PBMCs	3-MA and siRNA knockdown of ATG7	Down-regulated ↓	TNF-α ↓, IL-1β ↑, IL-6 ↑	Kleinnijenhuis *et al.* (2011)
Borrelia burgdorferi	*In vitro*	PBMCs	3-MA, wortmannin	Down-regulated ↓	IL-1β ↑, IL-6 ↑, IL-23 ↑, IL-17 ↑, IFN-γ ↑	Buffen *et al.* (2016)
	In vivo	Patellae of mice	*Atg7*$^{-/-}$ mice		IL-1β ↑, IL-6 ↑	
	In vitro	BMDMs	*Atg7*$^{-/-}$ mice		IL-1β ↑	Buffen *et al.* (2013)
Pseudomonas aeruginosa	*In vivo*	Lung macrophages of infected mice	*Anxa2*$^{-/-}$ mice	Down-regulated ↓	TNF-α ↑, IL-6 ↑, IL-1β ↑ IFNγ ↑	Li *et al.* (2015)
Adherent Invasive *E. coli* (AIEC)	*In vitro*	THP-1 macrophage-like cells	siRNA knockdown of ATG16L1	Down-regulated↓	TNF-α ↑, IL-6 ↑	Lapaquette *et al.* (2012)
Salmonella Typhimurium	*In vivo*	Epithelial cells in terminal ileum and cecum of mice	*Atg16l1*$^{-/-}$ mice	Down-regulated ↓	IL-6 ↑, IL-1β ↑	Conway *et al.* (2013)
Shigella flexneri	*In vitro*	Splenic macrophages of mice	Atg16L1 T300A knock-in mice	Down-regulated ↓	Il-1β ↑	Lassen *et al.* (2014)
Legionella pneumophila	*In vitro*	BMDMs	Δ*legS2*	Up-regulated ↑	IL-10 ↓, IL-6 ↓, TNFα ↓ Il-1β ↑	Abu Khweek *et al.* (2016)
Mycobacterium tuberculosis	*In vitro*	BMDMs	Δ*eis*	Up-regulated ↑	TNFα ↑ IL-6 ↑	Shin *et al.* (2010)
	In vivo	Lungs and spleen of infected mice	Δ*tlyA*	Up-regulated ↑	IL-12p40 ↑ IL-1β ↓, IL-10 ↓	Rahman *et al.* (2015)

(Table 4.1). These studies have shown that inhibition of autophagy by 3MA decreased TNF-α production but enhanced IL-1β and IL-6 production in *M. tuberculosis*-infected peripheral blood mononuclear cells (PBMCs). In contrast, induction of autophagy by starvation had the opposite effect (Kleinnijenhuis *et al.*, 2011). Similar results were obtained for 3MA inhibition of autophagy during infection of PBMCs with *Borrelia burgdorferi*, the causative agent of Lyme disease, as IL-1β and IL-6 increased while TNF-α remained unaltered (Buffen *et al.*, 2013). In a mouse model of *Pseudomonas aeruginosa* infection, Li and colleagues showed that Annexin A2 (AnxA2) regulates autophagosome formation through the mTORC1–ULK1 signalling pathway (Li *et al.*, 2015) as infected *Anxa2*$^{-/-}$ mice displayed reduced autophagy levels and a marked increase of cytokines, in particular IL-1β, TNF-α, IL-6 and IFN-γ, in bronchoalveolar lavage fluid (BALF). Given the body of work available, it seems plausible to affirm that autophagy reduction induces an increase in cytokine production during bacterial infection.

Crohn's disease is an inflammatory disorder characterized by an excessive immune response to intestinal microbiota. The Thr300Ala polymorphism on ATG16L1 is associated with Crohn's disease, a fact that inspired several studies on cytokines production on ATG16L1 down-regulation conditions (Conway *et al.*, 2013; Lapaquette *et al.*, 2012; Lassen *et al.*, 2014). One of the first studies showed that knockdown of *Atg16l1* in THP-1 macrophage-like cells led to decreased autophagy and increased production of TNF-α and IL-6 during infection with adherent invasive *Escherichia coli* (Lapaquette *et al.*, 2012). In a later study, Conway and colleagues showed that *S.* Typhimurium was unable to associate with autophagosomes in *Atg16l1*-deficient epithelial cells and showed significantly higher levels of IL-1β and IL-6 compared to WT mice in the terminal ileum and caecum (Conway *et al.*, 2013). But the most informative study was carried out using a knock-in mouse model expressing the ATG16L1 T300A variant (Lassen *et al.*, 2014). As autophagy was reduced in multiple cell types from T300A knock-in mice compared to WT mice, the authors explored several aspects of the immune response to *S. flexneri* infection. Their results showed that the T300A polymorphism was associated with decreased antibacterial autophagy and increased IL-1β production in primary cells and *in vivo*. Interestingly, *in vitro* bacterial infections of isolated splenic CD11b^{+} macrophages from ATG16L1 T300A mice, led also to higher levels of IL-1β. Collectively, this study shows that defective autophagy caused by a disease-associated polymorphism of an autophagy gene leads to increased IL-1β production. However, the mechanism linking ATG16L1 to IL-1β secretion was not elucidated (Lassen *et al.*, 2014).

Furthermore, a recent report showed that injection of *B. burgdorferi* into the knees of *Atg7*$^{-/-}$ mice increased joint swelling and cytokine levels (IL-1β and IL-6) when compared to WT mice, suggesting that *B. burgdorferi*-induced joint inflammation is controlled by autophagy (Buffen *et al.*, 2016). Autophagy inhibition by wortmannin led to an increase in the production of IL-1β and IL-23 cytokines by human PBMCs infected with *B. burgdorferi* suggesting that the production of these cytokines is controlled by autophagy during *B. burgdorferi* infection *in vitro*. Increased production of IL-1β led to the polarization of CD4^{+} T cells to IL-17-producing Th17 cells, a specific subtype of T cells that is commonly elevated in patients with confirmed neuroborreliosis (Henningsson *et al.*, 2011) and which seem to be involved in the pathogenesis of Lyme arthritis (Burchill *et al.*, 2003). Collectively, these results show that autophagy controls *B. burgdorferi*-induced secretion of cytokines such as IL-1β and IL-23, which in turn impact T-cell polarization during infection.

Thus, it seems that autophagy has a role limiting cytokine production during bacterial infection (especially IL-1β production). This resembles the cytokine production profiles observed during stimulation of cells with bacterial ligands in the presence of autophagy inhibitors discussed in previous sections.

Intracellular bacteria modulate autophagy in the infected cell

Infection by bacterial pathogens triggers autophagy in infected cells as a cell-autonomous defence mechanism aimed at degrading the invading pathogen. However, pathogenic bacteria evolved mechanisms to manipulate autophagy and counteract these host self defences. While certain intracellular bacteria induce and manipulate autophagy taking advantage of it, other bacteria inhibit autophagy in order to avoid xenophagy and lysosomal degradation (Escoll *et al.*, 2016).

Anaplasma phagocytophilum, *Yersinia pseudotuberculosis*, *Coxiella burnetii* and *F. tularensis* are intracellular pathogens that have evolved mechanisms to hijack autophagosomes during infection. These bacteria redirect the by-products of the autophagic degradation of cellular components for their own nutritional use, thereby promoting their replication (Steele *et al.*, 2015). They replicate within bacterial vacuoles decorated with autophagy components, such as LC3, and show defective replication in autophagy-deficient cells. Consequently, treatment of host cells with autophagy activators, increases bacterial replication rather than promoting bacterial clearance (Escoll *et al.*, 2016). *A. phagocytophilum* uses the secreted effector Ats-1 to promote autophagosome nucleation and utilization of the nutrients contained in the autophagosomes (Niu *et al.*, 2012). Similarly, it was elegantly shown by monitoring autophagy-derived radiolabelled amino acids that during *F. tularensis* infection a transfer from host proteins to invading bacteria takes place (Steele *et al.*, 2013).

Other bacteria, such as *M. tuberculosis* and S. Typhimurium, inhibit autophagy initiation upstream of autophagosome formation, thus evading xenophagy and pathogen degradation (Shin *et al.*, 2010; Tattoli *et al.*, 2012). *S. flexneri* evades autophagy recognition by masking the bacterial surface (Ogawa *et al.*, 2005).

An important observation is that some bacterial pathogens actively induce autophagy but, at the same time, block autophagosome maturation and fusion with the lysosome. *L. pneumophila* is one of these intracellular bacterial pathogens that uses this dual strategy. After phagocytosis, the establishment of the *Legionella*-containing vacuole (LCV) in the infected macrophage is accompanied by the acquisition of autophagy markers like LC3 at the LCV, showing that the LCVs rapidly become autophagosomes (Amer and Swanson, 2005). The secreted *L. pneumophila* effector LegA9 was shown to promote the recognition of the LCV by the autophagy machinery (Khweek *et al.*, 2013). In line with this observation, inhibition of autophagy in permissive A/J mouse macrophages reduces *L. pneumophila* survival at 2 hours post infection (Amer and Swanson, 2005; Amer *et al.*, 2005), supporting the idea that autophagy has a role in the promotion of the survival of *L. pneumophila* within the host cell early during infection (Amer *et al.*, 2005). Later during infection, *L. pneumophila* inhibits autophagy by secreting the effectors *Lp*SPL and RavZ (Choy *et al.*, 2012; Rolando *et al.*, 2016), which respectively inhibit autophagosome formation and maturation. This strategy delays the maturation of the LCV-containing autophagosome into autolysosomes, thus gaining precious time for multiplication of the pathogen (Amer and Swanson, 2005; Joshi and Swanson, 2011).

Autophagy subversion by pathogenic bacteria shapes host immunity

As previously discussed, autophagy and immunity are strongly linked and coordinated processes. Considering that many pathogenic bacteria manipulate autophagy, it is expected that this strategy has direct consequences on immunity. Regarding cell-autonomous immunity, bacteria-induced inhibition of autophagy promotes evasion from xenophagy, a benefit for the pathogen as they escape lysosomal degradation. However, autophagy is also linked to 'conventional' immune responses, i.e. innate and adaptive immunity (Kuballa *et al.*, 2012), but this relationship and the impact of bacterial modulation of autophagy on immune responses remains poorly understood.

Manipulation of autophagy by pathogenic bacteria may impact cytokine production by innate immune cells, as autophagy is interconnected to PRR signalling. This may represent an underappreciated effect on innate immunity induced by bacteria that modulate autophagy. Indeed, as listed in Table 4.1, some studies have pointed to this possibility. For example, the *M. tuberculosis* protein Eis is involved in bacterial survival within the host. Macrophages infected with a mutant lacking *eis* displayed markedly increased accumulation of autophagy vacuoles and formation of autophagosomes *in vitro* and *in vivo*, suggesting that Eis down-regulates autophagy in the host cell (Shin *et al.*, 2010). Interestingly, infection of macrophages with a Δ*eis* mutant strain increased the production of TNF-α and IL-6 as compared to those measured after infection with the WT strain, suggesting that the loss of the capacity of the *eis* mutant to inhibit autophagy leads to the increased TNF-α and IL-6 production (Shin *et al.*, 2010). In addition, the virulence factor TlyA also inhibits autophagy and significantly contributes to the pathogenesis of *M. tuberculosis*. DCs infected with a Δ*tlyA* strain displayed increased autophagy and showed increased IL-12p40 and reduced IL-1β and IL-10 cytokine responses, which clearly contrasts with the immune responses induced by the WT strain (Rahman *et al.*, 2015). Collectively, these two studies suggest that inhibition of autophagy induced by Eis and TlyA reduces production of TNF-α, IL-6 and IL-12p40 while it boosts the secretion of IL-1β and IL-10 in response to *M. tuberculosis* infection.

The bacterial effector *Lp*SPL secreted by *L. pneumophila* inhibits autophagy in human cells by modulating the sphingolipid metabolism of the host cell during infection (Rolando *et al.*, 2016). It was also shown, that a mutant lacking this effector induces higher secretion of IL-1β and lower secretion of TNF-α, IL-6 and IL-10 compared to the WT strain during infection of BMDMs (Abu Khweek *et al.*, 2016). As these studies were carried out in different host cells (human and mice) and none of them tested the influence of *L. pneumophila*-induced inhibition of autophagy on cytokine production of infected cells directly, it is difficult to draw conclusions about how manipulation of autophagy by *L. pneumophila* influences cytokine production of infected macrophages; however, it is tempting to assume that there is a link.

As autophagy is also involved in antigen presentation to T cells, autophagy modulation by bacterial pathogens may also impact adaptive immunity. Indeed, a recent report showed that autophagy modulation by *M. tuberculosis* impacts adaptive immunity. A genome-wide screening identified the protein PE_PGRS47 of *M. tuberculosis* as responsible for the inhibition of MHC class II antigen presentation of infected DCs as Δ*PE_PGRS47*-infected DCs showed increased MHC class II antigen presentation compared to the WT strain, and infection of mice with the Δ*PE_PGRS47* strain resulted in an increased number of bacterial-antigen-specific $CD4^+$ cells compared to WT-infected mice (Saini *et al.*, 2016). Interestingly, infection with a *M. tuberculosis* mutant lacking PE_PGRS47 showed increased autophagy when compared to the WT strain, suggesting that the bacteria use PE_PGRS47 to inhibit autophagy in the host cell. Moreover, the Δ*PE_PGRS47* strain was significantly attenuated *in vivo* and its defects in intracellular replication *in vitro* were restored to WT levels when autophagy was inhibited by 3MA treatment (Saini *et al.*, 2016). These results demonstrate that the *M. tuberculosis* protein PE_PGRS47 inhibits autophagy in the infected cells resulting in a reduction of MHC class II antigen presentation that impacts specific T-cell responses to infection.

Concluding remarks

Autophagy works as a cell-autonomous defence mechanism of immune and non-immune cells by removing invading pathogens immediately after infection. It is well established that bacteria-induced inhibition of autophagy promotes bacterial evasion from xenophagy and allows the pathogen to escape from lysosomal degradation (Huang and Brumell, 2014). Autophagy is also linked to conventional innate and adaptive immune responses but the impact of autophagy modulation by bacteria at these levels remains poorly understood.

Current data support a model where PRR signalling and autophagy cross-talk at different levels. In general, PRR activation through the recognition of bacterial ligands promotes autophagy (Delgado *et al.*, 2009). However, drugs or genetic approaches used to inhibit autophagy in infected cells or animals showed increased secretion of pro-inflammatory cytokines after stimulation with bacterial ligands or viable bacteria, suggesting that cells might also use autophagy to limit PRR-initiated immune responses as a negative-feedback loop during infection. Most of the available literature supports this model (Table 4.1), with data for inflammasome signalling and IL-1β secretion being very solid and reproducible through out the different studies.

One of the most compelling results is the role of ATG16L1 in Crohn's disease. Defects in ATG16L1, such as the T300A polymorphism carried by patients, lead to reduced autophagy in host cells during bacterial infection, reduced bacterial clearance and increased secretion of cytokines, mainly of IL-1β (Conway *et al.*, 2013; Lapaquette *et al.*, 2012; Lassen *et al.*, 2014). These studies highlight the existence of intrinsic mechanisms within host cells where autophagy and cytokine production are coordinated during host–pathogen interactions, suggesting that malfunction of autophagy might be one cause of inflammatory diseases with excessive cytokine production such as Crohn's disease. Finally, the investigation of the mechanisms used by bacteria to modulate autophagy identified some bacterial effectors inhibiting autophagy, thereby promoting bacterial survival during infection. However, only few studies addressed the question of how these bacterial-derived autophagy inhibitors directly impact immune responses during infection. Most is known for *M. tuberculosis*, in which the bacterial proteins Eis, TlyA and PE_PGRS47 have been shown to inhibit autophagy and to modulate cytokine production and MHC class II antigen presentation to T cells, shaping innate and adaptive immunity during *M. tuberculosis* infection.

Future work on this topic should be directed to uncover whether mechanisms exist that coordinate autophagy and cytokine production during infection. It would be important to investigate in parallel the impact of autophagy on the secretion of inflammasome-dependent/-independent cytokines (such as IL-1 and TNF-α, respectively) and also on the production of anti-inflammatory cytokines such as IL-10 or TGF-β. Additionally, an intriguing question yet to be answered is whether or not bacterial effectors that increase or reduce autophagy (such Ats-1 or *Lp*SPL, respectively) impact the innate and adaptive immune responses to infection. In depth knowledge on these questions will help to better understand bacterial infection and to better combat disease.

Acknowledgements

Work in CB laboratory is financed by the Institut Pasteur, the Institut Carnot-Pasteur MI, the French Region Ile de France (DIM Malinf) and the grant n°ANR-10-LABX-62-IBEID. FV was funded by the Infect-ERA project EUGENPATH (ANR-13-IFEC-0003-02), PE by the Fondation pour la Recherche Médicale, grant N° DEQ20120323697.

References

Abu Khweek, A., Kanneganti, A., Guttridge, D., D.C., and Amer, A.O. (2016). The Sphingosine-1-Phosphate Lyase (LegS2) Contributes to the Restriction of *Legionella pneumophila* in Murine Macrophages. PLOS ONE *11*, e0146410. https://doi.org/10.1371/journal.pone.0146410.

Aichinger, M., Wu, C., Nedjic, J., and Klein, L. (2013). Macroautophagy substrates are loaded onto MHC class II of medullary thymic epithelial cells for central tolerance. J. Exp. Med. *210*, 287–300. https://doi.org/10.1084/jem.20122149.

Amer, A.O., and Swanson, M.S. (2005). Autophagy is an immediate macrophage response to *Legionella pneumophila*. Cell. Microbiol. *7*, 765–778.

Amer, A.O., Byrne, B.G., and Swanson, M.S. (2005). Macrophages rapidly transfer pathogens from lipid raft vacuoles to autophagosomes. Autophagy *1*, 53–58.

Arsov, I., Adebayo, A., Kucerova-Levisohn, M., Haye, J., MacNeil, M., Papavasiliou, F.N., Yue, Z., and Ortiz, B.D. (2011). A role for autophagic protein beclin 1 early in lymphocyte development. J. Immunol. *186*, 2201–2209. https://doi.org/10.4049/jimmunol.1002223.

Buffen, K., Oosting, M., Li, Y., Kanneganti, T.D., Netea, M.G., and Joosten, L.A. (2016). Autophagy suppresses host adaptive immune responses toward Borrelia burgdorferi. J. Leukoc. Biol. *100*, 589–598. https://doi.org/10.1189/jlb.4A0715-331R.

Byrne, B.G., Dubuisson, J.F., Joshi, A.D., Persson, J.J., and Swanson, M.S. (2013). Inflammasome components coordinate autophagy and pyroptosis as macrophage responses to infection. MBio *4*, e00620–12. https://doi.org/10.1128/mBio.00620-12.

Choy, A., Dancourt, J., Mugo, B., O'Connor, T.J., Isberg, R.R., Melia, T.J., and Roy, C.R. (2012). The Legionella effector RavZ inhibits host autophagy through irreversible Atg8 deconjugation. Science *338*, 1072–1076. https://doi.org/10.1126/science.1227026.

Crotzer, V.L., and Blum, J.S. (2009). Autophagy and its role in MHC-mediated antigen presentation. J. Immunol. *182*, 3335–3341. https://doi.org/10.4049/jimmunol.0803458.

Dengjel, J., Schoor, O., Fischer, R., Reich, M., Kraus, M., Müller, M., Kreymborg, K., Altenberend, F., Brandenburg, J., Kalbacher, H., *et al.* (2005). Autophagy promotes MHC class II presentation of peptides from intracellular source proteins. Proc. Natl. Acad. Sci. U.S.A. *102*, 7922–7927.

Deretic, V., Saitoh, T., and Akira, S. (2013). Autophagy in infection, inflammation and immunity. Nat. Rev. Immunol. *13*, 722–737. https://doi.org/10.1038/nri3532.

Desai, M., Fang, R., and Sun, J. (2015). The role of autophagy in microbial infection and immunity. Immunotargets Ther. *4*, 13–26. https://doi.org/10.2147/ITT.S76720.

Escoll, P., Rolando, M., and Buchrieser, C. (2016). Modulation of Host Autophagy during Bacterial Infection: Sabotaging Host Munitions for Pathogen Nutrition. Front. Immunol. *7*, 81. https://doi.org/10.3389/fimmu.2016.00081.

Farfariello, V., Farfariello, V., Amantini, C., Amantini, C., Santoni, G., and Santoni, G. (2012). Transient receptor potential vanilloid 1 activation induces autophagy in thymocytes through ROS-regulated AMPK and Atg4C pathways. J. Leukoc. Biol. *92*, 421–431. https://doi.org/10.1189/jlb.0312123.

Harris, J., Hartman, M., Roche, C., Zeng, S.G., O'Shea, A., Sharp, F.A., Lambe, E.M., Creagh, E.M., Golenbock, D.T., Tschopp, J., *et al.* (2011). Autophagy controls IL-1beta secretion by targeting pro-IL-1beta for degradation. J. Biol. Chem. *286*, 9587–9597. https://doi.org/10.1074/jbc.M110.202911.

Huang, J., and Brumell, J.H. (2014). Bacteria-autophagy interplay: a battle for survival. Nat. Rev. Microbiol. *12*, 101–114. https://doi.org/10.1038/nrmicro3160.

Jagannath, C., Lindsey, D.R., Dhandayuthapani, S., Xu, Y., Hunter, R.L., and Eissa, N.T. (2009). Autophagy enhances the efficacy of BCG vaccine by increasing peptide presentation in mouse dendritic cells. Nat. Med. *15*, 267–276. https://doi.org/10.1038/nm.1928.

Joshi, A.D., and Swanson, M.S. (2011). Secrets of a successful pathogen: legionella resistance to progression along the autophagic pathway. Front. Microbiol. *2*, 138. https://doi.org/10.3389/fmicb.2011.00138.

Khweek, A.A., Caution, K., Akhter, A., Abdulrahman, B.A., Tazi, M., Hassan, H., Majumdar, N., Doran, A., Guirado, E., Schlesinger, L.S., *et al.* (2013). A bacterial protein promotes the recognition of the *Legionella pneumophila* vacuole by autophagy. Eur. J. Immunol. *43*, 1333–1344. https://doi.org/10.1002/eji.201242835.

Kovacs, J.R., Li, C., Yang, Q., Li, G., Garcia, I.G., Ju, S., Roodman, D.G., Windle, J.J., Zhang, X., and Lu, B. (2012). Autophagy promotes T-cell survival through degradation of proteins of the cell death machinery. Cell Death Differ. *19*, 144–152. https://doi.org/10.1038/cdd.2011.78.

Kuballa, P., Nolte, W.M., Castoreno, A.B., and Xavier, R.J. (2012). Autophagy and the immune system. Annu. Rev. Immunol. *30*, 611–646. https://doi.org/10.1146/annurev-immunol-020711-074948.

Latz, E., Xiao, T.S., and Stutz, A. (2013). Activation and regulation of the inflammasomes. Nat. Rev. Immunol. *13*, 397–411. https://doi.org/10.1038/nri3452.

Miller, B.C., Zhao, Z., Stephenson, L.M., Cadwell, K., Pua, H.H., Lee, H.K., Mizushima, N.N., Iwasaki, A., He, Y.W., Swat, W., *et al.* (2008). The autophagy gene ATG5 plays an essential role in B lymphocyte development. Autophagy *4*, 309–314. https://doi.org/10.4161/auto.5474.

Miller, C., and Celli, J. (2016). Avoidance and Subversion of Eukaryotic Homeostatic Autophagy Mechanisms by Bacterial Pathogens. J. Mol. Biol. *428*, 3387–3398. https://doi.org/10.1016/j.jmb.2016.07.007.

Molofsky, A.B., Byrne, B.G., Whitfield, N.N., Madigan, C.A., Fuse, E.T., Tateda, K., and Swanson, M.S. (2006). Cytosolic recognition of flagellin by mouse macrophages restricts *Legionella pneumophila* infection. J. Exp. Med. *203*, 1093–1104. https://doi.org/10.1084/jem.20051659.

Monroe, K.M., McWhirter, S.M., and Vance, R.E. (2009). Identification of host cytosolic sensors and bacterial factors regulating the type I interferon response to *Legionella pneumophila*. PLOS Pathog. *5*, e1000665. https://doi.org/10.1371/journal.ppat.1000665.

Mortensen, M., Soilleux, E.J., Djordjevic, G., Tripp, R., Lutteropp, M., Sadighi-Akha, E., Stranks, A.J., Glanville, J., Knight, S., Jacobsen, S.E., *et al.* (2011). The autophagy protein Atg7 is essential for hematopoietic stem cell maintenance. J. Exp. Med. *208*, 455–467. https://doi.org/10.1084/jem.20101145.

Münz, C. (2010). Antigen processing via autophagy – not only for MHC class II presentation anymore? Curr. Opin. Immunol. *22*, 89–93. https://doi.org/10.1016/j.coi.2010.01.016.

Nedjic, J., Aichinger, M., Emmerich, J., Mizushima, N., and Klein, L. (2008). Autophagy in thymic epithelium shapes the T-cell repertoire and is essential for tolerance. Nature *455*, 396–400. https://doi.org/10.1038/nature07208.

Nimmerjahn, F., Milosevic, S., Behrends, U., Jaffee, E.M., Pardoll, D.M., Bornkamm, G.W., and Mautner, J. (2003). Major histocompatibility complex class II-restricted presentation of a cytosolic antigen by autophagy. Eur. J. Immunol. *33*, 1250–1259. https://doi.org/10.1002/eji.200323730.

O'Sullivan, T.E., Geary, C.D., Weizman, O.-E., Geiger, T.L., Rapp, M., Dorn, G.W., Overholtzer, M., and Sun, J.C. (2016). Atg5 is essential for the development and survival of innate lymphocytes. Cell Rep. *15*, 1910–1919. https://doi.org/10.1016/j.celrep.2016.04.082.

Ohsumi, Y. (2014). Historical landmarks of autophagy research. Cell Res. *24*, 9–23. https://doi.org/10.1038/cr.2013.169.

Pengo, N., Scolari, M., Oliva, L., Milan, E., Mainoldi, F., Raimondi, A., Fagioli, C., Merlini, A., Mariani, E., Pasqualetto, E., *et al.* (2013). Plasma cells require autophagy for sustainable immunoglobulin production. Nat. Immunol. *14*, 298–305. https://doi.org/10.1038/ni.2524.

Pua, H.H., Guo, J., Komatsu, M., and He, Y.W. (2009). Autophagy is essential for mitochondrial clearance in mature T lymphocytes. J. Immunol. *182*, 4046–4055. https://doi.org/10.4049/jimmunol.0801143.

Pua, H.H., Dzhagalov, I., Chuck, M., Mizushima, N., and He, Y.W. (2007). A critical role for the autophagy gene Atg5 in T cell survival and proliferation. J. Exp. Med. *204*, 25–31.

Rahman, M.A., Sobia, P., Dwivedi, V.P., Bhawsar, A., Singh, D.K., Sharma, P., Moodley, P., Van Kaer, L., Bishai, W.R., and Das, G. (2015). *Mycobacterium tuberculosis* TlyA protein negatively regulates T helper (Th) 1 and Th17 differentiation and promotes tuberculosis pathogenesis. J. Biol. Chem. *290*, 14407–14417. https://doi.org/10.1074/jbc.M115.653600.

Rolando, M., Escoll, P., Nora, T., Botti, J., Boitez, V., Bedia, C., Daniels, C., Abraham, G., Stogios, P.J., Skarina, T., *et al.* (2016). *Legionella pneumophila* S1P-lyase targets host sphingolipid metabolism and restrains autophagy. Proc. Natl. Acad. Sci. U.S.A. *113*, 1901–1906. https://doi.org/10.1073/pnas.1522067113.

Rubinsztein, D.C., Codogno, P., and Levine, B. (2012). Autophagy modulation as a potential therapeutic target for diverse diseases. Nat. Rev. Drug Discov. *11*, 709–730. https://doi.org/10.1038/nrd3802.

Saini, N.K., Baena, A., Ng, T.W., Venkataswamy, M.M., Kennedy, S.C., Kunnath-Velayudhan, S., Carreño, L.J., Xu, J., Chan, J., Larsen, M.H., *et al.* (2016). Suppression of autophagy and antigen presentation by *Mycobacterium tuberculosis* PE_PGRS47. Nat. Microbiol. *1*, 16133. https://doi.org/10.1038/nmicrobiol.2016.133.

Schmid, D., Pypaert, M., and Münz, C. (2007). Antigen-loading compartments for Major Histocompatibility Complex Class II molecules continuously receive input from autophagosomes. Immunity *26*, 79–92. https://doi.org/10.1016/j.immuni.2006.10.018.

Shi, C.S., Shenderov, K., Huang, N.N., Kabat, J., Abu-Asab, M., Fitzgerald, K.A., Sher, A., and Kehrl, J.H. (2012). Activation of autophagy by inflammatory signals limits IL-1β production by targeting ubiquitinated inflammasomes for destruction. Nat. Immunol. *13*, 255–263. https://doi.org/10.1038/ni.2215.

Shibutani, S.T., Saitoh, T., Nowag, H., Münz, C., and Yoshimori, T. (2015). Autophagy and autophagy-related proteins in the immune system. Nat. Immunol. *16*, 1014–1024. https://doi.org/10.1038/ni.3273.

Sun, J.C., and Lanier, L.L. (2011). NK cell development, homeostasis and function: parallels with $CD8^+$ T cells. Nat. Rev. Immunol. *11*, 645–657. https://doi.org/10.1038/nri3044.

Walker, J.A., Barlow, J.L., and McKenzie, A.N. (2013). Innate lymphoid cells – how did we miss them? Nat. Rev. Immunol. *13*, 75–87. https://doi.org/10.1038/nri3349.

Wang, S., Xia, P., Huang, G., Zhu, P., Liu, J., Ye, B., Du, Y., and Fan, Z. (2016). FoxO1-mediated autophagy is required for NK cell development and innate immunity. Nat. Commun. *7*, 11023. https://doi.org/10.1038/ncomms11023.

Zamboni, D.S., Kobayashi, K.S., Kohlsdorf, T., Ogura, Y., Long, E.M., Vance, R.E., Kuida, K., Mariathasan, S., Dixit, V.M., Flavell, R.A., et al. (2006). The Birc1e cytosolic pattern-recognition receptor contributes to the detection and control of *Legionella pneumophila* infection. Nat. Immunol. *7*, 318–325. https://doi.org/10.1038/ni1305.

Zhou, D., Li, P., Lin, Y., Lott, J.M., Hislop, A.D., Canaday, D.H., Brutkiewicz, R.R., and Blum, J.S. (2005). Lamp-2a facilitates MHC class II presentation of cytoplasmic antigens. Immunity *22*, 571–581. https://doi.org/10.1016/j.immuni.2005.03.009

Inflammasome-dependent Mechanisms Involved in Sensing and Restriction of Bacterial Replication

5

Warrison A. Andrade and Dario S. Zamboni*

Department of Cell Biology, Ribeirão Preto Medical School, University of São Paulo (FMRP-USP), Ribeirão Preto, Brazil.

*Correspondence: dszamboni@fmrp.usp.br

http://doi.org/10.21775/9781910190692.05

Abstract

Inflammasomes are multiprotein platforms assembled in the cytosol in response to pathogens and cell stress. Inflammasomes are recognized by their important role on defences against bacterial infections and have been also implicated in a range of human inflammatory disorders. Intracellular sensors such as NLRP1, NLRP3, NLRC4, AIM2 and Pyrin induce assembly of inflammasomes, while caspase-11 induces the non-canonical pathway for activation of the NLRP3 inflammasome. The formation of the inflammasome leads to caspase-1 activation that triggers pyroptosis and activation of interleukin 1β (IL-1β) and IL-18. Pyroptotic cell death and cytokines production are involved in restriction of bacterial replication by limiting the replication niche of intracellular bacteria and by inducing inflammatory responses. In this review we focus on the mechanisms mediated by inflammasome activation that leads to inflammatory responses and restriction of bacterial infection.

Introduction

Invading pathogens have been interacting with their hosts during evolution. Vertebrates from fish to mammals have selected several defence strategies to protect against invading bacteria, parasites and viruses. The first line of host defence is the innate immune system, which relies on the recognition of conserved pathogen-associated molecular patterns (PAMPs) or danger-associated molecular patterns (DAMPs), by germline-encoded pattern-recognition receptors (PRRs) that survey the intracellular and extracellular spaces and initiate signalling cascades leading mainly to the induction of pro-inflammatory cytokines (Jack, 2015; Kawai and Akira, 2011).

PRRs are classified into five major families, including the Toll-like receptors (TLRs), retinoic acid-inducible (RIG-I)-like receptors (RLRs), absent of melanoma 2 (AIM2)-like receptors (ALRs), C-type lectin receptors (CLRs) and nucleotide-binding domain (NBD), leucine-rich repeat (LRR)-containing receptor (NLRs). TLRs and CLRs are bound to the plasma membrane to survey the extracellular space, or to the endosomal compartment. RLRs, ALRs and NLRs are intracellular receptors located in the cytosol, where they sense the presence of intracellular pathogens or danger signals (Broz and Dixit, 2016; Kawai and Akira, 2011).

The activation of NLRs usually results in the formation of a large multimeric protein complex, named inflammasome by Tschopp and co-workers (Martinon *et al.*, 2002; Tschopp *et al.*, 2003). After the recognition of the inflammatory stimulus by the NLRs, including ligands from pathogens (DNA, RNA, flagellin, muramyl dipeptide) self-antigens (ATP, amyloid β, cholesterol crystals) or environmental stimulus (silica, asbestos, alum), it leads to the NLR self oligomerization allowing the binding to the adaptor apoptosis-associated speck-like protein containing a CARD (ASC). Nucleation of ASC allows the recruitment of pro-caspase-1, and through proximity-induced auto processing, results in caspase-1 activation and inflammasome formation. Active caspase-1 can cleave pro-interleukin 1β (IL-1β) and pro-IL-18 into the mature cytokines, and induce a specific inflammatory cell death called pyroptosis (Fig. 5.1) (Latz *et al.*, 2013; Man and Kanneganti, 2015).

Thus, inflammasome activation provides two different outcomes, the release of mature cytokines and the removal of infected cells by pyroptosis, important features of the innate immune system that are involved in restriction of pathogen replication. Mutations on the NLRs are associated to human inflammatory diseases including gout and periodic fever syndromes, which include cryopyrin-associated periodic fever syndromes (CAPS) and Familial Mediterranean fever (FMF), both characterized by IL-1β overproduction (Broz and Dixit, 2016; Hoffman *et al.*, 2001; Rigante *et al.*, 2016).

NLR family members are characterized by a central NOD domain, a C-terminal leucine-rich domain (LRRs) and a N-terminal effector domain. The C-terminal domain is involved on the recognition of the ligand and sensing of the stimulus, while the N-terminal domain is involved on the interaction with other proteins from the complex (Sharma and Kanneganti, 2016). The NLR family members are grouped into two categories, NLRC and NLRP, depending on whether the N-terminal domain contains caspase activation and recruitment domain (CARD) or a pyrin domain, respectively. NLRP1, NLRP3, and NLRC4 are well studied NLRs that assemble the canonical inflammasomes, and several reports characterized their function on pathogen and stress recognition, which leads to cytokine production and pyroptosis. Caspase-11 was also demonstrated to play an important role on the secretion of mature IL-1β and pyroptosis induced by Gram-negative bacteria, forming the non-canonical inflammasome (Fig. 5.1) (de Vasconcelos *et al.*, 2016; Kayagaki *et al.*, 2011; Sharma and Kanneganti, 2016). In this review we provide an overview of inflammasome activation and mechanisms used by the host to control bacterial replication.

Canonical and non-canonical inflammasomes

NLRP1 inflammasome

NLRP1 was the first NLR described to its capacity to form a platform able to activate caspase-1 (Martinon *et al.*, 2002). The human proteome harbours only one NLRP1 that was described to be involved in Vitiligo, autoimmune Addison's disease, type I diabetes,

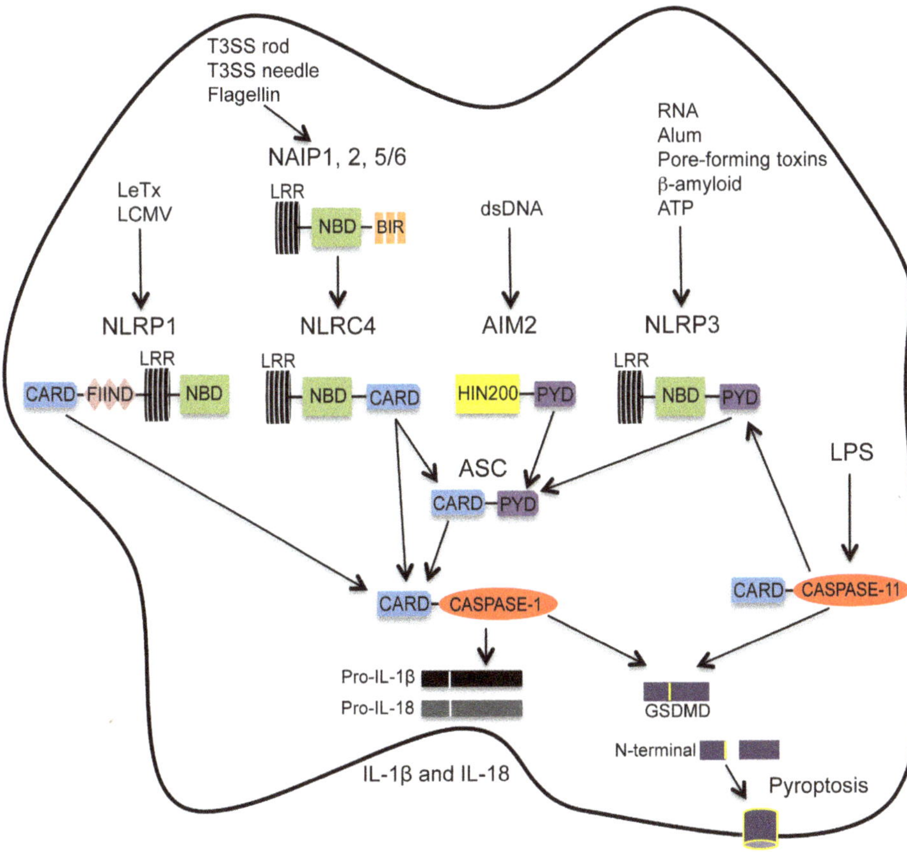

Figure 5.1 Inflammasome components and caspase-1 activation by canonical and non-canonical pathways. NOD-like receptors (NLRs) (NLRP1, NLRP3 and NLRC4) and AIM2-like receptors (ALRs) (AIM2) are depicted with individual domains and activate the canonical inflammasome pathway by sensing several intracellular bacterial ligands and cell stress. Caspase-11 directly senses LPS and activates NLRP3/caspase-1 through the non-cannonical inflammasome pathway. Activation of both canonical and non-cannonical pathways induce the secretion of mature IL-1β and IL-18, and pyroptosis by cleaving gasdermin D (GSDMD) leading to pore formation and cell death. For CARD-containing inflammasomes, pyroptosis and cytokine secretion occurs independently and dependently of ASC, respectively. Cytokine secretion by NLRP1 does not require ASC. CARD, caspase recruitment domain; NBD, nucleotide binding domain; PYD, pyrin domain; LRR, leucine-rich repeat domain; FIIND- function to find domain; HIN-HIN200 domain; BIR, baculovirus inhibitor of apoptosis protein repeat.

multiple self-healing palmoplantar carcinoma and familial keratosis lichenoides chronica (Jin *et al.*, 2007; Magitta *et al.*, 2009; Zhong *et al.*, 2016). On the other hand, mouse genome encodes three paralogues, NLRP1a, NLRP1b and NLRP1c. Human NLRP1 contains an amino terminal PYD, a NOD and LRR domains, a function-to-find and also a C-terminal CARD domain, whereas all murine NALP1 lack the PYD (Hiller *et al.*, 2003; Martinon *et al.*, 2001; Tschopp *et al.*, 2003).

Different studies have shown that NLRP1a and NLRP1b can form a functional inflammasome that activate caspase-1. NLRP1a is activated in progenitor cells infected by lymphocytic choriomeningitis virus (LCMV), and a mutation in the NLRP1a gene leads to a systemic inflammatory disease mediated by caspase-1 and IL-1β (Masters *et al.*, 2012). NLRP1b is activated by anthrax lethal toxin (LeTx) produced by *Bacillus anthrax*. LeTx consists of a protective antigen that generates pore in the membrane and lethal factor that enters the cell and cleave NLRP1b at the N-terminal site (Boyden and Dietrich, 2006; Chavarría-Smith and Vance, 2013). Macrophages from NLRP1b deficient mice are defective in caspase-1 activation, IL-1β secretion and pyroptosis in response to LeTx (Kovarova *et al.*, 2012), thus NLRP1b inflammasome is important to control *B. anthrax* infection in mice (Moayeri *et al.*, 2010). Interesting, NLRP1b is also involved in the control and protection against *Toxoplasma gondii* infection, although NLRP1b cleavage is not required (Ewald *et al.*, 2014; Gorfu *et al.*, 2014; Zamboni and Lima-Junior, 2015).

NLRP3 inflammasome

NLRP3 (cryopyrin) was first shown to be associated with autoinflammatory syndromes called CAPS, including familial cold autoinflammatory syndrome (FCAS) and Muckle-Wells syndrome (MWS), and also in gout, diabetes and atherosclerosis (Duewell *et al.*, 2010; Guo *et al.*, 2015; Hoffman *et al.*, 2001). NLRP3 is now implicated in the pathogenesis of several neuroinflammatory diseases, including Alzheimer's, Parkinson's and Prion diseases (Guo *et al.*, 2015; Heneka *et al.*, 2013; Shi *et al.*, 2015a).

NLRP3 respond to a wide range of stimulus, including ligand derived from pathogens (microbial cell-wall components, RNA, DNA and pore-forming toxins), environmental ligands (silica, asbestos and alum) and endogenous danger signals (ATP, β-amyloid and uric acid crystals) (Latz *et al.*, 2013; Man and Kanneganti, 2015). The lack of similarity between the diversity of stimulus, raise the possibility that NLRP3 senses an endogenous signal that is triggered after the cell gets in contact with the stimulus. In fact, several mechanisms have being proposed to be involved in NLRP3 activation. For example, the production of reactive oxygen species (ROS), oxidized mitochondrial DNA, potassium efflux, lysosomal release of cathepsins, changing in intracellular calcium levels (Cassel *et al.*, 2008; Cruz *et al.*, 2007; Hornung *et al.*, 2008; Lee *et al.*, 2012; Misawa *et al.*, 2013; Muñoz-Planillo *et al.*, 2013; Shimada *et al.*, 2012; Zhou *et al.*, 2010), but no unifying mechanism was identified. There are some evidences from several groups that potassium efflux is a common mechanism for NLRP3 activation, but it was also demonstrated to be involved in NLRP1b activation (Munñoz-Planillo *et al.*, 2013; Pétrilli *et al.*, 2007); thus this is not specific for NLRP3 activation. Despite this controversy, it is clear that NLRP3 activation requires two signals; the first stimulus is an NF-κB activator that is responsible for pro-IL1β expression and NLRP3 induction (Bauernfeind *et al.*, 2009), and a second stimulus that induces NLRP3 inflammasome assembly (Latz *et al.*, 2013). NLRP3 deubiquitination mediated by the k63-specific deubiquitinase BRCC3 is also required for NLRP3 assembly and activation (Juliana *et al.*, 2012; Lopez-Castejon *et al.*, 2013; Py *et al.*, 2013). Recently, a new report about NLRP3 in human monocytes is challenge the dogma for NLRP3 activation. Gaidt and co-workers identified a new alternative NLRP3 inflammasome activation pathway. LPS alone activates caspase-1 and induces IL-1β production through a TLR4/TRIF/RIP1/FADD/caspase-8-dependent pathway. This inflammasome activation leads to IL-1β secretion but no pyropstosis, and is independent of potassium efflux (Gaidt *et al.*, 2016). New insights into NLRP3 activation came from reports demonstrating an important role for NIMA-related kinase 7 (NEK7), a protein involved in cell-cycle progression, in NLRP3 activation. NEK7 interacts with the LRR domain of NLRP3 and it is important to its oligomerization and activation (He *et al.*, 2016; Schmid-Burgk *et al.*, 2016; Shi *et al.*, 2016). Under resting conditions, NLRP3 is localized at the endoplasmic reticulum (ER), and ASC is found in the nucleus, cytosol and mitochondria. Due to this different spatial localization, protein translocation is necessary for inflammasome assembly, and this is mediated by dynein that transports the mitochondria to the ER, bringing ASC and NLRP3 to the same location, demonstrating also a role for microtubules on the activation (Misawa *et al.*, 2013).

Due to the wide diversity of stimulus able to activate NLRP3, this is the best-studied inflammasome, and its role was described in a variety of infections and inflammatory diseases, but a lot of important information about NLRP3 activation and inhibition are still unclear.

NLRP12 inflammasome

Wang and co-workers described NLRP12 as a protein that could function as an inflammasome component (Wang *et al.*, 2002). It was first demonstrated that NLPR12 would act as NFκB signalling inhibitor (Williams *et al.*, 2005), and other studies suggested a role for NLRP12 as a negative regulator of colon inflammation and tumorigenesis in a DSS colitis model (Allen *et al.*, 2012; Zaki *et al.*, 2011). On the other hand, a pro-inflammatory role for NLRP12 was described on *Plasmodium* and *Yersinia pestis* infection, having an important role in IL-1β and IL-18 production and infection control (Ataide *et al.*, 2014; Vladimer *et al.*, 2012).

It is possible that NLRP12 can play a different role depending on the ligand source or on the cell type involved, but further studies are required to fully understand the mechanism of NLRP12 activation and signalling.

AIM2 inflammasome

The presence of DNA from pathogens or host into the cytosol is a danger signal sensed by the cell. It was observed that the presence of intracellular DNA induced caspase-1 activation in a ASC-dependent manner (Muruve *et al.*, 2008). The absent in melanoma 2 (AIM2) assembles a well-characterized inflammasome involved in the recognition of cytosolic double-stranded DNA (dsDNA) (Bürckstuümmer *et al.*, 2009; Fernandes-Alnemri *et al.*, 2009; Hornung *et al.*, 2009). AIM2 features a N-terminal PYD domain able to interact with the PYD of ASC and a C-terminal HIN domain involved in ligand binding and dsDNA recognition, that is sequence independent but requires about 80 pb length (Hornung *et al.*, 2009; Jin *et al.*, 2012; Roberts *et al.*, 2009).

AIM2 is involved in the recognition and host defence against several DNA viruses and bacteria, and its activation is important to infection control (Fernandes-Alnemri *et al.*, 2010; Jones *et al.*, 2010; Kim *et al.*, 2010; Rathinam *et al.*, 2010; Sauer *et al.*, 2010). It was suggested that AIM2 could be involved in some human autoimmune disorders, and increased AIM2 expression is associated with systemic lupus erythematosus, psoriasis and abdominal aortic aneurysm (Dihlmann *et al.*, 2014; Dombrowski *et al.*, 2011; Javierre *et al.*, 2010; Zhang *et al.*, 2013).

Pyrin

Pyrin was the latest PRR described as capable to assemble a canonical inflammasome. Pyrin was first described as an inflammasome involved in a mouse model of familial Mediterranean fever (FMF), an autoinflammatory disorder mediated by ASC and IL-1β (Chae *et al.*, 2011). Mutations in the C-terminal domain B30.2 from human pyrin are linked to FMF, and are associated to high levels of caspase-1 activation and IL-1β release (Chae *et al.*, 2011; Hesker *et al.*, 2012).

It was recently shown the definitive role of pyrin as an inflammasome platform. It was demonstrated that bacterial toxins produced by various species trigger pyrin inflammasome activation in response to Rho-GTPase modifications (Gavrilin *et al.*, 2012; Xu *et al.*, 2014).

NLRC4 inflammasome

NLRC4 contains a CARD domain on the N-terminal portion that mediates caspase-1 activation via CARD–CARD domain interaction. This probably explains why ASC is dispensable for pyroptosis induction by NLRC4. On the other hand, cytokine maturation and secretion by the NLRC4 inflammasome is dependent on ASC (Broz *et al.*, 2010b; Mariathasan *et al.*, 2004; Miao *et al.*, 2010a; Poyet *et al.*, 2001).

A number of bacteria induce caspase-1 activation via NLRC4 (Brodsky *et al.*, 2010; Miao *et al.*, 2008; Miao *et al.*, 2010b; Ren *et al.*, 2006; Zamboni *et al.*, 2006; Zhao *et al.*, 2011). NLRC4 is activated by three different bacterial components that enter the cytosol; flagellin, the monomeric subunit from bacteria flagellum (Franchi *et al.*, 2006; Mariathasan *et al.*, 2004; Miao *et al.*, 2006; Molofsky *et al.*, 2006); type III secretion (T3SS) rod proteins and T3SS needle proteins (Miao *et al.*, 2010b; Yang *et al.*, 2013a; Zhao *et al.*, 2011). Activation of NLRC4 is not direct mediated by the ligand; instead, a family of proteins called NAIPs (NLR family apoptosis inhibitory proteins) binds to the ligand and mediate NLRC4 activation (Cunha and Zamboni, 2013).

The NAIP family encodes seven NAIPs in mice, but only one NAIP in humans (Endrizzi *et al.*, 2000). Murine NAIP1 binds to the needle component of T3SS (Rayamajhi *et al.*, 2013; Yang *et al.*, 2013a), NAIP2 binds to the T3SS rod protein (Kofoed and Vance, 2011; Zhao *et al.*, 2011), flagellin binds to NAIP5 and NAIP6 and induces NLRC4 oligomerization (Ren *et al.*, 2006). The homologue human NAIP can sense the needle subunit of the T3SS (Yang *et al.*, 2013a), and the role of human NAIP in flagellin recognition is controversial (Kortmann *et al.*, 2015; Zhao *et al.*, 2011). However, a recent report elegantly demonstrated that a specific isoform of the human NAIP senses flagellin and trigger responses against flagellated intracellular bacteria (Kortmann *et al.*, 2015). Phosphorylation of NLRC4 on the Ser^{533} was also described to be important for NLRC4 activation (Qu *et al.*, 2012). NLRC4 is kept in an inactive state on the cytosol, and contains a winged-helix domain that stabilizes the protein in a closed conformation. A substantial reorganization is required for oligomerization and NLRC4 activation (Hu *et al.*, 2013; Hu *et al.*, 2015; Tenthorey *et al.*, 2014; Zhang *et al.*, 2015). Upon flagellin binding, NAIP5 recruits 11/12 NLRC4 molecules and forms a disk-like structure with a diameter of 28 nm. A rotation of the LRR domain is required for NAIP5/NLRC4 structure assembly (Diebolder *et al.*, 2015; Halff *et al.*, 2012).

Human genetic studies have shown gain-of-functions mutations on NLRC4 associated to autoinflammation (Canna *et al.*, 2014). One striking report showed that activation of NLRC4 inflammasome by intraperitoneal injection of flagellin induces high mortality independent of IL-1β and IL-18. The death was caused by the production of eicosanoids, including prostaglandins and leukotrienes, that led to a rapid vascular fluid loss (von Moltke *et al.*, 2012).

Caspase-11 and the non-canonical pathway for activation of the NLRP3 inflammasome

Caspase-11, an orthologue of human caspases-4 and caspase-5, was described as a binding partner and mediator of caspase-1 activation in a model of endotoxic shock by LPS (Wang *et al.*, 1998). The role of caspase-11 was masked in the mouse models used for years due to the lack of a functional caspase-11 on the strains engineered for caspase-1 deficiency (Kayagaki *et al.*, 2011). Recently, it was demonstrated that mouse caspase-11 is involved in a caspase-1-independent cell death in response to Gram-negative bacteria infection (Broz *et al.*, 2012; Case *et al.*, 2013; Hagar *et al.*, 2013; Rathinam *et al.*, 2012), but not Gram-positive bacteria (Rathinam *et al.*, 2012). Caspase-11 plays a dual role, acting as a sensor, by direct binding to LPS, and a pyroptosis inducer mediated by LPS (Kayagaki *et al.*, 2011; Shi *et al.*, 2014). Once caspase-11 activation is mediated by intracellular LPS, extracellular LPS present in Gram-negative bacteria need to gain access to the cytosol to activate caspase-11. Outer membrane vesicles (OMV) carry LPS to the cytosol by clathrin-mediated endocytosis, and LPS gains access to the cytosol from early endosomes (Vanaja *et al.*, 2016). Interestingly, LPS-mediated septic shock, first described to be TLR4 dependent, was demonstrated to be dependent on caspase-11. TLR4 was required only for type I interferon production that drives caspase-11 induction (Hagar *et al.*, 2013; Kayagaki *et al.*, 2011; Kayagaki *et al.*, 2013; Rathinam *et al.*, 2012; Shi *et al.*, 2014).

Pyroptosis induced via caspase-11 by Gram-negative bacteria is independent of ASC and NLRP3, whereas cytokine processing depends on both ASC and NLRP3 (Hagar *et al.*, 2013; Kayagaki *et al.*, 2011; Shi *et al.*, 2014). Until recently it was not known how activation of caspase-1 and caspase-11 lead to cell death. Three different groups identified gasdermin (GSDMD) as the link between caspase activation and cell death (He *et al.*, 2015; Kayagaki *et al.*, 2015; Shi *et al.*, 2015b). The inflammatory caspases 1, 4, 5 and 11 cleave GSDMD on the same site, and the N-terminal portion is required and sufficient to induce pyroptosis (Kayagaki *et al.*, 2015; Shi *et al.*, 2015b). Now, it was demonstrated that the GSDMD N-terminal portion binds to phosphatidylinositol phosphates, phosphatidylserine and cardiolipin present on the plasma membrane forming the pore and inducing pyroptosis (Aglietti *et al.*, 2016; Ding *et al.*, 2016; Liu *et al.*, 2016).

The mechanism by which caspase-11 induces NLRP3 activation is not described, and the lack of caspase-11 does not affect the canonical activation of NLRP3. Once NLRP3 inflammasome assembly is not fully understood, there are different hypotheses trying to explain how caspase-11 leads to NLRP3 activation. For example, caspase-11 activation leads to potassium efflux and drop in the intracellular levels, that could be responsible for NLRP3 activation (Rühl and Broz, 2015). Yang and collaborators showed that pannexin-1 channel is also cleaved by caspase-11, leading to ATP release and P2X7 activation, that ultimately leads to intracellular potassium levels drop and NLRP3 activation (Yang *et al.*, 2015). Another hypothesis that now can be postulated is the induction of pyroptosis and NLRP3 activation through the GSDMD pores, by which potassium could freely move to the extracellular space and induce NLRP3 activation. New studies will be required to confirm the link between caspase-11, intracellular potassium levels and NLRP3 activation.

Inflammasome activation during bacterial infections

Legionella

Legionella pneumophila is a Gram-negative intracellular bacterium that can be found in freshwater and soil, where they normally infect unicellular protozoa including several amoebae species (Newton *et al.*, 2010). *L. pneumophila* infection occurs by the inhalation of contaminated water and is normally controlled by the innate immune response, but if not controlled in the early stages have the potential to cause a form of pneumonia known as Legionnaires' disease (Parr *et al.*, 2015; Sabrià and Campins, 2003). In the lung, *L. pneumophila* initially infects alveolar macrophages using a type IV secretion system (T4SS), termed Dot/Icm, to inject bacterial proteins into the cytosol (Copenhaver *et al.*, 2014; Finsel and Hilbi, 2015; Hubber and Roy, 2010). These effectors block phagosomal maturation and fusion with lysosomes, thus preventing *L. pneumophila* degradation, and promoting the establishment of a *Legionella*-containing vacuole (LCV), a favourable niche for *Legionella* replication.

It was extensively demonstrated that *L. pneumophila* strongly activates different inflammasomes, including the canonical NLRC4 and NLRP3 (Case and Roy, 2011; Cerqueira *et al.*, 2015; Pereira *et al.*, 2011a; Pereira *et al.*, 2011b; Silveira and Zamboni, 2010) and non-canonical inflammasomes (Aachoui *et al.*, 2013; Akhter *et al.*, 2012; Case *et al.*, 2013; Casson *et al.*, 2013). *L. pneumophila* flagellin secretion is dependent on the Dot/Icm system that along with the effector proteins is secreted into the host cell cytosol. Flagellin recognition is mediated by NAIP5/NLRC4, which activates caspase-1 and triggers pyroptosis and IL-1β secretion in an ASC-independent and ASC-dependent manner, respectively (Amer *et al.*, 2006; Broz *et al.*, 2010b; Case *et al.*, 2009; Jamilloux *et al.*, 2013; Molofsky *et al.*, 2006). The recognition of flagellin and activation of NAIP5/NLRC4/caspase-1 is the main mechanism responsible for the control of *Legionella* infection both *in vivo* and *in vitro* (Coers *et al.*, 2007; Molofsky *et al.*, 2006; Pereira *et al.*, 2011b; Ren *et al.*, 2006). Caspase-7 also plays a role in restriction of *L. pneumophila* replication in murine macrophages. Caspase-7 activation downstream of NLRC4 was required for caspase-1 activation, and cells deficient for caspase-7 have an increased bacterial replication (Akhter *et al.*, 2009).

L. pneumophila deficient for the flagellin, encoded by the gene *flaA*, bypass NLRC4 inflammasome activation and replicates in macrophages (Molofsky *et al.*, 2006; Ren *et al.*, 2006). The use of the bacteria mutant for *flaA*, enabled the identification of NAIP5/NLRC4-independent pathways that are also involved in control of bacterial replication. Caspase-11 is activated by *L. pneumophila* and leads to pyroptosis independent of flagellin, but dependent on Dot/Icm (Case *et al.*, 2013), which is mediated by the release of LPS into the cytosol triggering caspase-11 activation (Kayagaki *et al.*, 2011; Shi *et al.*, 2014). Activation of caspase-11 leads to NLRP3/ASC/Caspase-1 activation and IL-1β secretion.

Activation of NAIP5/NLRC4 by flagellin induces caspase-1 activation. On this platform ASC is not present, and caspase-1 is present in an active state, but not cleaved. The dogma was that caspase-1 autoprocessing was required for its activity, but it was demonstrated that during the induction of pyroptosis by NAIP5/NLRC4/caspase-1, caspase-1 is not processed into the p10 and p20 forms (Broz *et al.*, 2010b). But ASC is required for IL-1β processing and cleavage of caspase-1. Until recently, the mechanism behind the control of *Legionella* replication was unknown. Miao and co-works demonstrated that in the case of *L. pneumophila*, caspase-1-induced pyroptosis, but not IL-1β and IL-18, is the main mechanism for restriction of bacterial replication, releasing bacteria to the extracellular space to be killed by neutrophils. Infection of IL-1/IL-18 double deficient mice did not recapitulate the phenotype from caspase-1 deficient mice, which was also observed for *Burkholderia thailandensis* (Miao *et al.*, 2010a).

L. pneumophila evolved in direct contact with unicellular protozoans and probably encountered almost no pressure to evade the recognition by the innate immune system, including inflammasome activation (Massis and Zamboni, 2011). Strains lacking SdhA are defective for intracellular replication and activate host cell death (Laguna *et al.*, 2006). The effector protein SdhA is responsible for maintaining the LCV architecture, and mutants lacking SdhA are unable to control vacuole integrity, leading to vacuole rupture and bacteria release to the cytosol, triggering host responses (Creasey and Isberg, 2012). *L. pneumophila* SdhA deficient is detected by caspase-11 (Aachoui *et al.*, 2013), and also by AIM2 through the detection of cytoplasmic bacteria DNA and induction of pyroptosis and IL-1β (Ge *et al.*, 2012).

After uptake, *L. pneumophila* creates the LCV that allows bacteria replication and survival. Through the T4SS, the bacteria secrete more than 300 effector proteins that are important to manipulate the host response, including delaying phagosome maturation (Hubber and Roy, 2010; Isberg *et al.*, 2009; Kagan and Roy, 2002). Rab proteins are involved in vesicular trafficking and phagosome maturation, and are regular target of intracellular pathogens to avoid phagosome acidification (Bhuin and Roy, 2014; Fratti *et al.*, 2003; Stein *et al.*, 2012; Via *et al.*, 1997). It was demonstrated that Rab1 is involved in phagosome acidification by *L. pneumophila;* Rab1 recruitment to the LCV avoids phagosome acidification and bacterial killing (Connor *et al.*, 2015). ER stress is a common response to infection by intracellular bacteria, and activation or inhibition of unfolded protein response (UPR) is used by pathogens to manipulate host defence (Lee *et al.*, 2008; Zhang and Kaufman, 2008). Recently two groups demonstrated that *L. pneumophila* manipulate UPR response by inhibiting the expression of immunoglobulin binding protein (BiP) and C/EBP homologous transcription factor protein (CHOP), and also by the inhibition of X-box binding protein 1 (XBP1) splicing, controlling host protein translation (Hempstead and Isberg, 2015; Treacy-Abarca and Mukherjee, 2015). While *L. pneumophila* triggers ER stress, inhibition of UPR responses may allow bacteria survival by limiting immune response and apoptosis.

Salmonella

Salmonella belongs to the Enterobacteriaceae family of Gram-negative bacteria, which consists of a large group with the ability to infect animal hosts (Baker and Dougan, 2007; Costa *et al.*, 2012). A prime example of an infectious agent is the pathogen *Salmonella enterica,* a facultative intracellular bacterium that is responsible for an estimated 90 million cases of human gastroenteritis (Majowicz *et al.*, 2010). The majority of clinical disease in animals and humans is caused by serovars within the *Salmonella enterica* subspecies and this can range from local gastroenteritis to a fatal disseminated disease. After oral ingestion, *Salmonella* invade intestinal epithelial cells in the distal ileum. *Salmonella* usually interacts first with epithelial cells, which can recognize pathogenic bacteria and initiate an inflammatory response (Jones *et al.*, 1994; Tam *et al.*, 2008).

Salmonella Typhimurium have two distinct pathogenic islands, SPI-1 and SPI-2 T3SS, which are expressed on different phase of the infection. During epithelial cells invasion, *Salmonella* utilizes the SPI-1 to invade cells. After invasion, bacteria reside in the *Salmonella*-containing vacuole (SCV) and expresses SPI-2 inside epithelial cells and macrophages, altering the vacuolar membrane and limiting lysosomal fusion (Drecktrah *et al.*, 2007; Figueira and Holden, 2012; Ibarra and Steele-Mortimer, 2009). The initial immune response to *Salmonella* involves the recruitment of neutrophils and monocytes, and TLRs and NLRs play a pivotal role (Broz *et al.*, 2010a; Hayashi *et al.*, 2001; Rydström and Wick, 2007; Sivick *et al.*, 2014). TLR4 and TLR5 play a role in the recognition of *Salmonella* by sensing LPS and flagellin, respectively (Feuillet *et al.*, 2006), and TLRs involved in the recognition of nucleic acids are also involved in *Salmonella* infection control (Sivick *et al.*, 2014). *Salmonella* is detected by NLRP3 and NLRC4, and activation of both NLRs is associated with host protection (Broz *et al.*, 2010a).

NLRP3 activation under *Salmonella* infection occurs on late time points, and the mechanism of activation is unknown. Activation of NLRP3 by some bacteria is dependent on T3SS (Brodsky *et al.*, 2010), but in *Salmonella* infection NLRP3 activation is independent of SPI-1 and SPI-2 T3SS (Broz *et al.*, 2010a). In a model of *Salmonella* infection, NLRP3 and ASC play a role in cytokine maturation, but bacterial counts and inflammation were not affected on the absence of NLRP3 and ASC (De Jong *et al.*, 2014).

NLRC4 is activated by the recognition of flagellin or T3SS and T4SS proteins from a variety of bacteria (Amer *et al.*, 2006; Franchi *et al.*, 2006; Miao *et al.*, 2006; Suzuki *et al.*, 2014; Zhao *et al.*, 2011). Macrophages can detect *Salmonella* flagellin and rod protein PrgJ translocated to the cytosol by SPI-1. *Salmonella* flagellin is recognized by NAIP5 and NAIP6, while PrgJ is sensed by NAIP2 (Kofoed and Vance, 2011; Zhao *et al.*, 2011). A mechanism of exclusion of *Salmonella*-infected enterocytes is used by the host to remove infected cells, expulsing them to the lumen. This response depends on NAIPs, NLRC4 and caspase1/11, but is independent of IL-1β and IL-18 (Sellin *et al.*, 2014). *Salmonella* is also detected by caspase-11, assembling the non-canonical inflammasome. SifA is a SPI-2 effector that is vital for SCV stabilization (Cirillo *et al.*, 1998). A *Salmonella* SifA mutant strongly induces pyroptosis dependent on caspase-11, but independent of NLRP3 and NLRC4 (Aachoui *et al.*, 2013). Caspase-11 recognizes *Salmonella* after it escapes from the SCV to the cytosol, but it can also detect when the vacuole is compromised (Broz *et al.*, 2012; Rathinam *et al.*, 2012). GBP2 expression induced by type I interferon and SCV destabilization mediated by GBP2 are required for caspase-11 activation in *Salmonella* infection (Meunier *et al.*, 2014). Caspase-11 and its human orthologue caspase-4 were also involved on the control of *S.* Typhimurium in intestinal epithelial cells by IL-18 secretion and cell death (Knodler *et al.*, 2014). Recently, a new report showed that growth of cytosolic *Salmonella* in a subpopulation of host cells is inhibited prior to pyroptosis, and requires caspase-1 and caspase-11 enzymatic activity (Thurston *et al.*, 2016). Macrophages are the major cell involved in *Salmonella* recognition and inflammasome activation, but neutrophils also play an important role. During acute *S.* Typhimurium infection, neutrophils are the main source of IL-1β, and different from macrophages, neutrophils do not undergo pyroptosis, sustaining IL-1β production on the site of infection (Chen *et al.*, 2014). On the other hand, NLRP6 and NLRP12 were also described in *Salmonella* infection, but in an opposite way. Mice deficient for NLRP6 and NLRP12 have a decreased bacterial burden and produce elevated levels of IL-6 and TNF-α, (Anand *et al.*, 2012; Zaki *et al.*, 2014). The mechanism of NLRP6 and NLRP12 activation by *Salmonella* still needs to be determined.

Caspase-1-deficient mice are highly susceptible to *S.* Typhimurium infection (Lara-Tejero *et al.*, 2006; Raupach *et al.*, 2006), and mice double deficient for NLRP3 and NLRC4 recapitulate this phenotype (Broz *et al.*, 2010a). However, IL-1β deficient mice have a mild phenotype. *Salmonella* control is largely mediated by IL-18, once IL-18 deficient mice are as susceptible as caspase-1 deficient mice (Raupach *et al.*, 2006). IL-18 is important to induce IFN-γ production by NK cells and T lymphocytes, a cytokine extremely important in the control of *Salmonella* infection (Eckmann and Kagnoff, 2001; Kupz *et al.*, 2014). A *S.* Typhimurium engineered to continuously express flagellin induces pyroptosis mediated by NLRC4 and caspase-1, and bacteria control was mediated by pyroptosis, removing the replicative niche and allowing extracellular bacteria to be killed by neutrophils. However, IL-18 and IL-1β did not play an important role in the control of this modified strain (Miao *et al.*, 2010a). Thus, IL-18 plays an important role in the control of WT *S.* Typhimurium, but in the case of a bacteria persistently expressing flagellin, pyroptosis is the main mechanism involved in the control. This discrepancy maybe is explained by the strong NLRC4 activation by this modified strain that leads to pyroptosis and bacteria killing by neutrophils. On the other hand, wild-type *S.* Typhimurium evades NLRC4 recognition by inhibiting flagellin expression and modifying T3SS rod protein (Miao *et al.*, 2010a; Miao and Rajan, 2011). Recently, another study from Ed Miao's group demonstrated that pore formation induced by *S.* Typhimurium infected cells leads to pyroptosis that traps live bacteria inside pyroptotic cell corpses, which ultimately are phagocytosed by neutrophils by efferocytose and killed via ROS. On this model, neutrophil recruitment to pyroptotic macrophages is dependent on IL-1β, IL-18 and eicosanoids (Jorgensen *et al.*, 2016a; Jorgensen *et al.*, 2016b).

As bona fide mammalian pathogen, *Salmonella* have developed strategies to dampen inflammasome activation and persist on the host. For example, it was identified two enzymes, aconitase and isocitrate dehydrogenase, involved on the bacterial tricarboxylic acid cycle. Mutations on these genes led to high intracellular citrate and elevated mitochondrial ROS, resulting in NLRP3 activation (Wynosky-Dolfi *et al.*, 2014). *Salmonella* also evades NLRC4 recognition by inhibiting flagellin expression in the intracellular environment when SPI-2 T3SS is active. *Salmonella* also expresses a T3SS rod protein termed Ssa1 that evades NAIP2 binding (Miao *et al.*, 2010b). Modifications on several amino acids on the Ssa1 protein inhibit NAIP2 recognition and NLRC4 activation (Jorgensen and Miao, 2015).

Francisella

Francisella tularensis is a Gram-negative intracellular bacterium and the causative agent tularemia, a life-threatening infectious disease of the respiratory tract. Depending on the route of infection, various organs and tissues can be colonized by *F. tularensis*, including lungs, liver, skin, and spleen (Carvalho *et al.*, 2014). Phagocytosis of *Francisella* by macrophages has been extensively studied and involves the engagement of different phagocytic receptors. Following uptake, *Francisella* resides within an initial vacuolar compartment on the endocytic pathway that would normally change to a bactericidal phagolysosome (Chong *et al.*, 2008; Clemens *et al.*, 2004, 2005; Santic *et al.*, 2006). However, *F. tularensis* escapes from the phagosome to the cytosol, and in contrast to other intracellular bacteria, it does not depend on T3SS or T4SS for virulence and secretion of effector proteins (Larsson *et al.*, 2005).

The role of inflammasome in the recognition and control of *F. novicida* in mouse infection was initially demonstrated by Denise Monack's group. Recognition of *F. novicida* in the cytosol induces a host response dependent on ASC and caspase-1, and is required for bacterial control and resistance to infection (Mariathasan *et al.*, 2005). Later on, three different groups reported the AIM2 inflammasome as responsible for *F. tularensis and F.*

novicida recognition and inflammasome activation *in vitro* and *in vivo*. AIM2 is involved on the recognition of *Francisella* DNA and is required for ASC oligomerization, caspase-1 activation, IL-1β maturation and cell death (Fernandes-Alnemri *et al.*, 2010; Jones *et al.*, 2010; Rathinam *et al.*, 2010). On the other hand, NLRP3 is dispensable for inflammasome activation by *F. novicida* (Fernandes-Alnemri *et al.*, 2010).

It was previously reported a role for type I Interferon in AIM2 activation by *F. novicida*. Macrophages knockout for type I IFN receptor or IRF3 are defective in caspase-1 activation, IL-1β and IL-18 secretion during *F. novicida* infection (Henry *et al.*, 2007). These data were confirmed by other results demonstrating that mice deficient for the DNA sensor cGAS and the adaptor STING are defective in both IFN-β and IL-1β production (Jones *et al.*, 2010; Storek *et al.*, 2015). Until recently, the mechanism by which type I IFN controls AIM2 inflammasome activation by *Francisella* was not described. Once bacterial DNA needs to be released to activate AIM2, and type I IFN is not required for AIM2 activation by synthetic dsDNA, it was proposed that type I IFN would be involved in bacteriolysis and DNA release. It was also reported that the IFN-inducible transcription factor IRF1 is involved in AIM2 activation by *F. novicida* downstream of IFN-β (Man *et al.*, 2015). IRF1 is required for the induction of guanylate-binding proteins Gbp2 and Gbp5, which are involved in intracellular bacteria killing and DNA release. Gbp2 and Gbp5 knockdown reduced AIM2 inflammasome activation by *F. novicida* infection (Man *et al.*, 2015; Meunier *et al.*, 2015). Recently, Man and collaborators identified IRGB10, together with GBPs, as essential for the disruption of *F. novicida* membrane, DNA release and AIM2 recognition. IRGB10 also plays a partial role on the activation of the non-canonical inflammasome by *Escherichia coli* and *Citrobacter rodentium*, supposedly by making LPS more available for caspase-11 recognition (Man *et al.*, 2016). On the other hand, *in vivo* infection by *F. novicida*, type I IFN has a detrimental role (Henry *et al.*, 2010), and IFN-γ is important for Gbps induction and infection control (Pierini *et al.*, 2013).

Cytosolic localization of *F. novicida* is also required for IL-1β release by human monocytes (Gavrilin *et al.*, 2006), and heat-killed bacteria is unable to induce IL-1β maturation (Li *et al.*, 2006). Several groups have described that AIM2, but not NLRP3, plays a role in *F. novicida* infection in murine cells. However, in human cells, both NLRP3 and AIM2 are involved in *F. novicida* recognition and IL-1β induction, once knockdown of NLRP3 or AIM2 reduces IL-1β secretion by THP-1 cells (Atianand *et al.*, 2011). NLRP3 also forms a speck with ASC in 293T cells infected by *F. novicida* (Bedoya *et al.*, 2007), but how the bacteria triggers NLRP3 activation is unknown.

Both caspase-1 activation, and IL-1β and IL-18 cytokines are involved in the control of *Francisella* infection. IL-18 is important for IFN-γ production by NK cells, which is essential for bacterial control. IL-18 deficient mice are susceptible to *F. tularensis* infection, but the treatment with recombinant IFN-γ rescues the phenotype. The same group also demonstrated that IL-1β is important for anti-LPS IgM production that increases opsonization and bacterial killing (del Barrio *et al.*, 2015). A nice set of experiments using mice deficient for caspase-1 and for both IL-1β and IL-18, showed a role for both cytokines and caspase-1 in the control of *F. novicida* infection. Mice double deficient for IL-1β and IL-18 have an intermediate phenotype when compared to WT and caspase-1 deficient (Henry and Monack, 2007). Similar results were reported for *S.* Typhimurium, *L. pneumophila* and *B. thailandensis* (Miao *et al.*, 2010a).

Bacteria from the genus *Francisella* express a modified LPS that does not engage TLR4 activation (Hajjar *et al.*, 2006). Despite the induction of caspase-11 by *F. novicida* (Akhter *et al.*, 2012), the bacterium is not able to activate caspase-11 mostly due to its tetra-acylated lipid A, evading the non-canonical inflammasome activation (Hagar *et al.*, 2013). Mutations in the gene FTL_0325 from *F. tularensis*, which encodes an OmpA-like protein, interferes with NF-κB signalling and reduces cytokine production (Mahawar *et al.*, 2012), and the same mutant also delays AIM2 inflammasome activation and cell death at the early stages of infection (Dotson *et al.*, 2013).

Listeria

Listeria monocytogenes, the causative agent of listeriosis, a life-threatening food-borne disease, is a Gram-positive facultative intracellular pathogen that survives in different environments and infects several hosts (Hernandez-Milian and Payeras-Cifre, 2014). Infection with *L. monocytogenes* is usually caused by ingestion of contaminated food such as dairy products, meat, vegetables and seafood. *L. monocytogenes* can infect and survive in a variety of cells, including myeloid cells through phagocytosis or through invasion of epithelial cells by its virulence factor internalin A (Mengaud *et al.*, 1996). After entering the cell, *L. monocytogenes* is initially present inside a phagocytic vacuole, and through the pore-forming toxin listeriolysin-O (LLO, encoded by the gene *hly*), *L. monocytogenes* escapes from the phagosome and gains access to the cytosol (Hamon *et al.*, 2012; Portnoy *et al.*, 1988). *L. monocytogenes* is detected *in vitro* by the NLRP3, NLRC4 and AIM2 inflammasomes. NLRP3 activation, caspase-1 activation and cytokine production are dependent on the adaptor ASC in *L. monocytogenes* infection (Kim *et al.*, 2010; Mariathasan *et al.*, 2006; Wu *et al.*, 2010). The induction of mature IL-1β is also dependent on TLR2, required for the first signal, transcription of pro-IL-1β (Ozören *et al.*, 2006). LLO-mediated pore formation is essential for NLRP3 activation through potassium efflux that induces its oligomerization (Hamon and Cossart, 2011; Kanneganti *et al.*, 2006; Meixenberger *et al.*, 2010). Vacuole escaping by *L. monocytogenes* mediated by LLO leads to cathepsin B release from the phagosome, and cathepsin B and phagosome disruption is also sensed by NLRP3. Infection by mutants lacking LLO or an avirulent *Listeria innocua*, had impaired IL-1β production (Hamon and Cossart, 2011; Meixenberger *et al.*, 2010). However, different reports demonstrated no role for NLRP3 in *L. monocytogenes* infection (Franchi *et al.*, 2007; Sauer *et al.*, 2010).

Another platform that senses *L. monocytogenes* is the AIM2 inflammasome (Kim *et al.*, 2010). After escaping from the phagosome, bacteriolysis occurs in the cytosol and the released DNA activates AIM2 (Sauer *et al.*, 2010). Macrophages AIM2 deficient have reduced caspase-1 activation and IL-1β production after *L. monocytogenes* infection (Rathinam *et al.*, 2010; Warren *et al.*, 2010), but the combined deficiency of both NLRP3 and AIM2 totally abrogates inflammasome activation (Kim *et al.*, 2010).

L. monocytogenes activates NLRC4 *in vitro*, but is a poor inducer *in vivo*. It was demonstrated an important role for NLRC4 in non-primed macrophages infected with *L. monocytogenes*. In cells NLRC4 deficient, caspase-1 activation and cell death are not induced after infection, but after LPS priming, NLRC4 is not required for cell death and caspase-1 activation (Wu *et al.*, 2010). Sauer and co-workers engineered a flagellin-expressing *L. monocytogenes* strain and investigated the role of this inflammasome *in vivo*. This strain activates NLRC4 and was highly attenuated *in vivo*, and restriction of bacterial replication was independent of IL-1β and IL-18 (Sauer *et al.*, 2011). *In vivo*, caspase-1/11 deficient mice have increased susceptibility to *L. monocytogenes* infection, but this feature is not corroborated by other groups (Sauer *et al.*, 2011; Tsuji *et al.*, 2004).

L. monocytogenes is able to evade the immune system through different mechanisms. It avoids the activation of NLRC4 inflammasome by suppressing flagellin expression at 37°C via the transcriptional regulator MogR (Grundling *et al.*, 2004; Shen and Higgins, 2006). NLRP3 activation is avoided by controlling LLO expression transcriptionally or post transcriptionally through the activity of the transcriptional factor PrfA (Schnupf *et al.*, 2006; Shen and Higgins, 2006), and also by the limited activity of LLO in acidic pH compartments, preventing the damage of infected cells (Glomski *et al.*, 2003; Glomski *et al.*, 2002).

Mycobacterium

Mycobacterium tuberculosis (Mtb) causes a chronic bacterial disease called tuberculosis (TB). The majority of people infected by Mtb do not display symptoms. Many individuals contain Mtb confined in structures of immune cells termed granulomas (Nowag and Hartmann, 2016). The innate immune response is critical for host resistance against TB, where pro-inflammatory cytokines as IL-12, IFN-γ and TNF-α plays an important role (Dorhoi and Kaufmann, 2014; Sia *et al.*, 2015). Murine models of TB have demonstrated an important role for these cytokines in defence against Mtb. Cytokines from the IL-1 family have emerged as key players in the Mtb control. IL-1R-deficient mice infected by Mtb succumbed to the infection (Di Paolo *et al.*, 2015; Fremond *et al.*, 2007).

One important characteristic of virulent strains of Mtb is the presence of type VII secretion system (T7SS), among them ESX-1, that exports bacterial proteins into the host and is involved in the escape from the phagosome to the cytosol (de Jonge *et al.*, 2007; Houben *et al.*, 2012; Simeone *et al.*, 2012). Mutants lacking ESX-1 are severely attenuated in murine infection models (Hsu *et al.*, 2003). ESX-1 is important to activate PRR including the inflammasome. Mtb mutant strains that lack ESX-1 are poor IL-1β inducers in both human and mouse macrophages (Dorhoi *et al.*, 2012; Koo *et al.*, 2008; Mishra *et al.*, 2010).

The role of NLRP3, ASC and caspase-1 in *Mycobacterium* infection is controversial. Some reports show a role for NLRP3 in Mtb IL-1β induction and ESX-1 plays a key role, but other reports showed no role for NLRP3 and caspase-1 in IL-1β production (Dorhoi *et al.*, 2012; Koo *et al.*, 2008; Kurenuma *et al.*, 2009; Netea *et al.*, 2015). A human study showed that individuals carrying a gain-of-function mutation on the *NLRP3* gene, in combination with mutations on *CARD8*, have an improved Mtb control by macrophages (Eklund *et al.*, 2014). *In vivo* Mtb murine models found that both ASC and caspase-1 are not required for IL-1β production, and almost no increase in bacteria load was observed (Dorhoi *et al.*, 2012; McElvania Tekippe *et al.*, 2010). IL-1β and IL-1R deficient mice are more susceptible than ASC and caspase-1 knockout mice, and caspase-1-independent IL-1β cleavage was already described (Mayer-Barber *et al.*, 2010; Netea *et al.*, 2015). Recognition of *Mycobacterium leprae* by Dectin-1 in DC leads to IL-1β production mediated by caspase-8 and ASC (Gringhuis *et al.*, 2012), and macrophages could also induce IL-1β secretion independent of caspase-1 through RIPK3 (Vince *et al.*, 2012). It was also recently described that *Mycobacterium bovis* induces IL-1β secretion to a mechanism dependent on NLRP7 that also leads to pyroptosis (Zhou *et al.*, 2016).

The AIM2 inflammasome have also been described as involved in Mtb recognition. DNA from *Mycobacterium* can reach the cytosol through ESX-1, where it can activate AIM2 and induce IL-1β. Mice AIM2 knockout are more susceptible to Mtb infection (Saiga *et al.*, 2012; Yang *et al.*, 2013b).

Mtb can modulate inflammasome activation to escape from this response. *Mycobacterium smegmatis* induces AIM2 inflammasome activation dependent on IFN-β production; in contrast Mtb inhibits AIM2 activation by blocking type I interferon induction (Shah *et al.*, 2013). ZMP-1, an effector molecule produced by Mtb can diminish inflammasome activation, IL-1β production and phagosome maturation, increasing Mtb survival (Master *et al.*, 2008).

Yersinia

The *Yersinia* genus consists of Gram-negative coccobacilli or rod-shaped bacteria, and three species within the genus *Yersinia* are pathogenic for humans, *Yersinia pestis, Yersinia pseudotuberculosis*, and *Yersinia enterocolitica. Y. pestis* is the causative agent of the plague, and the other two species are enteric pathogens that cause a self-limiting gastroenteritis. They share a common virulence plasmid that encodes the T3SS (Fällman and Gustavsson, 2005). Translocation of *Yersinia* effector proteins (YopE, YopM, YpkA/YopO, YopH, YopT and YopP/J) into the cytosol of host cells results in immune response modulation and cell death. Both epithelial cells and macrophages have been shown to respond to *Yersinia* infection, and massive of immune cells is hallmark of infection by *Yersinia* species (Cornelis, 2002; Trosky *et al.*, 2008; Wren, 2003). It is believed that *Yersinia* primarily replicates extracellularly and evades phagocytosis in lymphoid tissues, but intracellular replication was already described (Grabenstein *et al.*, 2004). Distinct types of cell death including apoptosis, pyroptosis and even necrosis were described in *Yersinia* infection, but their contribution to pathogenesis or bacteria control is still debatable, especially *in vivo* (Monack *et al.*, 1998; Philip *et al.*, 2014; Ruckdeschel *et al.*, 1998; Ruckdeschel *et al.*, 1997; Zheng *et al.*, 2012). Cell death was described as a mechanism used by *Yersinia* for phagocyte elimination (Monack *et al.*, 1998), but some reports suggested that *Yersinia* induced cell death could trigger immune responses against the bacteria (Bergman *et al.*, 2009; Zauberman *et al.*, 2009).

The effector YopJ, called YopP in *Y. enterocolitica*, induces caspase-1-dependent apoptosis during infection (Brodsky *et al.*, 2010; Zheng *et al.*, 2011). YopJ is a MAPK and NFκB signalling inhibitor (Ruckdeschel *et al.*, 1998; Ruckdeschel *et al.*, 2001), and was also described as an ubiquitin-like protein protease and also as a deubiquitinase (Orth *et al.*, 2000; Sweet *et al.*, 2007; Zhou *et al.*, 2005). TLR activation and signalling through MAPK and NFκB induces up-regulation of pro-survival genes that limit caspase-8-dependent apoptosis, thus inhibition of MAPK and NFκB by YopJ induces caspase-8/caspase-1 activation and cell death. Recently, a caspase-independent cell death, mediated by Receptor Interacting Protein Kinases 1 and 3 (Ripk1 and Ripk3) was also described in *Yersinia* infection (Philip *et al.*, 2014; Weng *et al.*, 2014).

Despite the mechanism of inflammasome activation by *Yersinia* is still unknown, the induction of IL-1β and IL-18 by YopJ requires NLRP3, ASC and caspase-1, but caspase-1 activation is independent of NLRP3 and ASC. Moreover, T3SS recognition induces caspase-1 activation and IL-1β secretion dependent on NLRP3, NLRC4 and ASC, but the mechanism is unclear (Brodsky *et al.*, 2010; Zheng *et al.*, 2011). NLRP12 was also demonstrated to play a role in *Y. pestis* infection, and both NLRP12 and NLRP3 were involved in host defence against *Yersinia in vivo*, probably via caspase-1-dependent IL-1β and IL-18 induction (Vladimer *et al.*, 2012). In contrast to other Gram-negative bacteria such as *Salmonella*, *Yersinia* do not strongly activate the NLRC4 inflammasome, possibly because bacteria inhibits flagellin expression (Brodsky *et al.*, 2010; Minnich and Rohde, 2007).

Although the precise role of cell death in *Yersinia* infections is not entirely clear, evasion of inflammasome activation might be beneficial to the bacteria in some circumstances. YopM was originally described to play a role in IL-10 induction and immune response suppression (McPhee *et al.*, 2010; McPhee *et al.*, 2012; Nemeth and Straley, 1997). One isoform of YopM from *Y. pestis* is able to bind and inhibits caspase-1. YopM can bind directly to caspase-1 acting as a pseudosubstrate, or to interact with the small GTPase IQGAPI to inhibit inflammasome activation. The decreased virulence of YopM mutant was reversed in caspase-1 -deficient mice (Chung *et al.*, 2014; LaRock and Cookson, 2012), but the mechanism was unclear. A new report demonstrated a role for YopM in the inhibition of pyrin inflammasome in *Yersinia* infection. *In vivo* infection by *Yersinia* activates pyrin inflammasome that is important for bacteria control and host survival, and *Yersinia* YopM mutant was unable to inhibit pyrin inflammasome. These data suggest that YopM promotes virulence by inhibiting pyrin through its phosphorylation by protein kinase C-related kinases, PRK1 and PRK2 (Chung *et al.*, 2016). Recently, two reports showed that YopJ, that induces inflammasome activation, is also involved in inflammasome inhibition by acting together with YopM (Ratner *et al.*, 2016; Schoberle *et al.*, 2016).

A secreted effector protein called YopK prevents inflammasome activation. YopK deficient bacteria inject increased amount of proteins (YopB and YopD) into the host cell and activates the inflammasome (Zwack *et al.*, 2015). YopE and YopT, which are involved in Rho GTPase inactivation (Black and Bliska, 2000; Shao *et al.*, 2003), are also involved in IL-1β inhibition in *Y. enterocolitica* infection (Schotte *et al.*, 2004). Recently it was demonstrated that *Yersinia* modifies its own LPS from the hexa-acylated form to a tetra-acylated form when grew at 37°C. This hypoacylated LPS evades TLR4 recognition (Matsuura, 2013), and probably caspase-11 activation (Kayagaki *et al.*, 2011; Shi *et al.*, 2014).

Shigella

Shigella is a Gram-negative intracellular pathogen and is the causative agent of the acute recto-colitis shigellosis, otherwise known as bacillary dysentery. Children living in endemic areas of poor countries are the most affected by this disease (Raqib *et al.*, 2000). Strong acute inflammation is a characteristic of the host innate immune response to *Shigella* infection, marked by a rapid influx of neutrophils that leads to tissue destruction (Parsot and Sansonetti, 1996; Perdomo *et al.*, 1994). The pathogenesis of *Shigella* relies on the expression of a T3SS, through which the delivery of bacterial effectors into host cell cytosol promotes pathogenesis (Ogawa *et al.*, 2008; Parsot, 2009; Schroeder and Hilbi, 2008). The main response against intracellular pathogens, including *Salmonella*, *Legionella* and also *Shigella* is mediated by the activation of NLRC4 inflammasome. As such, *S. flexneri* is detected by NLRC4 and induces caspase-1 activation and IL-1β and IL-18 secretion (Suzuki *et al.*, 2014). However, *S. flexneri* is non-flagellated bacteria, thus NLRC4 do not respond to flagellin during *Shigella* infection. The bacterial T3SS rod protein MxiI and the needle component MxiH are detected by NAIP2 and NAIP1 respectively (Miao *et al.*, 2010b; Yang *et al.*, 2013a; Zhao *et al.*, 2011). The T3SS effector protein invasion plasmid antigen B (IpaB) that is required for virulence also induces pyroptosis and IL-1β secretion dependent on caspase-1 (Chen *et al.*, 1996; Hilbi *et al.*, 1998). IpaB oligomerizes and forms an ion channel allowing potassium influx and endolysosomal destabilization, leading to caspase-1 activation (Senerovic *et al.*, 2012). An NLRC5/NLRP3 inflammasome was also involved in the recognition and caspase-1 activation in *S. flexneri* infection. Cells knockdown for NLRC5 or NLRP3 have a decreased caspase-1 activation and IL-1β secretion (Davis *et al.*, 2011). Inflammasome activation by *S. flexineri* leads to caspase-1 activation and Il-1β and IL-18 secretion. Experiments using caspase-1 deficient mice, and also IL-1β and IL-18 deficient mice demonstrated an important role for both cytokines in inducing inflammation and infection control (Sansonetti *et al.*, 2000).

Recently, a guinea pig model of *S. flexneri* infection reported a caspase-4-dependent cell death, suggesting a role for human caspase-4 in *S. flexineri* infection. The *S. flexneri* effector OspC3 inhibits caspase-4 activation by interacting with the p19 subunit preventing its heterodimerization. Bacteria deficient in OspC3 induced higher cell death and showed decreased bacterial burden (Kobayashi *et al.*, 2013). Another mechanism described for *S. flexineri* to dampen IL-1β production is the modification of its own LPS structure. *S. flexneri* expresses a hypoacylated LPS that differs from the LPS of bacteria growing in laboratory. This hypoacylated LPS induces lower levels of IL-1β and has reduced priming capacity (Paciello *et al.*, 2013).

Concluding remarks

In the past few years we have vastly increased our knowledge about the mechanisms involved in inflammasome activation and suppression by intracellular bacteria (Table 5.1), but a number of unanswered questions still remains. For example, the molecular mechanisms responsible for NLRP3 activation, the role of pyroptotic cell death in restriction of pathogen replication *in vivo*, and the mechanism by which IL-1β is secreted still need to be determined. Inflammasome activation and downstream innate immune responses present a strong host immune defence against pathogens, and successful pathogens evolved mechanisms to avoid inflammasome activation, or in some cases to induce cell death. The identification and characterization of the pyrin inflammasome and the non-canonical inflammasome gave new insights on how the innate immune system recognize and respond to intracellular pathogens.

Inflammasomes mediate host defence by two major mechanisms. First, the induction and secretion of IL-1β and IL-18, which are important to trigger cellular responses mediated mostly by macrophages and neutrophils. Second, pyroptosis is important for the removal of the bacterial replicative niche and bacteria release to the extracellular milieu and killing mostly by neutrophils.

In addition, inflammasome biology has been extensively studied in murine models, facilitated by new techniques for genetic manipulation, allowing the study of hundreds of genes. But some of these discoveries cannot be easily translated to human inflammasome biology, and there is a need for more investigation on inflammasome biology in human cells. Future studies shall determine the precise role of cytokines and pyroptosis, and also other mechanisms operating downstream of the inflammasome for the control of bacterial infections. On the other hand, new strategies used by pathogens to escape and inhibit inflammasome activation will also be identified.

Table 5.1 Putative inflammasome activators and suppressors in bacterial infection

Pathogen	Inflammasome sensor	Putative inflammasome activator	Role *in vivo*	Putative inflammasome suppressor	References
Legionella	NLRC4	Flagellin	Yes	SdhA	Ren *et al*. (2006)
	NLRP3	ND	No		Case *et al*. (2013)
	Caspase-11	LPS	No		Silveira and Zamboni (2010)
	AIM2	DNA	ND		Ge *et al*. (2012)
Salmonella	NLRP3/NLRC4	T3SS	Yes	Ssa1	Broz *et al*. (2010a)
	NLRC4	Flagellin, PrgJ	Yes		Zhao *et al*. (2011)
	Caspase-11	LPS	Yes		Aachoui *et al*. (2013)
	NLRP6	ND	No		Anand *et al*. (2012)
	NLRP12	ND	No		Zaki *et al*. (2014)
Francisella	AIM2	DNA	Yes	FTL_0325	Fernandes-Alnemri *et al*. (2010)
	NLRP3	ND	No	Tetra-acylated lipid A	Rathinam *et al*. (2010)
					Atianand *et al*. (2011)
Listeria	NLRP3	LLO	ND	PrfA	Kanneganti *et al*. (2006)
	NLRC4	Flagellin	No	MogR	Kim *et al*. (2010)
	AIM2	DNA	Yes		Wu *et al*. (2010)
Mycobacterium	NLRP3	ESX-1	No	ZMP-1	Dorhoi *et al*. (2012)
	AIM2	DNA	Yes		Saiga *et al*. (2012)
	NLRP7	ND	ND		Zhou *et al*. (2016)
Yersinia	NLRP3	YopJ	Yes	YopK	Brodsky *et al*. (2010)
	Naip/NLRC4	T3SS	No	YopT	Zheng *et al*. (2011)
	NLRP12	T3SS, YopJ	Yes	YopE	Vladimer *et al*. (2012)
			Yes	YopM	Ratner *et al*. (2016)
				Tetra-acylated lipid A	Matsuura (2013)
Shigella	NLRC4	MxiI, MxiH, IpaB	Yes	OspC3	Suzuki *et al*. (2014)
	NLRC5/NLRP3	ND	ND		Yang *et al*. (2013a)
					Davis *et al*. (2011)

ND, not determined.

Acknowledgements

The work in our laboratory is supported by grants from the Fundação de Amparo à Pesquisa do Estado de São Paulo (FAPESP; grants 2013/08216-2, 2014/50268-2, and 2014/04684-4), Conselho Nacional do Desenvolvimento Cientifico e Tecnológico (CNPq), Coordenação de Aperfeiçoamento de Pessoal de Nível Superior (CAPES) and Fundação de Amparo ao Ensino, Pesquisa e Assistência do Hospital das Clínicas da FMRP/USP (FAEPA). D.S.Z. is a research fellow from CNPq, Brazil.

References

Aachoui, Y., Leaf, I.A., Hagar, J.A., Fontana, M.F., Campos, C.G., Zak, D.E., Tan, M.H., Cotter, P.A., Vance, R.E., Aderem, A., *et al.* (2013). Caspase-11 protects against bacteria that escape the vacuole. Science *339*, 975–978. https://doi.org/10.1126/science.1230751.

Aglietti, R.A., Estevez, A., Gupta, A., Ramirez, M.G., Liu, P.S., Kayagaki, N., Ciferri, C., Dixit, V.M., and Dueber, E.C. (2016). GsdmD p30 elicited by caspase-11 during pyroptosis forms pores in membranes. Proc. Natl. Acad. Sci. U.S.A. *113*, 7858–7863. https://doi.org/10.1073/pnas.1607769113.

Akhter, A., Caution, K., Abu Khweek, A., Tazi, M., Abdulrahman, B.A., Abdelaziz, D.H., Voss, O.H., Doseff, A.I., Hassan, H., Azad, A.K., *et al.* (2012). Caspase-11 promotes the fusion of phagosomes harboring pathogenic bacteria with lysosomes by modulating actin polymerization. Immunity *37*, 35-47.

Akhter, A., Gavrilin, M.A., Frantz, L., Washington, S., Ditty, C., Limoli, D., Day, C., Sarkar, A., Newland, C., Butchar, J., *et al.* (2009). Caspase-7 activation by the Nlrc4/Ipaf inflammasome restricts *Legionella pneumophila* infection. PLOS Pathog. *5*, e1000361. https://doi.org/10.1371/journal.ppat.1000361.

Allen, I.C., Wilson, J.E., Schneider, M., Lich, J.D., Roberts, R.A., Arthur, J.C., Woodford, R.M., Davis, B.K., Uronis, J.M., Herfarth, H.H., *et al.* (2012). NLRP12 suppresses colon inflammation and tumorigenesis through the negative regulation of noncanonical NF-kappaB signaling. Immunity *36*, 742-754.

Amer, A., Franchi, L., Kanneganti, T.D., Body-Malapel, M., Ozören, N., Brady, G., Meshinchi, S., Jagirdar, R., Gewirtz, A., Akira, S., *et al.* (2006). Regulation of Legionella phagosome maturation and infection through flagellin and host Ipaf. J. Biol. Chem. *281*, 35217–35223.

Anand, P.K., Malireddi, R.K., Lukens, J.R., Vogel, P., Bertin, J., Lamkanfi, M., and Kanneganti, T.D. (2012). NLRP6 negatively regulates innate immunity and host defence against bacterial pathogens. Nature *488*, 389–393. https://doi.org/10.1038/nature11250.

Ataide, M.A., Andrade, W.A., Zamboni, D.S., Wang, D., Souza, M.d.o.C., Franklin, B.S., Elian, S., Martins, F.S., Pereira, D., Reed, G., *et al.* (2014). Malaria-induced NLRP12/NLRP3-dependent caspase-1 activation mediates inflammation and hypersensitivity to bacterial superinfection. PLOS Pathog. *10*, e1003885. https://doi.org/10.1371/journal.ppat.1003885.

Atianand, M.K., Duffy, E.B., Shah, A., Kar, S., Malik, M., and Harton, J.A. (2011). Francisella tularensis reveals a disparity between human and mouse NLRP3 inflammasome activation. J Biol Chem *286*, 39033-39042.

Baker, S., and Dougan, G. (2007). The genome of *Salmonella enterica* serovar Typhi. Clin. Infect. Dis. 45 (Suppl. 1), S29–33.

Bauernfeind, F.G., Horvath, G., Stutz, A., Alnemri, E.S., MacDonald, K., Speert, D., Fernandes-Alnemri, T., Wu, J., Monks, B.G., Fitzgerald, K.A., *et al.* (2009). Cutting edge: NF-kappaB activating pattern recognition and cytokine receptors license NLRP3 inflammasome activation by regulating NLRP3 expression. J Immunol *183*, 787-791.

Bedoya, F., Sandler, L.L., and Harton, J.A. (2007). Pyrin-only protein 2 modulates NF-kappaB and disrupts ASC:CLR interactions. J. Immunol. *178*, 3837–3845.

Bergman, M.A., Loomis, W.P., Mecsas, J., Starnbach, M.N., and Isberg, R.R. (2009). CD8(+) T cells restrict Yersinia pseudotuberculosis infection: bypass of anti-phagocytosis by targeting antigen-presenting cells. PLOS Pathog *5*, e1000573.

Bhuin, T., and Roy, J.K. (2014). Rab proteins: the key regulators of intracellular vesicle transport. Exp. Cell Res. *328*, 1–19. https://doi.org/10.1016/j.yexcr.2014.07.027.

Black, D.S., and Bliska, J.B. (2000). The RhoGAP activity of the Yersinia pseudotuberculosis cytotoxin YopE is required for antiphagocytic function and virulence. Mol. Microbiol. 37, 515–527.

Boyden, E.D., and Dietrich, W.F. (2006). Nalp1b controls mouse macrophage susceptibility to anthrax lethal toxin. Nat. Genet. *38*, 240–244.

Brodsky, I.E., Palm, N.W., Sadanand, S., Ryndak, M.B., Sutterwala, F.S., Flavell, R.A., Bliska, J.B., and Medzhitov, R. (2010). A Yersinia effector protein promotes virulence by preventing inflammasome recognition of the type III secretion system. Cell Host Microbe 7, 376-387.

Broz, P., and Dixit, V.M. (2016). Inflammasomes: mechanism of assembly, regulation and signalling. Nat. Rev. Immunol. *16*, 407–420. https://doi.org/10.1038/nri.2016.58.

Broz, P., Newton, K., Lamkanfi, M., Mariathasan, S., Dixit, V.M., and Monack, D.M. (2010a). Redundant roles for inflammasome receptors NLRP3 and NLRC4 in host defense against Salmonella. J. Exp. Med. *207*, 1745–1755. https://doi.org/10.1084/jem.20100257.

Broz, P., Ruby, T., Belhocine, K., Bouley, D.M., Kayagaki, N., Dixit, V.M., and Monack, D.M. (2012). Caspase-11 increases susceptibility to Salmonella infection in the absence of caspase-1. Nature *490*, 288–291. https://doi.org/10.1038/nature11419.

Broz, P., von Moltke, J., Jones, J.W., Vance, R.E., and Monack, D.M. (2010b). Differential requirement for Caspase-1 autoproteolysis in pathogen-induced cell death and cytokine processing. Cell Host Microbe *8*, 471–483. https://doi.org/10.1016/j.chom.2010.11.007.

Bürckstümmer, T., Baumann, C., Blüml, S., Dixit, E., Dürnberger, G., Jahn, H., Planyavsky, M., Bilban, M., Colinge, J., Bennett, K.L., *et al.* (2009). An orthogonal proteomic-genomic screen identifies AIM2 as a cytoplasmic DNA sensor for the inflammasome. Nat. Immunol. *10*, 266–272. https://doi.org/10.1038/ni.1702.

Canna, S.W., de Jesus, A.A., Gouni, S., Brooks, S.R., Marrero, B., Liu, Y., DiMattia, M.A., Zaal, K.J., Sanchez, G.A., Kim, H., *et al.* (2014). An activating NLRC4 inflammasome mutation causes autoinflammation with recurrent macrophage activation syndrome. Nat Genet 46, 1140-1146.

Carvalho, C.L., Lopes de Carvalho, I., Zé-Zé, L., Núncio, M.S., and Duarte, E.L. (2014). Tularaemia: a challenging zoonosis. Comp. Immunol. Microbiol. Infect. Dis. *37*, 85–96. https://doi.org/10.1016/j.cimid.2014.01.002.

Case, C.L., Kohler, L.J., Lima, J.B., Strowig, T., de Zoete, M.R., Flavell, R.A., Zamboni, D.S., and Roy, C.R. (2013). Caspase-11 stimulates rapid flagellin-independent pyroptosis in response to *Legionella pneumophila*. Proc. Natl. Acad. Sci. U.S.A. *110*, 1851–1856. https://doi.org/10.1073/pnas.1211521110.

Case, C.L., and Roy, C.R. (2011). Asc modulates the function of NLRC4 in response to infection of macrophages by *Legionella pneumophila*. MBio *2*, e00117–11. https://doi.org/10.1128/mBio.00117-11.

Case, C.L., Shin, S., and Roy, C.R. (2009). Asc and Ipaf Inflammasomes direct distinct pathways for caspase-1 activation in response to *Legionella pneumophila*. Infect. Immun. *77*, 1981–1991. https://doi.org/10.1128/IAI.01382-08.

Cassel, S.L., Eisenbarth, S.C., Iyer, S.S., Sadler, J.J., Colegio, O.R., Tephly, L.A., Carter, A.B., Rothman, P.B., Flavell, R.A., and Sutterwala, F.S. (2008). The Nalp3 inflammasome is essential for the development of silicosis. Proc. Natl. Acad. Sci. U.S.A. *105*, 9035–9040. https://doi.org/10.1073/pnas.0803933105.

Casson, C.N., Copenhaver, A.M., Zwack, E.E., Nguyen, H.T., Strowig, T., Javdan, B., Bradley, W.P., Fung, T.C., Flavell, R.A., Brodsky, I.E., *et al.* (2013). Caspase-11 activation in response to bacterial secretion systems that access the host cytosol. PLOS Pathog. *9*, e1003400. https://doi.org/10.1371/journal.ppat.1003400.

Cerqueira, D.M., Pereira, M.S., Silva, A.L., Cunha, L.D., and Zamboni, D.S. (2015). Caspase-1 but not caspase-11 is required for NLRC4-mediated pyroptosis and restriction of infection by flagellated *Legionella* species in mouse macrophages and in vivo. J. Immunol. *195*, 2303–2311. https://doi.org/10.4049/jimmunol.1501223.

Chae, J.J., Cho, Y.H., Lee, G.S., Cheng, J., Liu, P.P., Feigenbaum, L., Katz, S.I., and Kastner, D.L. (2011). Gain-of-function Pyrin mutations induce NLRP3 protein-independent interleukin-1beta activation and severe autoinflammation in mice. Immunity *34*, 755-768.

Chavarría-Smith, J., and Vance, R.E. (2013). Direct proteolytic cleavage of NLRP1B is necessary and sufficient for inflammasome activation by anthrax lethal factor. PLOS Pathog. *9*, e1003452. https://doi.org/10.1371/journal.ppat.1003452.

Chen, K.W., Gross, C.J., Sotomayor, F.V., Stacey, K.J., Tschopp, J., Sweet, M.J., and Schroder, K. (2014). The neutrophil NLRC4 inflammasome selectively promotes IL-1beta maturation without pyroptosis during acute *Salmonella* challenge. Cell Rep *8*, 570-582.

Chen, Y., Smith, M.R., Thirumalai, K., and Zychlinsky, A. (1996). A bacterial invasin induces macrophage apoptosis by binding directly to ICE. EMBO J. *15*, 3853–3860.

Chong, A., Wehrly, T.D., Nair, V., Fischer, E.R., Barker, J.R., Klose, K.E., and Celli, J. (2008). The early phagosomal stage of Francisella tularensis determines optimal phagosomal escape and Francisella pathogenicity island protein expression. Infect. Immun. *76*, 5488-5499.

Chung, L.K., Park, Y.H., Zheng, Y., Brodsky, I.E., Hearing, P., Kastner, D.L., Chae, J.J., and Bliska, J.B. (2016). The *Yersinia* virulence factor YopM hijacks host kinases to inhibit Type III effector-triggered activation of the pyrin INFLAMMASOME. Cell Host Microbe *20*, 296–306. https://doi.org/10.1016/j.chom.2016.07.018.

Chung, L.K., Philip, N.H., Schmidt, V.A., Koller, A., Strowig, T., Flavell, R.A., Brodsky, I.E., and Bliska, J.B. (2014). IQGAP1 is important for activation of caspase-1 in macrophages and is targeted by *Yersinia pestis* type III effector YopM. MBio *5*, e01402–14. https://doi.org/10.1128/mBio.01402-14.

Cirillo, D.M., Valdivia, R.H., Monack, D.M., and Falkow, S. (1998). Macrophage-dependent induction of the *Salmonella* pathogenicity island 2 type III secretion system and its role in intracellular survival. Mol Microbiol *30*, 175-188.

Clemens, D.L., Lee, B.Y., and Horwitz, M.A. (2004). Virulent and avirulent strains of Francisella tularensis prevent acidification and maturation of their phagosomes and escape into the cytoplasm in human macrophages. Infect Immun *72*, 3204-3217.

Clemens, D.L., Lee, B.Y., and Horwitz, M.A. (2005). Francisella tularensis enters macrophages via a novel process involving pseudopod loops. Infect. Immun. *73*, 5892–5902.

Coers, J., Vance, R.E., Fontana, M.F., and Dietrich, W.F. (2007). Restriction of *Legionella pneumophila* growth in macrophages requires the concerted action of cytokine and Naip5/Ipaf signalling pathways. Cell. Microbiol. *9*, 2344–2357.

Connor, M.G., Pulsifer, A.R., Price, C.T., Abu Kwaik, Y., and Lawrenz, M.B. (2015). *Yersinia pestis* Requires Host Rab1b for Survival in Macrophages. PLOS Pathog. *11*, e1005241. https://doi.org/10.1371/journal.ppat.1005241.

Copenhaver, A.M., Casson, C.N., Nguyen, H.T., Fung, T.C., Duda, M.M., Roy, C.R., and Shin, S. (2014). Alveolar macrophages and neutrophils are the primary reservoirs for *Legionella pneumophila* and mediate cytosolic surveillance of type IV secretion. Infect Immun *82*, 4325-4336.

Cornelis, G.R. (2002). Yersinia type III secretion: send in the effectors. J. Cell Biol. *158*, 401–408. https://doi.org/10.1083/jcb.200205077.

Costa, L.F., Paixão, T.A., Tsolis, R.M., Bäumler, A.J., and Santos, R.L. (2012). Salmonellosis in cattle: advantages of being an experimental model. Res. Vet. Sci. *93*, 1–6. https://doi.org/10.1016/j.rvsc.2012.03.002.

Creasey, E.A., and Isberg, R.R. (2012). The protein SdhA maintains the integrity of the Legionella-containing vacuole. Proc. Natl. Acad. Sci. U.S.A. *109*, 3481–3486. https://doi.org/10.1073/pnas.1121286109.

Cruz, C.M., Rinna, A., Forman, H.J., Ventura, A.L., Persechini, P.M., and Ojcius, D.M. (2007). ATP activates a reactive oxygen species-dependent oxidative stress response and secretion of proinflammatory cytokines in macrophages. J Biol Chem *282*, 2871-2879.

Cunha, L.D., and Zamboni, D.S. (2013). Subversion of inflammasome activation and pyroptosis by pathogenic bacteria. Front. Cell. Infect. Microbiol. *3*, 76. https://doi.org/10.3389/fcimb.2013.00076.

Davis, B.K., Roberts, R.A., Huang, M.T., Willingham, S.B., Conti, B.J., Brickey, W.J., Barker, B.R., Kwan, M., Taxman, D.J., Accavitti-Loper, M.A., *et al.* (2011). Cutting edge: NLRC5-dependent activation of the inflammasome. J. Immunol. *186*, 1333–1337. https://doi.org/10.4049/jimmunol.1003111.

De Jong, H.K., Koh, G.C., van Lieshout, M.H., Roelofs, J.J., van Dissel, J.T., van der Poll, T., and Wiersinga, W.J. (2014). Limited role for ASC and NLRP3 during in vivo Salmonella Typhimurium infection. BMC Immunol. *15*, 30. https://doi.org/10.1186/s12865-014-0030-7.

de Jonge, M.I., Pehau-Arnaudet, G., Fretz, M.M., Romain, F., Bottai, D., Brodin, P., Honore, N., Marchal, G., Jiskoot, W., England, P., *et al.* (2007). ESAT-6 from *Mycobacterium tuberculosis* dissociates from its putative chaperone CFP-10 under acidic conditions and exhibits membrane-lysing activity. J Bacteriol *189*, 6028-6034.

de Vasconcelos, N.M., Van Opdenbosch, N., and Lamkanfi, M. (2016). Inflammasomes as polyvalent cell death platforms. Cell. Mol. Life Sci. *73*, 2335–2347. https://doi.org/10.1007/s00018-016-2204-3.

del Barrio, L., Sahoo, M., Lantier, L., Reynolds, J.M., Ceballos-Olvera, I., and Re, F. (2015). Production of anti-LPS IgM by B1a B cells depends on IL-1β and is protective against lung infection with Francisella tularensis LVS. PLOS Pathog. *11*, e1004706. https://doi.org/10.1371/journal.ppat.1004706.

Di Paolo, N.C., Shafiani, S., Day, T., Papayannopoulou, T., Russell, D.W., Iwakura, Y., Sherman, D., Urdahl, K., and Shayakhmetov, D.M. (2015). Interdependence between interleukin-1 and tumor necrosis factor regulates TNF-dependent control of *Mycobacterium tuberculosis* Infection. Immunity *43*, 1125-1136.

Diebolder, C.A., Halff, E.F., Koster, A.J., Huizinga, E.G., and Koning, R.I. (2015). Cryoelectron Tomography of the NAIP5/NLRC4 Inflammasome: Implications for NLR Activation. Structure *23*, 2349–2357. https://doi.org/10.1016/j.str.2015.10.001.

Dihlmann, S., Erhart, P., Mehrabi, A., Nickkholgh, A., Lasitschka, F., Bockler, D., and Hakimi, M. (2014). Increased expression and activation of absent in melanoma 2 inflammasome components in lymphocytic infiltrates of abdominal aortic aneurysms. Mol Med *20*, 230-237.

Ding, J., Wang, K., Liu, W., She, Y., Sun, Q., Shi, J., Sun, H., Wang, D.C., and Shao, F. (2016). Pore-forming activity and structural autoinhibition of the gasdermin family. Nature *535*, 111-116.

Dombrowski, Y., Peric, M., Koglin, S., Kammerbauer, C., Göss, C., Anz, D., Simanski, M., Gläser, R., Harder, J., Hornung, V., *et al.* (2011). Cytosolic DNA triggers inflammasome activation in keratinocytes in psoriatic lesions. Sci. Transl. Med. *3*, 82ra38. https://doi.org/10.1126/scitranslmed.3002001.

Dorhoi, A., and Kaufmann, S.H. (2014). Perspectives on host adaptation in response to *Mycobacterium tuberculosis*: modulation of inflammation. Semin Immunol *26*, 533-542.

Dorhoi, A., Nouailles, G., Jörg, S., Hagens, K., Heinemann, E., Pradl, L., Oberbeck-Müller, D., Duque-Correa, M.A., Reece, S.T., Ruland, J., *et al.* (2012). Activation of the NLRP3 inflammasome by *Mycobacterium tuberculosis* is uncoupled from susceptibility to active tuberculosis. Eur. J. Immunol. *42*, 374–384. https://doi.org/10.1002/eji.201141548.

Dotson, R.J., Rabadi, S.M., Westcott, E.L., Bradley, S., Catlett, S.V., Banik, S., Harton, J.A., Bakshi, C.S., and Malik, M. (2013). Repression of inflammasome by Francisella tularensis during early stages of infection. J. Biol. Chem. *288*, 23844–23857. https://doi.org/10.1074/jbc.M113.490086.

Drecktrah, D., Knodler, L.A., Howe, D., and Steele-Mortimer, O. (2007). Salmonella trafficking is defined by continuous dynamic interactions with the endolysosomal system. Traffic *8*, 212–225.

Duewell, P., Kono, H., Rayner, K.J., Sirois, C.M., Vladimer, G., Bauernfeind, F.G., Abela, G.S., Franchi, L., Nuñez, G., Schnurr, M., *et al.* (2010). NLRP3 inflammasomes are required for atherogenesis and activated by cholesterol crystals. Nature *464*, 1357–1361. https://doi.org/10.1038/nature08938.

Eckmann, L., and Kagnoff, M.F. (2001). Cytokines in host defense against Salmonella. Microbes Infect. *3*, 1191–1200.

Eklund, D., Welin, A., Andersson, H., Verma, D., Soderkvist, P., Stendahl, O., Sarndahl, E., and Lerm, M. (2014). Human gene variants linked to enhanced NLRP3 activity limit intramacrophage growth of *Mycobacterium tuberculosis*. J. Infect. Dis. *209*, 749-753.

Endrizzi, M.G., Hadinoto, V., Growney, J.D., Miller, W., and Dietrich, W.F. (2000). Genomic sequence analysis of the mouse Naip gene array. Genome Res. *10*, 1095–1102.

Ewald, S.E., Chavarria-Smith, J., and Boothroyd, J.C. (2014). NLRP1 is an inflammasome sensor for *Toxoplasma gondii*. Infect. Immun. *82*, 460–468. https://doi.org/10.1128/IAI.01170-13.

Fällman, M., and Gustavsson, A. (2005). Cellular mechanisms of bacterial internalization counteracted by *Yersinia*. Int. Rev. Cytol. *246*, 135–188.

Fernandes-Alnemri, T., Yu, J.W., Datta, P., Wu, J., and Alnemri, E.S. (2009). AIM2 activates the inflammasome and cell death in response to cytoplasmic DNA. Nature *458*, 509–513. https://doi.org/10.1038/nature07710.

Fernandes-Alnemri, T., Yu, J.W., Juliana, C., Solorzano, L., Kang, S., Wu, J., Datta, P., McCormick, M., Huang, L., McDermott, E., *et al.* (2010). The AIM2 inflammasome is critical for innate immunity to Francisella tularensis. Nat. Immunol. *11*, 385–393. https://doi.org/10.1038/ni.1859.

Feuillet, V., Medjane, S., Mondor, I., Demaria, O., Pagni, P.P., Galán, J.E., Flavell, R.A., and Alexopoulou, L. (2006). Involvement of Toll-like receptor 5 in the recognition of flagellated bacteria. Proc. Natl. Acad. Sci. U.S.A. *103*, 12487–12492.

Figueira, R., and Holden, D.W. (2012). Functions of the Salmonella pathogenicity island 2 (SPI-2) type III secretion system effectors. Microbiology *158*, 1147-1161.

Finsel, I., and Hilbi, H. (2015). Formation of a pathogen vacuole according to *Legionella pneumophila*: how to kill one bird with many stones. Cell. Microbiol. *17*, 935–950. https://doi.org/10.1111/cmi.12450.

Franchi, L., Amer, A., Body-Malapel, M., Kanneganti, T.D., Ozören, N., Jagirdar, R., Inohara, N., Vandenabeele, P., Bertin, J., Coyle, A., *et al.* (2006). Cytosolic flagellin requires Ipaf for activation of caspase-1 and interleukin 1beta in salmonella-infected macrophages. Nat. Immunol. *7*, 576–582.

Franchi, L., Kanneganti, T.D., Dubyak, G.R., and Núñez, G. (2007). Differential requirement of P2X7 receptor and intracellular K+ for caspase-1 activation induced by intracellular and extracellular bacteria. J. Biol. Chem. *282*, 18810–18818.

Fratti, R.A., Chua, J., Vergne, I., and Deretic, V. (2003). *Mycobacterium tuberculosis* glycosylated phosphatidylinositol causes phagosome maturation arrest. Proc Natl Acad Sci U S A *100*, 5437-5442.

Fremond, C.M., Togbe, D., Doz, E., Rose, S., Vasseur, V., Maillet, I., Jacobs, M., Ryffel, B., and Quesniaux, V.F. (2007). IL-1 receptor-mediated signal is an essential component of MyD88-dependent innate response to *Mycobacterium tuberculosis* infection. J. Immunol. *179*, 1178-1189.

Gaidt, M.M., Ebert, T.S., Chauhan, D., Schmidt, T., Schmid-Burgk, J.L., Rapino, F., Robertson, A.A., Cooper, M.A., Graf, T., and Hornung, V. (2016). Human Monocytes Engage an Alternative Inflammasome Pathway. Immunity *44*, 833–846. https://doi.org/10.1016/j.immuni.2016.01.012.

Gavrilin, M.A., Abdelaziz, D.H., Mostafa, M., Abdulrahman, B.A., Grandhi, J., Akhter, A., Abu Khweek, A., Aubert, D.F., Valvano, M.A., Wewers, M.D., *et al.* (2012). Activation of the pyrin inflammasome by intracellular Burkholderia cenocepacia. J. Immunol. *188*, 3469–3477. https://doi.org/10.4049/jimmunol.1102272.

Gavrilin, M.A., Bouakl, I.J., Knatz, N.L., Duncan, M.D., Hall, M.W., Gunn, J.S., and Wewers, M.D. (2006). Internalization and phagosome escape required for Francisella to induce human monocyte IL-1beta processing and release. Proc. Natl. Acad. Sci. U.S.A. *103*, 141–146.

Ge, J., Gong, Y.N., Xu, Y., and Shao, F. (2012). Preventing bacterial DNA release and absent in melanoma 2 inflammasome activation by a Legionella effector functioning in membrane trafficking. Proc. Natl. Acad. Sci. U.S.A. *109*, 6193–6198. https://doi.org/10.1073/pnas.1117490109.

Glomski, I.J., Decatur, A.L., and Portnoy, D.A. (2003). *Listeria monocytogenes* mutants that fail to compartmentalize listerolysin O activity are cytotoxic, avirulent, and unable to evade host extracellular defenses. Infect Immun *71*, 6754-6765.

Glomski, I.J., Gedde, M.M., Tsang, A.W., Swanson, J.A., and Portnoy, D.A. (2002). The *Listeria monocytogenes* hemolysin has an acidic pH optimum to compartmentalize activity and prevent damage to infected host cells. J. Cell Biol. *156*, 1029–1038. https://doi.org/10.1083/jcb.200201081.

Gorfu, G., Cirelli, K.M., Melo, M.B., Mayer-Barber, K., Crown, D., Koller, B.H., Masters, S., Sher, A., Leppla, S.H., Moayeri, M., *et al.* (2014). Dual role for inflammasome sensors NLRP1 and NLRP3 in murine resistance to Toxoplasma gondii. MBio *5*, e01117–13. https://doi.org/10.1128/mBio.01117-13.

Grabenstein, J.P., Marceau, M., Pujol, C., Simonet, M., and Bliska, J.B. (2004). The response regulator PhoP of Yersinia pseudotuberculosis is important for replication in macrophages and for virulence. Infect. Immun. *72*, 4973–4984. https://doi.org/10.1128/IAI.72.9.4973-4984.2004.

Gringhuis, S.I., Kaptein, T.M., Wevers, B.A., Theelen, B., van der Vlist, M., Boekhout, T., and Geijtenbeek, T.B. (2012). Dectin-1 is an extracellular pathogen sensor for the induction and processing of IL-1β via a noncanonical caspase-8 inflammasome. Nat. Immunol. *13*, 246–254. https://doi.org/10.1038/ni.2222.

Grundling, A., Burrack, L.S., Bouwer, H.G., and Higgins, D.E. (2004). *Listeria monocytogenes* regulates flagellar motility gene expression through MogR, a transcriptional repressor required for virulence. Proc Natl Acad Sci U S A *101*, 12318-12323.

Guo, W., Wang, P., Liu, Z., Yang, P., and Ye, P. (2015). The activation of pyrin domain-containing-3 inflammasome depends on lipopolysaccharide from Porphyromonas gingivalis and extracellular adenosine triphosphate in cultured oral epithelial cells. BMC Oral Health *15*, 133.

Hagar, J.A., Powell, D.A., Aachoui, Y., Ernst, R.K., and Miao, E.A. (2013). Cytoplasmic LPS activates caspase-11: implications in TLR4-independent endotoxic shock. Science *341*, 1250–1253. https://doi.org/10.1126/science.1240988.

Hajjar, A.M., Harvey, M.D., Shaffer, S.A., Goodlett, D.R., Sjostedt, A., Edebro, H., Forsman, M., Bystrom, M., Pelletier, M., Wilson, C.B., *et al.* (2006). Lack of in vitro and in vivo recognition of Francisella tularensis subspecies lipopolysaccharide by Toll-like receptors. Infect. Immun. *74*, 6730-6738.

Halff, E.F., Diebolder, C.A., Versteeg, M., Schouten, A., Brondijk, T.H., and Huizinga, E.G. (2012). Formation and structure of a NAIP5-NLRC4 inflammasome induced by direct interactions with conserved N- and C-terminal regions of flagellin. J. Biol. Chem. *287*, 38460–38472. https://doi.org/10.1074/jbc.M112.393512.

Hamon, M.A., and Cossart, P. (2011). K+ efflux is required for histone H3 dephosphorylation by *Listeria monocytogenes* listeriolysin O and other pore-forming toxins. Infect. Immun. *79*, 2839–2846. https://doi.org/10.1128/IAI.01243-10.

Hamon, M.A., Ribet, D., Stavru, F., and Cossart, P. (2012). Listeriolysin O: the Swiss army knife of Listeria. Trends Microbiol. *20*, 360–368. https://doi.org/10.1016/j.tim.2012.04.006.

Hayashi, F., Smith, K.D., Ozinsky, A., Hawn, T.R., Yi, E.C., Goodlett, D.R., Eng, J.K., Akira, S., Underhill, D.M., and Aderem, A. (2001). The innate immune response to bacterial flagellin is mediated by Toll-like receptor 5. Nature *410*, 1099–1103. https://doi.org/10.1038/35074106.

He, W.T., Wan, H., Hu, L., Chen, P., Wang, X., Huang, Z., Yang, Z.H., Zhong, C.Q., and Han, J. (2015). Gasdermin D is an executor of pyroptosis and required for interleukin-1β secretion. Cell Res. *25*, 1285–1298. https://doi.org/10.1038/cr.2015.139.

He, Y., Zeng, M.Y., Yang, D., Motro, B., and Núñez, G. (2016). NEK7 is an essential mediator of NLRP3 activation downstream of potassium efflux. Nature *530*, 354–357. https://doi.org/10.1038/nature16959.

Hempstead, A.D., and Isberg, R.R. (2015). Inhibition of host cell translation elongation by *Legionella pneumophila* blocks the host cell unfolded protein response. Proc. Natl. Acad. Sci. U.S.A. *112*, E6790–7. https://doi.org/10.1073/pnas.1508716112.

Heneka, M.T., Kummer, M.P., Stutz, A., Delekate, A., Schwartz, S., Vieira-Saecker, A., Griep, A., Axt, D., Remus, A., Tzeng, T.C., *et al.* (2013). NLRP3 is activated in Alzheimer's disease and contributes to pathology in APP/PS1 mice. Nature *493*, 674–678. https://doi.org/10.1038/nature11729.

Henry, T., Brotcke, A., Weiss, D.S., Thompson, L.J., and Monack, D.M. (2007). Type I interferon signaling is required for activation of the inflammasome during Francisella infection. J. Exp. Med. *204*, 987–994.

Henry, T., Kirimanjeswara, G.S., Ruby, T., Jones, J.W., Peng, K., Perret, M., Ho, L., Sauer, J.D., Iwakura, Y., Metzger, D.W., *et al.* (2010). Type I IFN signaling constrains IL-17A/F secretion by gammadelta T cells during bacterial infections. J. Immunol. *184*, 3755–3767. https://doi.org/10.4049/jimmunol.0902065.

Henry, T., and Monack, D.M. (2007). Activation of the inflammasome upon Francisella tularensis infection: interplay of innate immune pathways and virulence factors. Cell. Microbiol. *9*, 2543–2551.

Hernandez-Milian, A., and Payeras-Cifre, A. (2014). What is new in listeriosis? Biomed. Res. Int. *2014*, 358051. https://doi.org/10.1155/2014/358051.

Hesker, P.R., Nguyen, M., Kovarova, M., Ting, J.P., and Koller, B.H. (2012). Genetic loss of murine pyrin, the Familial Mediterranean Fever protein, increases interleukin-1β levels. PLOS ONE *7*, e51105. https://doi.org/10.1371/journal.pone.0051105.

Hilbi, H., Moss, J.E., Hersh, D., Chen, Y., Arondel, J., Banerjee, S., Flavell, R.A., Yuan, J., Sansonetti, P.J., and Zychlinsky, A. (1998). Shigella-induced apoptosis is dependent on caspase-1 which binds to IpaB. J. Biol. Chem. *273*, 32895–32900.

Hiller, S., Kohl, A., Fiorito, F., Herrmann, T., Wider, G., Tschopp, J., Grütter, M.G., and Wüthrich, K. (2003). NMR structure of the apoptosis- and inflammation-related NALP1 pyrin domain. Structure *11*, 1199–1205.

Hoffman, H.M., Mueller, J.L., Broide, D.H., Wanderer, A.A., and Kolodner, R.D. (2001). Mutation of a new gene encoding a putative pyrin-like protein causes familial cold autoinflammatory syndrome and Muckle-Wells syndrome. Nat. Genet. *29*, 301–305. https://doi.org/10.1038/ng756.

Hornung, V., Ablasser, A., Charrel-Dennis, M., Bauernfeind, F., Horvath, G., Caffrey, D.R., Latz, E., and Fitzgerald, K.A. (2009). AIM2 recognizes cytosolic dsDNA and forms a caspase-1-activating inflammasome with ASC. Nature *458*, 514–518. https://doi.org/10.1038/nature07725.

Hornung, V., Bauernfeind, F., Halle, A., Samstad, E.O., Kono, H., Rock, K.L., Fitzgerald, K.A., and Latz, E. (2008). Silica crystals and aluminum salts activate the NALP3 inflammasome through phagosomal destabilization. Nat. Immunol. *9*, 847-856.

Houben, D., Demangel, C., van Ingen, J., Perez, J., Baldeón, L., Abdallah, A.M., Caleechurn, L., Bottai, D., van Zon, M., de Punder, K., *et al.* (2012). ESX-1-mediated translocation to the cytosol controls virulence of mycobacteria. Cell. Microbiol. *14*, 1287–1298. https://doi.org/10.1111/j.1462-5822.2012.01799.x.

Hsu, T., Hingley-Wilson, S.M., Chen, B., Chen, M., Dai, A.Z., Morin, P.M., Marks, C.B., Padiyar, J., Goulding, C., Gingery, M., *et al.* (2003). The primary mechanism of attenuation of bacillus Calmette-Guerin is a loss of secreted lytic function required for invasion of lung interstitial tissue. Proc Natl Acad Sci U S A *100*, 12420-12425.

Hu, Z., Yan, C., Liu, P., Huang, Z., Ma, R., Zhang, C., Wang, R., Zhang, Y., Martinon, F., Miao, D., *et al.* (2013). Crystal structure of NLRC4 reveals its autoinhibition mechanism. Science *341*, 172–175. https://doi.org/10.1126/science.1236381.

Hu, Z., Zhou, Q., Zhang, C., Fan, S., Cheng, W., Zhao, Y., Shao, F., Wang, H.W., Sui, S.F., and Chai, J. (2015). Structural and biochemical basis for induced self-propagation of NLRC4. Science *350*, 399–404. https://doi.org/10.1126/science.aac5489.

Hubber, A., and Roy, C.R. (2010). Modulation of host cell function by *Legionella pneumophila* type IV effectors. Annu. Rev. Cell Dev. Biol. *26*, 261–283. https://doi.org/10.1146/annurev-cellbio-100109-104034.

Ibarra, J.A., and Steele-Mortimer, O. (2009). Salmonella--the ultimate insider. Salmonella virulence factors that modulate intracellular survival. Cell Microbiol *11*, 1579-1586.

Isberg, R.R., O'Connor, T.J., and Heidtman, M. (2009). The *Legionella pneumophila* replication vacuole: making a cosy niche inside host cells. Nat. Rev. Microbiol. *7*, 13–24. https://doi.org/10.1038/nrmicro1967.

Jack, R.S. (2015). Evolution of Immunity and Pathogens. Results Probl. Cell. Differ. *57*, 1–20. https://doi.org/10.1007/978-3-319-20819-0_1.

Jamilloux, Y., Pierini, R., Querenet, M., Juruj, C., Fauchais, A.L., Jauberteau, M.O., Jarraud, S., Lina, G., Etienne, J., Roy, C.R., *et al.* (2013). Inflammasome activation restricts *Legionella pneumophila* replication in primary microglial cells through flagellin detection. Glia *61*, 539-549.

Javierre, B.M., Fernandez, A.F., Richter, J., Al-Shahrour, F., Martin-Subero, J.I., Rodriguez-Ubreva, J., Berdasco, M., Fraga, M.F., O'Hanlon, T.P., Rider, L.G., *et al.* (2010). Changes in the pattern of DNA methylation associate with twin discordance in systemic lupus erythematosus. Genome Res. *20*, 170-179.

Jin, T., Perry, A., Jiang, J., Smith, P., Curry, J.A., Unterholzner, L., Jiang, Z., Horvath, G., Rathinam, V.A., Johnstone, R.W., *et al.* (2012). Structures of the HIN domain:DNA complexes reveal ligand binding and activation mechanisms of the AIM2 inflammasome and IFI16 receptor. Immunity *36*, 561–571. https://doi.org/10.1016/j.immuni.2012.02.014.

Jin, Y., Mailloux, C.M., Gowan, K., Riccardi, S.L., LaBerge, G., Bennett, D.C., Fain, P.R., and Spritz, R.A. (2007). NALP1 in vitiligo-associated multiple autoimmune disease. N. Engl. J. Med. *356*, 1216–1225.

Jones, B.D., Ghori, N., and Falkow, S. (1994). Salmonella typhimurium initiates murine infection by penetrating and destroying the specialized epithelial M cells of the Peyer's patches. J. Exp. Med. *180*, 15-23.

Jones, J.W., Kayagaki, N., Broz, P., Henry, T., Newton, K., O'Rourke, K., Chan, S., Dong, J., Qu, Y., Roose-Girma, M., *et al.* (2010). Absent in melanoma 2 is required for innate immune recognition of *Francisella tularensis*. Proc. Natl. Acad. Sci. U.S.A. *107*, 9771–9776. https://doi.org/10.1073/pnas.1003738107.

Jorgensen, I., Lopez, J.P., Laufer, S.A., and Miao, E.A. (2016a). IL-1beta, IL-18, and eicosanoids promote neutrophil recruitment to pore-induced intracellular traps following pyroptosis. Eur J Immunol.

Jorgensen, I., and Miao, E.A. (2015). Pyroptotic cell death defends against intracellular pathogens. Immunol. Rev. *265*, 130–142. https://doi.org/10.1111/imr.12287.

Jorgensen, I., Zhang, Y., Krantz, B.A., and Miao, E.A. (2016b). Pyroptosis triggers pore-induced intracellular traps (PITs) that capture bacteria and lead to their clearance by efferocytosis. J. Exp. Med. *213*, 2113–2128. https://doi.org/10.1084/jem.20151613.

Juliana, C., Fernandes-Alnemri, T., Kang, S., Farias, A., Qin, F., and Alnemri, E.S. (2012). Non-transcriptional priming and deubiquitination regulate NLRP3 inflammasome activation. J. Biol. Chem. *287*, 36617–36622. https://doi.org/10.1074/jbc.M112.407130.

Kagan, J.C., and Roy, C.R. (2002). Legionella phagosomes intercept vesicular traffic from endoplasmic reticulum exit sites. Nat. Cell Biol. *4*, 945–954. https://doi.org/10.1038/ncb883.

Kanneganti, T.D., Ozören, N., Body-Malapel, M., Amer, A., Park, J.H., Franchi, L., Whitfield, J., Barchet, W., Colonna, M., Vandenabeele, P., *et al.* (2006). Bacterial RNA and small antiviral compounds activate caspase-1 through cryopyrin/Nalp3. Nature *440*, 233–236.

Kawai, T., and Akira, S. (2011). Toll-like receptors and their cross-talk with other innate receptors in infection and immunity. Immunity *34*, 637–650. https://doi.org/10.1016/j.immuni.2011.05.006.

Kayagaki, N., Stowe, I.B., Lee, B.L., O'Rourke, K., Anderson, K., Warming, S., Cuellar, T., Haley, B., Roose-Girma, M., Phung, Q.T., *et al.* (2015). Caspase-11 cleaves gasdermin D for non-canonical inflammasome signalling. Nature *526*, 666–671. https://doi.org/10.1038/nature15541.

Kayagaki, N., Warming, S., Lamkanfi, M., Vande Walle, L., Louie, S., Dong, J., Newton, K., Qu, Y., Liu, J., Heldens, S., *et al.* (2011). Non-canonical inflammasome activation targets caspase-11. Nature *479*, 117–121. https://doi.org/10.1038/nature10558.

Kayagaki, N., Wong, M.T., Stowe, I.B., Ramani, S.R., Gonzalez, L.C., Akashi-Takamura, S., Miyake, K., Zhang, J., Lee, W.P., Muszyński, A., *et al.* (2013). Noncanonical inflammasome activation by intracellular LPS independent of TLR4. Science *341*, 1246–1249. https://doi.org/10.1126/science.1240248.

Kim, S., Bauernfeind, F., Ablasser, A., Hartmann, G., Fitzgerald, K.A., Latz, E., and Hornung, V. (2010). *Listeria monocytogenes* is sensed by the NLRP3 and AIM2 inflammasome. Eur. J. Immunol. *40*, 1545–1551. https://doi.org/10.1002/eji.201040425.

Knodler, L.A., Crowley, S.M., Sham, H.P., Yang, H., Wrande, M., Ma, C., Ernst, R.K., Steele-Mortimer, O., Celli, J., and Vallance, B.A. (2014). Noncanonical inflammasome activation of caspase-4/caspase-11 mediates epithelial defenses against enteric bacterial pathogens. Cell Host Microbe *16*, 249-256.

Kobayashi, T., Ogawa, M., Sanada, T., Mimuro, H., Kim, M., Ashida, H., Akakura, R., Yoshida, M., Kawalec, M., Reichhart, J.M., *et al.* (2013). The Shigella OspC3 effector inhibits caspase-4, antagonizes inflammatory cell death, and promotes epithelial infection. Cell Host Microbe *13*, 570-583.

Kofoed, E.M., and Vance, R.E. (2011). Innate immune recognition of bacterial ligands by NAIPs determines inflammasome specificity. Nature *477*, 592-595.

Koo, I.C., Wang, C., Raghavan, S., Morisaki, J.H., Cox, J.S., and Brown, E.J. (2008). ESX-1-dependent cytolysis in lysosome secretion and inflammasome activation during mycobacterial infection. Cell. Microbiol. *10*, 1866-1878.

Kortmann, J., Brubaker, S.W., and Monack, D.M. (2015). Cutting edge: inflammasome activation in primary human macrophages is dependent on flagellin. J. Immunol. *195*, 815–819. https://doi.org/10.4049/jimmunol.1403100.

Kovarova, M., Hesker, P.R., Jania, L., Nguyen, M., Snouwaert, J.N., Xiang, Z., Lommatzsch, S.E., Huang, M.T., Ting, J.P., and Koller, B.H. (2012). NLRP1-dependent pyroptosis leads to acute lung injury and morbidity in mice. J. Immunol. *189*, 2006–2016. https://doi.org/10.4049/jimmunol.1201065.

Kupz, A., Curtiss, R., Bedoui, S., and Strugnell, R.A. (2014). In vivo IFN-γ secretion by NK cells in response to Salmonella typhimurium requires NLRC4 inflammasomes. PLOS ONE *9*, e97418. https://doi.org/10.1371/journal.pone.0097418.

Kurenuma, T., Kawamura, I., Hara, H., Uchiyama, R., Daim, S., Dewamitta, S.R., Sakai, S., Tsuchiya, K., Nomura, T., and Mitsuyama, M. (2009). The RD1 locus in the *Mycobacterium tuberculosis* genome contributes to activation of caspase-1 via induction of potassium ion efflux in infected macrophages. Infect Immun 77, 3992-4001.

Laguna, R.K., Creasey, E.A., Li, Z., Valtz, N., and Isberg, R.R. (2006). A *Legionella pneumophila*-translocated substrate that is required for growth within macrophages and protection from host cell death. Proc. Natl. Acad. Sci. U.S.A. *103*, 18745–18750.

Lara-Tejero, M., Sutterwala, F.S., Ogura, Y., Grant, E.P., Bertin, J., Coyle, A.J., Flavell, R.A., and Galán, J.E. (2006). Role of the caspase-1 inflammasome in Salmonella typhimurium pathogenesis. J. Exp. Med. *203*, 1407–1412.

LaRock, C.N., and Cookson, B.T. (2012). The Yersinia virulence effector YopM binds caspase-1 to arrest inflammasome assembly and processing. Cell Host Microbe *12*, 799–805. https://doi.org/10.1016/j.chom.2012.10.020.

Larsson, P., Oyston, P.C., Chain, P., Chu, M.C., Duffield, M., Fuxelius, H.H., Garcia, E., Hälltorp, G., Johansson, D., Isherwood, K.E., *et al.* (2005). The complete genome sequence of Francisella tularensis, the causative agent of tularemia. Nat. Genet. *37*, 153–159.

Latz, E., Xiao, T.S., and Stutz, A. (2013). Activation and regulation of the inflammasomes. Nat. Rev. Immunol. *13*, 397–411. https://doi.org/10.1038/nri3452.

Lee, G.S., Subramanian, N., Kim, A.I., Aksentijevich, I., Goldbach-Mansky, R., Sacks, D.B., Germain, R.N., Kastner, D.L., and Chae, J.J. (2012). The calcium-sensing receptor regulates the NLRP3 inflammasome through Ca^{2+} and cAMP. Nature *492*, 123–127. https://doi.org/10.1038/nature11588.

Lee, S.Y., Lee, M.S., Cherla, R.P., and Tesh, V.L. (2008). Shiga toxin 1 induces apoptosis through the endoplasmic reticulum stress response in human monocytic cells. Cell Microbiol *10*, 770-780.

Li, H., Nookala, S., Bina, X.R., Bina, J.E., and Re, F. (2006). Innate immune response to *Francisella tularensis* is mediated by TLR2 and caspase-1 activation. J. Leukoc. Biol. *80*, 766–773.

Liu, X., Zhang, Z., Ruan, J., Pan, Y., Magupalli, V.G., Wu, H., and Lieberman, J. (2016). Inflammasome-activated gasdermin D causes pyroptosis by forming membrane pores. Nature *535*, 153–158. https://doi.org/10.1038/nature18629.

Lopez-Castejon, G., Luheshi, N.M., Compan, V., High, S., Whitehead, R.C., Flitsch, S., Kirov, A., Prudovsky, I., Swanton, E., and Brough, D. (2013). Deubiquitinases regulate the activity of caspase-1 and interleukin-1β secretion via assembly of the inflammasome. J. Biol. Chem. *288*, 2721–2733. https://doi.org/10.1074/jbc.M112.422238.

Magitta, N.F., Bøe Wolff, A.S., Johansson, S., Skinningsrud, B., Lie, B.A., Myhr, K.M., Undlien, D.E., Joner, G., Njølstad, P.R., Kvien, T.K., *et al.* (2009). A coding polymorphism in NALP1 confers risk for autoimmune Addison's disease and type 1 diabetes. Genes Immun. *10*, 120–124. https://doi.org/10.1038/gene.2008.85.

Mahawar, M., Atianand, M.K., Dotson, R.J., Mora, V., Rabadi, S.M., Metzger, D.W., Huntley, J.F., Harton, J.A., Malik, M., and Bakshi, C.S. (2012). Identification of a novel Francisella tularensis factor required for intramacrophage survival and subversion of innate immune response. J Biol Chem *287*, 25216-25229.

Majowicz, S.E., Musto, J., Scallan, E., Angulo, F.J., Kirk, M., O'Brien, S.J., Jones, T.F., Fazil, A., Hoekstra, R.M., and International Collaboration on Enteric Disease 'Burden of Illness, S. (2010). The global burden of nontyphoidal *Salmonella* gastroenteritis. Clin Infect Dis *50*, 882-889.

Man, S.M., and Kanneganti, T.D. (2015). Regulation of inflammasome activation. Immunol. Rev. *265*, 6–21. https://doi.org/10.1111/imr.12296.

Man, S.M., Karki, R., Malireddi, R.K., Neale, G., Vogel, P., Yamamoto, M., Lamkanfi, M., and Kanneganti, T.D. (2015). The transcription factor IRF1 and guanylate-binding proteins target activation of the AIM2 inflammasome by *Francisella* infection. Nat Immunol *16*, 467-475.

Man, S.M., Karki, R., Sasai, M., Place, D.E., Kesavardhana, S., Temirov, J., Frase, S., Zhu, Q., Malireddi, R.K., Kuriakose, T., *et al.* (2016). IRGB10 liberates bacterial ligands for sensing by the AIM2 and caspase-11-NLRP3 Inflammasomes. Cell *167*, 382–396.e17.

Mariathasan, S., Newton, K., Monack, D.M., Vucic, D., French, D.M., Lee, W.P., Roose-Girma, M., Erickson, S., and Dixit, V.M. (2004). Differential activation of the inflammasome by caspase-1 adaptors ASC and Ipaf. Nature *430*, 213–218. https://doi.org/10.1038/nature02664.

Mariathasan, S., Weiss, D.S., Dixit, V.M., and Monack, D.M. (2005). Innate immunity against *Francisella tularensis* is dependent on the ASC/caspase-1 axis. J. Exp. Med. *202*, 1043–1049.

Mariathasan, S., Weiss, D.S., Newton, K., McBride, J., O'Rourke, K., Roose-Girma, M., Lee, W.P., Weinrauch, Y., Monack, D.M., and Dixit, V.M. (2006). Cryopyrin activates the inflammasome in response to toxins and ATP. Nature *440*, 228–232.

Martinon, F., Burns, K., and Tschopp, J. (2002). The inflammasome: a molecular platform triggering activation of inflammatory caspases and processing of proIL-beta. Mol. Cell *10*, 417–426.

Martinon, F., Hofmann, K., and Tschopp, J. (2001). The pyrin domain: a possible member of the death domain-fold family implicated in apoptosis and inflammation. Curr. Biol. *11*, R118–20.

Massis, L.M., and Zamboni, D.S. (2011). Innate immunity to *Legionella pneumophila*. Front. Microbiol. *2*, 109. https://doi.org/10.3389/fmicb.2011.00109.

Master, S.S., Rampini, S.K., Davis, A.S., Keller, C., Ehlers, S., Springer, B., Timmins, G.S., Sander, P., and Deretic, V. (2008). *Mycobacterium tuberculosis* prevents inflammasome activation. Cell Host Microbe *3*, 224-232.

Masters, S.L., Gerlic, M., Metcalf, D., Preston, S., Pellegrini, M., O'Donnell, J.A., McArthur, K., Baldwin, T.M., Chevrier, S., Nowell, C.J., *et al.* (2012). NLRP1 inflammasome activation induces pyroptosis of hematopoietic progenitor cells. Immunity *37*, 1009–1023. https://doi.org/10.1016/j.immuni.2012.08.027.

Matsuura, M. (2013). Structural modifications of bacterial lipopolysaccharide that facilitate Gram-negative bacteria evasion of host innate immunity. Front Immunol *4*, 109.

Mayer-Barber, K.D., Barber, D.L., Shenderov, K., White, S.D., Wilson, M.S., Cheever, A., Kugler, D., Hieny, S., Caspar, P., Núñez, G., *et al.* (2010). Caspase-1 independent IL-1beta production is critical for host resistance to *Mycobacterium tuberculosis* and does not require TLR signaling in vivo. J. Immunol. *184*, 3326–3330. https://doi.org/10.4049/jimmunol.0904189.

McElvania Tekippe, E., Allen, I.C., Hulseberg, P.D., Sullivan, J.T., McCann, J.R., Sandor, M., Braunstein, M., and Ting, J.P. (2010). Granuloma formation and host defense in chronic *Mycobacterium tuberculosis* infection requires PYCARD/ASC but not NLRP3 or caspase-1. PLOS ONE *5*, e12320. https://doi.org/10.1371/journal.pone.0012320.

McPhee, J.B., Mena, P., and Bliska, J.B. (2010). Delineation of regions of the Yersinia YopM protein required for interaction with the RSK1 and PRK2 host kinases and their requirement for interleukin-10 production and virulence. Infect. Immun. *78*, 3529–3539. https://doi.org/10.1128/IAI.00269-10.

McPhee, J.B., Mena, P., Zhang, Y., and Bliska, J.B. (2012). Interleukin-10 induction is an important virulence function of the Yersinia pseudotuberculosis type III effector YopM. Infect. Immun. *80*, 2519–2527. https://doi.org/10.1128/IAI.06364-11.

Meixenberger, K., Pache, F., Eitel, J., Schmeck, B., Hippenstiel, S., Slevogt, H., N'Guessan, P., Witzenrath, M., Netea, M.G., Chakraborty, T., *et al.* (2010). *Listeria monocytogenes*-infected human peripheral blood mononuclear cells produce IL-1beta, depending on listeriolysin O and NLRP3. J Immunol *184*, 922-930.

Mengaud, J., Ohayon, H., Gounon, P., Mege, R.M., and Cossart, P. (1996). E-cadherin is the receptor for internalin, a surface protein required for entry of L. monocytogenes into epithelial cells. Cell *84*, 923-932.

Meunier, E., Dick, M.S., Dreier, R.F., Schürmann, N., Kenzelmann Broz, D., Warming, S., Roose-Girma, M., Bumann, D., Kayagaki, N., Takeda, K., *et al.* (2014). Caspase-11 activation requires lysis of pathogen-containing vacuoles by IFN-induced GTPases. Nature *509*, 366–370. https://doi.org/10.1038/nature13157.

Meunier, E., Wallet, P., Dreier, R.F., Costanzo, S., Anton, L., Ruhl, S., Dussurgey, S., Dick, M.S., Kistner, A., Rigard, M., *et al.* (2015). Guanylate-binding proteins promote activation of the AIM2 inflammasome during infection with *Francisella novicida*. Nat Immunol *16*, 476-484.

Miao, E.A., Alpuche-Aranda, C.M., Dors, M., Clark, A.E., Bader, M.W., Miller, S.I., and Aderem, A. (2006). Cytoplasmic flagellin activates caspase-1 and secretion of interleukin 1beta via Ipaf. Nat. Immunol. *7*, 569–575.
Miao, E.A., Ernst, R.K., Dors, M., Mao, D.P., and Aderem, A. (2008). *Pseudomonas aeruginosa* activates caspase 1 through Ipaf. Proc. Natl. Acad. Sci. U.S.A. *105*, 2562–2567. https://doi.org/10.1073/pnas.0712183105.
Miao, E.A., Leaf, I.A., Treuting, P.M., Mao, D.P., Dors, M., Sarkar, A., Warren, S.E., Wewers, M.D., and Aderem, A. (2010a). Caspase-1-induced pyroptosis is an innate immune effector mechanism against intracellular bacteria. Nat. Immunol. *11*, 1136–1142. https://doi.org/10.1038/ni.1960.
Miao, E.A., Mao, D.P., Yudkovsky, N., Bonneau, R., Lorang, C.G., Warren, S.E., Leaf, I.A., and Aderem, A. (2010b). Innate immune detection of the type III secretion apparatus through the NLRC4 inflammasome. Proc. Natl. Acad. Sci. U.S.A. *107*, 3076–3080. https://doi.org/10.1073/pnas.0913087107.
Miao, E.A., and Rajan, J.V. (2011). Salmonella and Caspase-1: A complex Interplay of Detection and Evasion. Front. Microbiol. *2*, 85. https://doi.org/10.3389/fmicb.2011.00085.
Minnich, S.A., and Rohde, H.N. (2007). A rationale for repression and/or loss of motility by pathogenic Yersinia in the mammalian host. Adv. Exp. Med. Biol. *603*, 298–310. https://doi.org/10.1007/978-0-387-72124-8_27.
Misawa, T., Takahama, M., Kozaki, T., Lee, H., Zou, J., Saitoh, T., and Akira, S. (2013). Microtubule-driven spatial arrangement of mitochondria promotes activation of the NLRP3 inflammasome. Nat. Immunol. *14*, 454–460. https://doi.org/10.1038/ni.2550.
Mishra, B.B., Moura-Alves, P., Sonawane, A., Hacohen, N., Griffiths, G., Moita, L.F., and Anes, E. (2010). *Mycobacterium tuberculosis* protein ESAT-6 is a potent activator of the NLRP3/ASC inflammasome. Cell. Microbiol. *12*, 1046–1063. https://doi.org/10.1111/j.1462-5822.2010.01450.x.
Moayeri, M., Crown, D., Newman, Z.L., Okugawa, S., Eckhaus, M., Cataisson, C., Liu, S., Sastalla, I., and Leppla, S.H. (2010). Inflammasome sensor Nlrp1b-dependent resistance to anthrax is mediated by caspase-1, IL-1 signaling and neutrophil recruitment. PLOS Pathog. *6*, e1001222. https://doi.org/10.1371/journal.ppat.1001222.
Molofsky, A.B., Byrne, B.G., Whitfield, N.N., Madigan, C.A., Fuse, E.T., Tateda, K., and Swanson, M.S. (2006). Cytosolic recognition of flagellin by mouse macrophages restricts *Legionella pneumophila* infection. J Exp Med *203*, 1093-1104.
Monack, D.M., Mecsas, J., Bouley, D., and Falkow, S. (1998). Yersinia-induced apoptosis in vivo aids in the establishment of a systemic infection of mice. J. Exp. Med. *188*, 2127–2137.
Muñoz-Planillo, R., Kuffa, P., Martínez-Colón, G., Smith, B.L., Rajendiran, T.M., and Núñez, G. (2013). K^+ efflux is the common trigger of NLRP3 inflammasome activation by bacterial toxins and particulate matter. Immunity *38*, 1142–1153. https://doi.org/10.1016/j.immuni.2013.05.016.
Muruve, D.A., Pétrilli, V., Zaiss, A.K., White, L.R., Clark, S.A., Ross, P.J., Parks, R.J., and Tschopp, J. (2008). The inflammasome recognizes cytosolic microbial and host DNA and triggers an innate immune response. Nature *452*, 103–107. https://doi.org/10.1038/nature06664.
Nemeth, J., and Straley, S.C. (1997). Effect of *Yersinia pestis* YopM on experimental plague. Infect. Immun. *65*, 924–930.
Netea, M.G., van de Veerdonk, F.L., van der Meer, J.W., Dinarello, C.A., and Joosten, L.A. (2015). Inflammasome-independent regulation of IL-1-family cytokines. Annu. Rev. Immunol. *33*, 49–77. https://doi.org/10.1146/annurev-immunol-032414-112306.
Newton, H.J., Ang, D.K., van Driel, I.R., and Hartland, E.L. (2010). Molecular pathogenesis of infections caused by *Legionella pneumophila*. Clin. Microbiol. Rev. *23*, 274–298. https://doi.org/10.1128/CMR.00052-09.
Nowag, A., and Hartmann, P. (2016). [Immune response to *Mycobacterium tuberculosis*.] Internist *57*, 107–116. https://doi.org/10.1007/s00108-015-0016-4.
Ogawa, M., Handa, Y., Ashida, H., Suzuki, M., and Sasakawa, C. (2008). The versatility of Shigella effectors. Nat. Rev. Microbiol. *6*, 11–16.
Orth, K., Xu, Z., Mudgett, M.B., Bao, Z.Q., Palmer, L.E., Bliska, J.B., Mangel, W.F., Staskawicz, B., and Dixon, J.E. (2000). Disruption of signaling by Yersinia effector YopJ, a ubiquitin-like protein protease. Science *290*, 1594–1597.
Ozören, N., Masumoto, J., Franchi, L., Kanneganti, T.D., Body-Malapel, M., Ertürk, I., Jagirdar, R., Zhu, L., Inohara, N., Bertin, J., *et al.* (2006). Distinct roles of TLR2 and the adaptor ASC in IL-1beta/IL-18 secretion in response to *Listeria monocytogenes*. J. Immunol. *176*, 4337–4342.

Paciello, I., Silipo, A., Lembo-Fazio, L., Curcuru, L., Zumsteg, A., Noel, G., Ciancarella, V., Sturiale, L., Molinaro, A., and Bernardini, M.L. (2013). Intracellular Shigella remodels its LPS to dampen the innate immune recognition and evade inflammasome activation. Proc Natl Acad Sci U S A *110*, E4345-4354.

Parr, A., Whitney, E.A., and Berkelman, R.L. (2015). Legionellosis on the rise: a review of guidelines for prevention in the United States. J. Public Health Manag. Pract. *21*, E17–26. https://doi.org/10.1097/PHH.0000000000000123.

Parsot, C. (2009). Shigella type III secretion effectors: how, where, when, for what purposes? Curr. Opin. Microbiol. *12*, 110–116. https://doi.org/10.1016/j.mib.2008.12.002.

Parsot, C., and Sansonetti, P.J. (1996). Invasion and the pathogenesis of Shigella infections. Curr. Top. Microbiol. Immunol. *209*, 25–42.

Perdomo, O.J., Cavaillon, J.M., Huerre, M., Ohayon, H., Gounon, P., and Sansonetti, P.J. (1994). Acute inflammation causes epithelial invasion and mucosal destruction in experimental shigellosis. J. Exp. Med. *180*, 1307–1319.

Pereira, M.S., Marques, G.G., Dellama, J.E., and Zamboni, D.S. (2011a). The Nlrc4 inflammasome contributes to restriction of pulmonary infection by flagellated *Legionella* spp. that trigger pyroptosis. Front. Microbiol. *2*, 33. https://doi.org/10.3389/fmicb.2011.00033.

Pereira, M.S., Morgantetti, G.F., Massis, L.M., Horta, C.V., Hori, J.I., and Zamboni, D.S. (2011b). Activation of NLRC4 by flagellated bacteria triggers caspase-1-dependent and -independent responses to restrict *Legionella pneumophila* replication in macrophages and in vivo. J Immunol *187*, 6447-6455.

Pétrilli, V., Papin, S., Dostert, C., Mayor, A., Martinon, F., and Tschopp, J. (2007). Activation of the NALP3 inflammasome is triggered by low intracellular potassium concentration. Cell Death Differ. *14*, 1583–1589.

Philip, N.H., Dillon, C.P., Snyder, A.G., Fitzgerald, P., Wynosky-Dolfi, M.A., Zwack, E.E., Hu, B., Fitzgerald, L., Mauldin, E.A., Copenhaver, A.M., *et al.* (2014). Caspase-8 mediates caspase-1 processing and innate immune defense in response to bacterial blockade of NF-κB and MAPK signaling. Proc. Natl. Acad. Sci. U.S.A. *111*, 7385–7390. https://doi.org/10.1073/pnas.1403252111.

Pierini, R., Perret, M., Djebali, S., Juruj, C., Michallet, M.C., Förster, I., Marvel, J., Walzer, T., and Henry, T. (2013). ASC controls IFN-γ levels in an IL-18-dependent manner in caspase-1-deficient mice infected with Francisella novicida. J. Immunol. *191*, 3847–3857. https://doi.org/10.4049/jimmunol.1203326.

Portnoy, D.A., Jacks, P.S., and Hinrichs, D.J. (1988). Role of hemolysin for the intracellular growth of *Listeria monocytogenes*. J. Exp. Med. *167*, 1459–1471.

Poyet, J.L., Srinivasula, S.M., Tnani, M., Razmara, M., Fernandes-Alnemri, T., and Alnemri, E.S. (2001). Identification of Ipaf, a human caspase-1-activating protein related to Apaf-1. J. Biol. Chem. *276*, 28309–28313. https://doi.org/10.1074/jbc.C100250200.

Py, B.F., Kim, M.S., Vakifahmetoglu-Norberg, H., and Yuan, J. (2013). Deubiquitination of NLRP3 by BRCC3 critically regulates inflammasome activity. Mol. Cell *49*, 331–338. https://doi.org/10.1016/j.molcel.2012.11.009.

Qu, Y., Misaghi, S., Izrael-Tomasevic, A., Newton, K., Gilmour, L.L., Lamkanfi, M., Louie, S., Kayagaki, N., Liu, J., Kömüves, L., *et al.* (2012). Phosphorylation of NLRC4 is critical for inflammasome activation. Nature *490*, 539–542. https://doi.org/10.1038/nature11429.

Raqib, R., Mia, S.M., Qadri, F., Alam, T.I., Alam, N.H., Chowdhury, A.K., Mathan, M.M., and Andersson, J. (2000). Innate immune responses in children and adults with Shigellosis. Infect. Immun. *68*, 3620–3629.

Rathinam, V.A., Jiang, Z., Waggoner, S.N., Sharma, S., Cole, L.E., Waggoner, L., Vanaja, S.K., Monks, B.G., Ganesan, S., Latz, E., *et al.* (2010). The AIM2 inflammasome is essential for host defense against cytosolic bacteria and DNA viruses. Nat. Immunol. *11*, 395–402. https://doi.org/10.1038/ni.1864.

Rathinam, V.A., Vanaja, S.K., Waggoner, L., Sokolovska, A., Becker, C., Stuart, L.M., Leong, J.M., and Fitzgerald, K.A. (2012). TRIF licenses caspase-11-dependent NLRP3 inflammasome activation by gram-negative bacteria. Cell *150*, 606–619. https://doi.org/10.1016/j.cell.2012.07.007.

Ratner, D., Orning, M.P., Starheim, K.K., Marty-Roix, R., Proulx, M.K., Goguen, J.D., and Lien, E. (2016). Manipulation of interleukin-1beta and interleukin-18 production by *Yersinia pestis* effectors YopJ and YopM and redundant impact on virulence. J Biol Chem *291*, 16417.

Raupach, B., Peuschel, S.K., Monack, D.M., and Zychlinsky, A. (2006). Caspase-1-mediated activation of interleukin-1beta (IL-1beta) and IL-18 contributes to innate immune defenses against *Salmonella enterica* serovar Typhimurium infection. Infect Immun 74, 4922-4926.

Rayamajhi, M., Zak, D.E., Chavarria-Smith, J., Vance, R.E., and Miao, E.A. (2013). Cutting edge: Mouse NAIP1 detects the type III secretion system needle protein. J. Immunol. *191*, 3986–3989. https://doi.org/10.4049/jimmunol.1301549.

Ren, T., Zamboni, D.S., Roy, C.R., Dietrich, W.F., and Vance, R.E. (2006). Flagellin-deficient Legionella mutants evade caspase-1- and Naip5-mediated macrophage immunity. PLOS Pathog. 2, e18. https://doi.org/10.1371/journal.ppat.0020018.

Rigante, D., Frediani, B., and Cantarini, L. (2016). A comprehensive overview of the hereditary periodic fever syndromes. Clin. Rev. Allergy Immunol. [Epub ahead of print]. https://doi.org/10.1007/s12016-016-8537-8.

Roberts, T.L., Idris, A., Dunn, J.A., Kelly, G.M., Burnton, C.M., Hodgson, S., Hardy, L.L., Garceau, V., Sweet, M.J., Ross, I.L., *et al.* (2009). HIN-200 proteins regulate caspase activation in response to foreign cytoplasmic DNA. Science *323*, 1057–1060. https://doi.org/10.1126/science.1169841.

Ruckdeschel, K., Harb, S., Roggenkamp, A., Hornef, M., Zumbihl, R., Kohler, S., Heesemann, J., and Rouot, B. (1998). Yersinia enterocolitica impairs activation of transcription factor NF-kappaB: involvement in the induction of programmed cell death and in the suppression of the macrophage tumor necrosis factor alpha production. J. Exp. Med. *187*, 1069-1079.

Ruckdeschel, K., Mannel, O., Richter, K., Jacobi, C.A., Trulzsch, K., Rouot, B., and Heesemann, J. (2001). Yersinia outer protein P of Yersinia enterocolitica simultaneously blocks the nuclear factor-kappa B pathway and exploits lipopolysaccharide signaling to trigger apoptosis in macrophages. J. Immunol. *166*, 1823-1831.

Ruckdeschel, K., Roggenkamp, A., Lafont, V., Mangeat, P., Heesemann, J., and Rouot, B. (1997). Interaction of Yersinia enterocolitica with macrophages leads to macrophage cell death through apoptosis. Infect. Immun. *65*, 4813–4821.

Rühl, S., and Broz, P. (2015). Caspase-11 activates a canonical NLRP3 inflammasome by promoting K(+) efflux. Eur. J. Immunol. *45*, 2927–2936. https://doi.org/10.1002/eji.201545772.

Rydström, A., and Wick, M.J. (2007). Monocyte recruitment, activation, and function in the gut-associated lymphoid tissue during oral Salmonella infection. J. Immunol. *178*, 5789–5801.

Sabrià, M., and Campins, M. (2003). Legionnaires' disease: update on epidemiology and management options. Am. J. Respir. Med. *2*, 235–243.

Saiga, H., Kitada, S., Shimada, Y., Kamiyama, N., Okuyama, M., Makino, M., Yamamoto, M., and Takeda, K. (2012). Critical role of AIM2 in *Mycobacterium tuberculosis* infection. Int. Immunol. *24*, 637–644.

Sansonetti, P.J., Phalipon, A., Arondel, J., Thirumalai, K., Banerjee, S., Akira, S., Takeda, K., and Zychlinsky, A. (2000). Caspase-1 activation of IL-1beta and IL-18 are essential for *Shigella flexneri*-induced inflammation. Immunity *12*, 581–590.

Santic, M., Molmeret, M., Klose, K.E., and Abu Kwaik, Y. (2006). *Francisella tularensis* travels a novel, twisted road within macrophages. Trends Microbiol. *14*, 37–44.

Sauer, J.D., Pereyre, S., Archer, K.A., Burke, T.P., Hanson, B., Lauer, P., and Portnoy, D.A. (2011). *Listeria monocytogenes* engineered to activate the Nlrc4 inflammasome are severely attenuated and are poor inducers of protective immunity. Proc. Natl. Acad. Sci. U.S.A. *108*, 12419–12424. https://doi.org/10.1073/pnas.1019041108.

Sauer, J.D., Witte, C.E., Zemansky, J., Hanson, B., Lauer, P., and Portnoy, D.A. (2010). *Listeria monocytogenes* triggers AIM2-mediated pyroptosis upon infrequent bacteriolysis in the macrophage cytosol. Cell Host Microbe 7, 412–419. https://doi.org/10.1016/j.chom.2010.04.004.

Schmid-Burgk, J.L., Chauhan, D., Schmidt, T., Ebert, T.S., Reinhardt, J., Endl, E., and Hornung, V. (2016). A Genome-wide CRISPR (clustered regularly interspaced short palindromic repeats) screen identifies NEK7 as an essential component of NLRP3 inflammasome activation. J. Biol. Chem. *291*, 103–109.

Schnupf, P., Hofmann, J., Norseen, J., Glomski, I.J., Schwartzstein, H., and Decatur, A.L. (2006). Regulated translation of listeriolysin O controls virulence of *Listeria monocytogenes*. Mol. Microbiol. *61*, 999–1012.

Schoberle, T.J., Chung, L.K., McPhee, J.B., Bogin, B., and Bliska, J.B. (2016). Uncovering an important role for YopJ in the inhibition of Caspase-1 in activated macrophages and promoting *Yersinia pseudotuberculosis* virulence. Infect Immun *84*, 1062-1072.

Schotte, P., Denecker, G., Van Den Broeke, A., Vandenabeele, P., Cornelis, G.R., and Beyaert, R. (2004). Targeting Rac1 by the *Yersinia* effector protein YopE inhibits caspase-1-mediated maturation and release of interleukin-1beta. J. Biol. Chem. *279*, 25134–25142. https://doi.org/10.1074/jbc.M401245200.

Schroeder, G.N., and Hilbi, H. (2008). Molecular pathogenesis of Shigella spp.: controlling host cell signaling, invasion, and death by type III secretion. Clin. Microbiol. Rev. *21*, 134–156. https://doi.org/10.1128/CMR.00032-07.

Sellin, M.E., Muller, A.A., Felmy, B., Dolowschiak, T., Diard, M., Tardivel, A., Maslowski, K.M., and Hardt, W.D. (2014). Epithelium-intrinsic NAIP/NLRC4 inflammasome drives infected enterocyte expulsion to restrict *Salmonella* replication in the intestinal mucosa. Cell Host Microbe *16*, 237-248.

Senerovic, L., Tsunoda, S.P., Goosmann, C., Brinkmann, V., Zychlinsky, A., Meissner, F., and Kolbe, M. (2012). Spontaneous formation of IpaB ion channels in host cell membranes reveals how *Shigella* induces pyroptosis in macrophages. Cell Death Dis. *3*, e384. https://doi.org/10.1038/cddis.2012.124.
Shah, S., Bohsali, A., Ahlbrand, S.E., Srinivasan, L., Rathinam, V.A., Vogel, S.N., Fitzgerald, K.A., Sutterwala, F.S., and Briken, V. (2013). Cutting edge: *Mycobacterium tuberculosis* but not nonvirulent mycobacteria inhibits IFN-beta and AIM2 inflammasome-dependent IL-1beta production via its ESX-1 secretion system. J Immunol *191*, 3514-3518.
Shao, F., Vacratsis, P.O., Bao, Z., Bowers, K.E., Fierke, C.A., and Dixon, J.E. (2003). Biochemical characterization of the *Yersinia* YopT protease: cleavage site and recognition elements in Rho GTPases. Proc. Natl. Acad. Sci. U.S.A. *100*, 904–909. https://doi.org/10.1073/pnas.252770599.
Sharma, D., and Kanneganti, T.D. (2016). The cell biology of inflammasomes: Mechanisms of inflammasome activation and regulation. J. Cell Biol. *213*, 617–629. https://doi.org/10.1083/jcb.201602089.
Shen, A., and Higgins, D.E. (2006). The MogR transcriptional repressor regulates nonhierarchal expression of flagellar motility genes and virulence in *Listeria monocytogenes*. PLOS Pathog *2*, e30.
Shi, F., Kouadir, M., and Yang, Y. (2015a). NALP3 inflammasome activation in protein misfolding diseases. Life Sci. *135*, 9–14. https://doi.org/10.1016/j.lfs.2015.05.011.
Shi, H., Wang, Y., Li, X., Zhan, X., Tang, M., Fina, M., Su, L., Pratt, D., Bu, C.H., Hildebrand, S., *et al.* (2016). NLRP3 activation and mitosis are mutually exclusive events coordinated by NEK7, a new inflammasome component. Nat. Immunol. *17*, 250–258. https://doi.org/10.1038/ni.3333.
Shi, J., Zhao, Y., Wang, K., Shi, X., Wang, Y., Huang, H., Zhuang, Y., Cai, T., Wang, F., and Shao, F. (2015b). Cleavage of GSDMD by inflammatory caspases determines pyroptotic cell death. Nature *526*, 660–665. https://doi.org/10.1038/nature15514.
Shi, J., Zhao, Y., Wang, Y., Gao, W., Ding, J., Li, P., Hu, L., and Shao, F. (2014). Inflammatory caspases are innate immune receptors for intracellular LPS. Nature *514*, 187–192. https://doi.org/10.1038/nature13683.
Shimada, K., Crother, T.R., Karlin, J., Dagvadorj, J., Chiba, N., Chen, S., Ramanujan, V.K., Wolf, A.J., Vergnes, L., Ojcius, D.M., *et al.* (2012). Oxidized mitochondrial DNA activates the NLRP3 inflammasome during apoptosis. Immunity *36*, 401–414. https://doi.org/10.1016/j.immuni.2012.01.009.
Sia, J.K., Georgieva, M., and Rengarajan, J. (2015). Innate immune defenses in human tuberculosis: an overview of the interactions between *Mycobacterium tuberculosis* and innate immune cells. J Immunol Res *2015*, 747543.
Silveira, T.N., and Zamboni, D.S. (2010). Pore formation triggered by Legionella spp. is an Nlrc4 inflammasome-dependent host cell response that precedes pyroptosis. Infect. Immun. *78*, 1403–1413. https://doi.org/10.1128/IAI.00905-09.
Simeone, R., Bobard, A., Lippmann, J., Bitter, W., Majlessi, L., Brosch, R., and Enninga, J. (2012). Phagosomal rupture by *Mycobacterium tuberculosis* results in toxicity and host cell death. PLOS Pathog. *8*, e1002507. https://doi.org/10.1371/journal.ppat.1002507.
Sivick, K.E., Arpaia, N., Reiner, G.L., Lee, B.L., Russell, B.R., and Barton, G.M. (2014). Toll-like receptor-deficient mice reveal how innate immune signaling influences Salmonella virulence strategies. Cell Host Microbe *15*, 203-213.
Stein, M.P., Müller, M.P., and Wandinger-Ness, A. (2012). Bacterial pathogens commandeer Rab GTPases to establish intracellular niches. Traffic *13*, 1565–1588. https://doi.org/10.1111/tra.12000.
Storek, K.M., Gertsvolf, N.A., Ohlson, M.B., and Monack, D.M. (2015). cGAS and Ifi204 cooperate to produce type I IFNs in response to Francisella infection. J. Immunol. *194*, 3236–3245. https://doi.org/10.4049/jimmunol.1402764.
Suzuki, S., Franchi, L., He, Y., Muñoz-Planillo, R., Mimuro, H., Suzuki, T., Sasakawa, C., and Núñez, G. (2014). Shigella type III secretion protein MxiI is recognized by Naip2 to induce Nlrc4 inflammasome activation independently of Pkcδ. PLOS Pathog. *10*, e1003926. https://doi.org/10.1371/journal.ppat.1003926.
Sweet, C.R., Conlon, J., Golenbock, D.T., Goguen, J., and Silverman, N. (2007). YopJ targets TRAF proteins to inhibit TLR-mediated NF-kappaB, MAPK and IRF3 signal transduction. Cell. Microbiol. *9*, 2700–2715.
Tam, M.A., Rydström, A., Sundquist, M., and Wick, M.J. (2008). Early cellular responses to *Salmonella* infection: dendritic cells, monocytes, and more. Immunol. Rev. *225*, 140–162. https://doi.org/10.1111/j.1600-065X.2008.00679.x.
Tenthorey, J.L., Kofoed, E.M., Daugherty, M.D., Malik, H.S., and Vance, R.E. (2014). Molecular basis for specific recognition of bacterial ligands by NAIP/NLRC4 inflammasomes. Mol. Cell *54*, 17–29. https://doi.org/10.1016/j.molcel.2014.02.018.

Thurston, T.L., Matthews, S.A., Jennings, E., Alix, E., Shao, F., Shenoy, A.R., Birrell, M.A., and Holden, D.W. (2016). Growth inhibition of cytosolic Salmonella by caspase-1 and caspase-11 precedes host cell death. Nat. Commun. 7, 13292. https://doi.org/10.1038/ncomms13292.

Treacy-Abarca, S., and Mukherjee, S. (2015). *Legionella* suppresses the host unfolded protein response via multiple mechanisms. Nat. Commun. *6*, 7887. https://doi.org/10.1038/ncomms8887.

Trosky, J.E., Liverman, A.D., and Orth, K. (2008). *Yersinia* outer proteins: Yops. Cell. Microbiol. *10*, 557–565.

Tschopp, J., Martinon, F., and Burns, K. (2003). NALPs: a novel protein family involved in inflammation. Nat. Rev. Mol. Cell Biol. *4*, 95–104. https://doi.org/10.1038/nrm1019.

Tsuji, N.M., Tsutsui, H., Seki, E., Kuida, K., Okamura, H., Nakanishi, K., and Flavell, R.A. (2004). Roles of caspase-1 in Listeria infection in mice. Int. Immunol. *16*, 335–343.

Vanaja, S.K., Russo, A.J., Behl, B., Banerjee, I., Yankova, M., Deshmukh, S.D., and Rathinam, V.A. (2016). Bacterial outer membrane vesicles mediate cytosolic localization of LPS and caspase-11 activation. Cell *165*, 1106–1119. https://doi.org/10.1016/j.cell.2016.04.015.

Via, L.E., Deretic, D., Ulmer, R.J., Hibler, N.S., Huber, L.A., and Deretic, V. (1997). Arrest of mycobacterial phagosome maturation is caused by a block in vesicle fusion between stages controlled by rab5 and rab7. J. Biol. Chem. *272*, 13326–13331.

Vince, J.E., Wong, W.W., Gentle, I., Lawlor, K.E., Allam, R., O'Reilly, L., Mason, K., Gross, O., Ma, S., Guarda, G., *et al.* (2012). Inhibitor of apoptosis proteins limit RIP3 kinase-dependent interleukin-1 activation. Immunity *36*, 215–227. https://doi.org/10.1016/j.immuni.2012.01.012.

Vladimer, G.I., Weng, D., Paquette, S.W., Vanaja, S.K., Rathinam, V.A., Aune, M.H., Conlon, J.E., Burbage, J.J., Proulx, M.K., Liu, Q., *et al.* (2012). The NLRP12 inflammasome recognizes *Yersinia pestis*. Immunity *37*, 96–107. https://doi.org/10.1016/j.immuni.2012.07.006.

von Moltke, J., Trinidad, N.J., Moayeri, M., Kintzer, A.F., Wang, S.B., van Rooijen, N., Brown, C.R., Krantz, B.A., Leppla, S.H., Gronert, K., *et al.* (2012). Rapid induction of inflammatory lipid mediators by the inflammasome in vivo. Nature *490*, 107–111. https://doi.org/10.1038/nature11351.

Wang, L., Manji, G.A., Grenier, J.M., Al-Garawi, A., Merriam, S., Lora, J.M., Geddes, B.J., Briskin, M., DiStefano, P.S., and Bertin, J. (2002). PYPAF7, a novel PYRIN-containing Apaf1-like protein that regulates activation of NF-kappa B and caspase-1-dependent cytokine processing. J Biol Chem *277*, 29874-29880.

Wang, S., Miura, M., Jung, Y.K., Zhu, H., Li, E., and Yuan, J. (1998). Murine caspase-11, an ICE-interacting protease, is essential for the activation of ICE. Cell 92, 501–509.

Warren, S.E., Armstrong, A., Hamilton, M.K., Mao, D.P., Leaf, I.A., Miao, E.A., and Aderem, A. (2010). Cutting edge: Cytosolic bacterial DNA activates the inflammasome via Aim2. J. Immunol. *185*, 818–821. https://doi.org/10.4049/jimmunol.1000724.

Weng, D., Marty-Roix, R., Ganesan, S., Proulx, M.K., Vladimer, G.I., Kaiser, W.J., Mocarski, E.S., Pouliot, K., Chan, F.K., Kelliher, M.A., *et al.* (2014). Caspase-8 and RIP kinases regulate bacteria-induced innate immune responses and cell death. Proc. Natl. Acad. Sci. U.S.A. *111*, 7391–7396. https://doi.org/10.1073/pnas.1403477111.

Williams, K.L., Lich, J.D., Duncan, J.A., Reed, W., Rallabhandi, P., Moore, C., Kurtz, S., Coffield, V.M., Accavitti-Loper, M.A., Su, L., *et al.* (2005). The CATERPILLER protein monarch-1 is an antagonist of toll-like receptor-, tumor necrosis factor alpha-, and *Mycobacterium tuberculosis*-induced pro-inflammatory signals. J Biol Chem *280*, 39914-39924.

Wren, B.W. (2003). The yersiniae – a model genus to study the rapid evolution of bacterial pathogens. Nat. Rev. Microbiol. *1*, 55–64. https://doi.org/10.1038/nrmicro730.

Wu, J., Fernandes-Alnemri, T., and Alnemri, E.S. (2010). Involvement of the AIM2, NLRC4, and NLRP3 inflammasomes in caspase-1 activation by *Listeria monocytogenes*. J. Clin. Immunol. *30*, 693–702. https://doi.org/10.1007/s10875-010-9425-2.

Wynosky-Dolfi, M.A., Snyder, A.G., Philip, N.H., Doonan, P.J., Poffenberger, M.C., Avizonis, D., Zwack, E.E., Riblett, A.M., Hu, B., Strowig, T., *et al.* (2014). Oxidative metabolism enables Salmonella evasion of the NLRP3 inflammasome. J. Exp. Med. *211*, 653–668. https://doi.org/10.1084/jem.20130627.

Xu, H., Yang, J., Gao, W., Li, L., Li, P., Zhang, L., Gong, Y.N., Peng, X., Xi, J.J., Chen, S., *et al.* (2014). Innate immune sensing of bacterial modifications of Rho GTPases by the Pyrin inflammasome. Nature *513*, 237–241. https://doi.org/10.1038/nature13449.

Yang, D., He, Y., Muñoz-Planillo, R., Liu, Q., and Núñez, G. (2015). Caspase-11 Requires the Pannexin-1 Channel and the Purinergic P2X7 Pore to Mediate Pyroptosis and Endotoxic Shock. Immunity *43*, 923–932. https://doi.org/10.1016/j.immuni.2015.10.009.

Yang, J., Zhao, Y., Shi, J., and Shao, F. (2013a). Human NAIP and mouse NAIP1 recognize bacterial type III secretion needle protein for inflammasome activation. Proc. Natl. Acad. Sci. U.S.A. *110*, 14408–14413. https://doi.org/10.1073/pnas.1306376110.

Yang, Y., Zhou, X., Kouadir, M., Shi, F., Ding, T., Liu, C., Liu, J., Wang, M., Yang, L., Yin, X., *et al.* (2013b). the AIM2 inflammasome is involved in macrophage activation during infection with virulent *Mycobacterium bovis* strain. J. Infect. Dis. *208*, 1849–1858. https://doi.org/10.1093/infdis/jit347.

Zaki, M.H., Man, S.M., Vogel, P., Lamkanfi, M., and Kanneganti, T.D. (2014). Salmonella exploits NLRP12-dependent innate immune signaling to suppress host defenses during infection. Proc. Natl. Acad. Sci. U.S.A. *111*, 385–390. https://doi.org/10.1073/pnas.1317643111.

Zaki, M.H., Vogel, P., Malireddi, R.K., Body-Malapel, M., Anand, P.K., Bertin, J., Green, D.R., Lamkanfi, M., and Kanneganti, T.D. (2011). The NOD-like receptor NLRP12 attenuates colon inflammation and tumorigenesis. Cancer Cell. *20*, 649–660. https://doi.org/10.1016/j.ccr.2011.10.022.

Zamboni, D.S., Kobayashi, K.S., Kohlsdorf, T., Ogura, Y., Long, E.M., Vance, R.E., Kuida, K., Mariathasan, S., Dixit, V.M., Flavell, R.A., *et al.* (2006). The Birc1e cytosolic pattern-recognition receptor contributes to the detection and control of *Legionella pneumophila* infection. Nat Immunol 7, 318-325.

Zamboni, D.S., and Lima-Junior, D.S. (2015). Inflammasomes in host response to protozoan parasites. Immunol. Rev. *265*, 156–171. https://doi.org/10.1111/imr.12291.

Zauberman, A., Tidhar, A., Levy, Y., Bar-Haim, E., Halperin, G., Flashner, Y., Cohen, S., Shafferman, A., and Mamroud, E. (2009). *Yersinia pestis* endowed with increased cytotoxicity is avirulent in a bubonic plague model and induces rapid protection against pneumonic plague. PLOS ONE *4*, e5938.

Zhang, K., and Kaufman, R.J. (2008). Identification and characterization of endoplasmic reticulum stress-induced apoptosis in vivo. Meth. Enzymol. *442*, 395–419. https://doi.org/10.1016/S0076-6879(08)01420-1.

Zhang, L., Chen, S., Ruan, J., Wu, J., Tong, A.B., Yin, Q., Li, Y., David, L., Lu, A., Wang, W.L., *et al.* (2015). Cryo-EM structure of the activated NAIP2-NLRC4 inflammasome reveals nucleated polymerization. Science *350*, 404–409. https://doi.org/10.1126/science.aac5789.

Zhang, W., Cai, Y., Xu, W., Yin, Z., Gao, X., and Xiong, S. (2013). AIM2 facilitates the apoptotic DNA-induced systemic lupus erythematosus via arbitrating macrophage functional maturation. J. Clin. Immunol. *33*, 925–937. https://doi.org/10.1007/s10875-013-9881-6.

Zhao, Y., Yang, J., Shi, J., Gong, Y.N., Lu, Q., Xu, H., Liu, L., and Shao, F. (2011). The NLRC4 inflammasome receptors for bacterial flagellin and type III secretion apparatus. Nature *477*, 596–600. https://doi.org/10.1038/nature10510.

Zheng, Y., Lilo, S., Brodsky, I.E., Zhang, Y., Medzhitov, R., Marcu, K.B., and Bliska, J.B. (2011). A *Yersinia* effector with enhanced inhibitory activity on the NF-κB pathway activates the NLRP3/ASC/caspase-1 inflammasome in macrophages. PLOS Pathog. *7*, e1002026. https://doi.org/10.1371/journal.ppat.1002026.

Zheng, Y., Lilo, S., Mena, P., and Bliska, J.B. (2012). YopJ-induced caspase-1 activation in *Yersinia*-infected macrophages: independent of apoptosis, linked to necrosis, dispensable for innate host defense. PLOS ONE *7*, e36019. https://doi.org/10.1371/journal.pone.0036019.

Zhong, F.L., Mamaï, O., Sborgi, L., Boussofara, L., Hopkins, R., Robinson, K., Szeverényi, I., Takeichi, T., Balaji, R., Lau, A., *et al.* (2016). Germline NLRP1 mutations cause skin inflammatory and cancer susceptibility syndromes via inflammasome activation. Cell *167*, 187–202.e17. https://doi.org/10.1016/j.cell.2016.09.001.

Zhou, H., Monack, D.M., Kayagaki, N., Wertz, I., Yin, J., Wolf, B., and Dixit, V.M. (2005). *Yersinia* virulence factor YopJ acts as a deubiquitinase to inhibit NF-kappa B activation. J. Exp. Med. *202*, 1327–1332.

Zhou, R., Tardivel, A., Thorens, B., Choi, I., and Tschopp, J. (2010). Thioredoxin-interacting protein links oxidative stress to inflammasome activation. Nat. Immunol. *11*, 136–140. https://doi.org/10.1038/ni.1831.

Zhou, Y., Shah, S.Z., Yang, L., Zhang, Z., Zhou, X., and Zhao, D. (2016). Virulent *Mycobacterium bovis* Beijing Strain activates the NLRP7 inflammasome in THP-1 macrophages. PLOS ONE *11*, e0152853. https://doi.org/10.1371/journal.pone.0152853.

Zwack, E.E., Snyder, A.G., Wynosky-Dolfi, M.A., Ruthel, G., Philip, N.H., Marketon, M.M., Francis, M.S., Bliska, J.B., and Brodsky, I.E. (2015). Inflammasome activation in response to the *Yersinia* type III secretion system requires hyperinjection of translocon proteins YopB and YopD. MBio *6*, e02095-02014.

Molecular Mechanisms Used by *Salmonella* to Evade the Immune System

6

Joaquín Bernal-Bayard[1]* and Francisco Ramos-Morales[2]

[1]Département de Microbiologie, Institut Pasteur, Paris, France.
[2]Departamento de Genética, Universidad de Sevilla, Seville, Spain.

*Correspondence: jbernal@pasteur.fr

http://doi.org/10.21775/9781910190692.06

Abstract

Human and animal pathogens are able to circumvent, at least temporarily, the sophisticated immune defences of their hosts. Several serovars of the Gram-negative bacterium *Salmonella enterica* have been used as models for the study of pathogen–host interactions. In this review we discuss the strategies used by *Salmonella* to evade or manipulate three levels of host immune defences: physical barriers, innate immunity and adaptive immunity. During its passage through the digestive system, *Salmonella* has to face the acidic pH of the stomach, bile and antimicrobial peptides in the intestine, as well as the competition with resident microbiota. After host cell invasion, *Salmonella* manipulates inflammatory pathways and the autophagy process. Finally, *Salmonella* evades the adaptive immune system by interacting with dendritic cells, and T and B lymphocytes. Mechanisms allowing the establishment of persistent infections are also discussed.

Introduction

Salmonella spp. are pathogenic Gram-negative, rod-shaped, motile bacteria belonging to the family Enterobacteriaceae. The genus *Salmonella* includes the species *S. bongori* and *S. enterica*. *S. enterica* is further divided into several subspecies and more than 2500 serovars (Tindall *et al.*, 2005). *Salmonella* can cause gastroenteritis, typhoid fever, abortion, and bacteraemia, depending on the serovar and the host. For instance, *S. enterica* subspecies *enterica* serovar Typhi (*S.* Typhi) is responsible for typhoid fever in humans, whereas *S. enterica* subspecies *enterica* serovar Typhimurium (*S.* Typhimurium) causes gastrointestinal inflammation in humans and a systemic typhoid-like disease in mice (Garai *et al.*, 2012). *Salmonella* is a facultative intracellular pathogen that usually resides in a modified phagolysosome known as *Salmonella*-containing vacuole (SCV) during host infection. However, recent data indicate that this facultative intracellular pathogen has a distinct bimodal lifestyle in epithelial cells, where, there are subpopulations of vacuolar and cytosolic *Salmonella* (Knodler *et al.*, 2010). Target cells include M cells, gut epithelial cells, dendritic cells (DCs), macrophages, monocytes, neutrophils, B cells, and T cells. During its passage through the host, *Salmonella* must resist or evade multiple levels of immune defence. Many virulence genes contribute to the survival of these bacteria inside the host. Some of them are clustered in horizontally acquired genomic regions known as *Salmonella* pathogenicity islands (SPIs) (Gerlach and Hensel, 2007). The largest and best known are SPI1, which is present in all the members of the genus, and SPI2, which is present in *S. enterica* but not in *S. bongori*. These SPIs encode two distinct type III secretion systems, type III secretion system 1 (T3SS1) and type III secretion system 2 (T3SS2), respectively (Galán and Curtiss, 1989; Ochman *et al.*, 1996; Shea *et al.*, 1996). These systems are able to inject directly into the eukaryotic host cells a number of bacterial proteins known as effectors (Ramos-Morales, 2012a). Collectively, these virulence factors contribute to invade host cells, interfere with host cellular functions, subvert immunity, establish an intracellular niche, and promote pathogen proliferation.

In this review we summarize the mechanisms of immunity that pathogens in general have to face and the specific strategies that *S. enterica* uses to deal with the different levels of defence of the host.

Overview of mechanisms of immune defence

Humans, like any other animal are confronted every day to a myriad of microorganisms. However, these potentially infectious agents do not cause disease regularly due to a series of defence mechanisms that have been shaped over millions of years of evolution.

Three levels of immune defence can be recognized (Delves *et al.*, 2011; Kenneth, 2011): (i) physical barriers, (ii) the innate immune system and (iii) the adaptive immune system.

Physical barriers

The best way to avoid infections is by keeping potentially infectious agents out of the body. Physical barriers are very effective for this purpose. They are represented by the skin and other epithelia covering the gastrointestinal, respiratory and urogenital tracts. These surfaces provide, not only mechanical, but also chemical and microbiological barriers.

The primary mechanical protection common to all epithelia consists of epithelial cells linked by tight junctions. The internal epithelia, or mucosal epithelia, are covered with mucous secretions. Mucus blocks the adherence of bacteria to epithelial cells. Microorganisms trapped within it can be removed by the movement of cilia, coughing or sneezing. In the intestinal mucosa, mucus is secreted by specialized goblet cells and creates a physical barrier against microbial pathogens (Liévin-Le Moal and Servin, 2006). Other general mechanical factors that contribute to this first level of defence are the flow of air or other fluids, like the tears in the eyes.

In addition to these physical obstacles, a chemical barrier to infections is provided by substances that kill microbes or that inhibit their growth. These chemical substances include a group of antibacterial enzymes present in body fluids, like lactoperoxidase in milk, lysozyme and secretory phospholipase A2 in tears, nasal secretions and saliva. Lysozyme breaks the peptidoglycan (PG) in the bacterial cell wall of Gram-positive and Gram-negative bacteria. Phospholipase A2 kills bacteria by hydrolysing phospholipids in the cell membrane. Antimicrobial peptides, like defensins, cathelicidins and histatins, represent another important group of antimicrobial agents associated to epithelia. All of them are amphipathic peptides activated by proteolysis. Defensins and cathelicidins disrupt membranes and are toxic to many microorganisms, whereas histatins act against pathogenic fungi. Additional chemical barriers that are found by potential food pathogens include the acidity of the stomach pH and the bile in the intestine.

Microbiota is the microbial community composed of bacteria, protozoa, archaea, viruses and fungi that reside in different body niches. This microbiota competes with pathogenic microorganisms for nutrients and attachment sites. Therefore, the normal microbiota that is present in the intestine provides a microbiological physical defence against potential microbial pathogens that, like *Salmonella*, usually invade through the gut. In addition, some microbiota-produced metabolites affect the growth and virulence of pathogens. These metabolites include short-chain fatty acids, succinate, mucin O-glycans, molecular hydrogen, secondary bile acids, and the quorum sensing autoinducer AI-2 (Vogt *et al.*, 2015).

Innate immune defence

The second level of defence is the innate immune response. This response occurs immediately after entry of a foreign agent into the body, but its effectiveness does not improve upon secondary encounter with the same agent. Microorganisms are recognized and killed by two main classes of phagocytic cells: polymorphonuclear neutrophils and macrophages. These cells carry pattern recognition receptors (PRRs) on their cytoplasmic membranes that adhere to pathogen-associated molecular patterns (PAMPs) on the microbe surface. PRRs include Toll-like receptors (TLRs), C-type lectin receptors (CTLRs), NOD-like receptors (NLRs), RIG-like receptors (RLRs) and scavenger receptors. PAMPs that activate TLRs include lipopolysaccharide (LPS), PG, lipoproteins, flagellin and other pathogen-derived ligands. Binding of a TLR to its ligand leads to activation of nuclear factor κB (NF-κB) and several members of the interferon-regulated factor (IRF) family of transcription factors. CTLRs are very diverse transmembrane proteins. RLRs are found in the cytoplasm and activate NF-κB and IRF3/4 to induce antiviral type I interferons in response to double-stranded RNA. Scavenger receptors recognize modified low-density proteins, LPS and other ligands, and some of them can cooperate with TLRs (Murshid *et al.*, 2016). NLRs are soluble cytosolic proteins that usually are composed of three domains: an N-terminal domain that recruits proteases or kinases, a central oligomerization domain and C-terminal leucine-rich repeats (LRRs) that recognize PAMPs. NLRs exist in an autoinhibited conformation that is released upon binding to a PAMP, allowing oligomerization and recruitment of different host ligands depending on the particular NLR (Kim *et al.*, 2016). The NLRs NOD1 and NOD2 induce autophagy to remove pathogens by recruiting ATG16L1 to the plasma membrane at the site of bacterial entry. Autophagy is a conserved intracellular degradation pathway that involves the formation of double-membrane vesicles, known as autophagosomes, which deliver cytosolic components to the lysosome for degradation. Microbe-induced autophagy, or xenophagy, is important to restrict the growth of a number of intracellular pathogens, including *S.* Typhimurium (Gomes and Dikic, 2014). NOD1 and NOD2 activate NF-κB and mitogen-activated protein kinase (MAPK) signalling pathways, whereas NLRP2 and NLRP4 are negative regulators of the NF-κB pathway. Some NLRs, as well as absent in melanoma 2-like receptors (ALRs) and pyrin, act as inflammasome sensors (Sharma and Kanneganti, 2016). The inflammasome is mainly activated in myeloid cells as an essential component of innate immunity that promotes caspase-1-induced conversion of procytokines into active IL-1β and IL-18. It can result in an inflammatory form of cell death known as pyroptosis. Inflammasomes are cytosolic protein complexes that usually contain a specific sensor, an adaptor molecule (ASC) and pro-caspase-1. Canonical inflammasomes contain NLRP1, NLRP3, NLRC4, AIM2 or pyrin as sensors. NLRP3 is also involved in a non-canonical inflammasome that activates pro-caspase-11 in response to Gram-negative bacteria. This inflammasome is primed by recognition of LPS through a TLR and type I interferon signalling (Rathinam and Fitzgerald, 2016).

The general consequences of binding to PRRs include the activation of macrophages and neutrophils to increase their phagocytic activity, and the release of cytokines and chemokines that amplify the immune response. Once inside the phagocytic cell, the microorganism can be killed by reactive oxygen or nitrogen intermediates, by preformed antimicrobials like the defensins mentioned above or a neutral protease. This takes place with the assistance of other factors like low pH, lysozyme and lactoferrin. Other important cells in the innate immune system are natural killer cells, which identify host cells expressing abnormal patterns of proteins and induce apoptosis of these cells.

In addition to the cellular components of innate immunity, the complement system and other soluble circulating defensive proteins integrate the humoral innate immunity. The complement system is a multicomponent enzyme cascade that facilitates phagocytosis and lysis of microorganisms. This system can be activated by the lectin, classical or alternative pathways. All of the aforementioned generate a C3 convertase that leaves C3b bound to the microbial surface and releases C3a. C3b acts as an opsonin that increases the ability of phagocytes to ingest bacteria. C3a, together with C5a, recruit phagocytic cells to the site of infection and promote inflammation. Complement activation also leads to the formation of a membrane-attack complex that causes cell lysis. In addition, complement activation together with products of activated mast cells promotes inflammation, a response that leads to local swelling, redness, pain and temperature elevation.

Adaptive immune response

Unlike innate responses, acquired or adaptive immunity is able to recognize specific antigens and develop a response against these pathogens. In contrast to innate immunity, adaptive immunity is different in each individual of a species and responds to specific antigenic challenges, displaying then a flexible spectrum of action, as it is able to recognize millions of different antigenic molecules. The two major varieties of lymphocytes, B-lymphocytes and T-lymphocytes, are essential to generate humoral and cellular immunity, respectively. B-lymphocytes are able to make antibodies, specific antigen-recognition molecules, known as immunoglobulins, which deal with extracellular infections. T-lymphocytes are involved in the control of intracellular infections. Their receptors recognize processed antigens in association with molecules of the major histocompatibility complex (MHC). T cells are classified in two categories: CD4 (helpers) and CD8 (cytotoxic) T cells, which are activated upon recognition of an antigen presented by major histocompatibility complex class II (MHC-II) and major histocompatibility complex class I (MHC-I), respectively by the TCR. While MHC-I is present on most cell surfaces, MHC-II is expressed on antigen presenting cells (APC).

A connection between the innate and adaptive immune systems at the level of cellular immunity is provided by DCs, which are relevant APCs. Stimulation of DCs by PAMPs triggers their maturation. Mature DCs present antigens to T cells via MHC molecules and also provide co-stimulatory signals via B7 family ligands, promoting the activation of naive T cells. The interconnection between innate and adaptive immunity is also important for humoral immunity since the classical pathway to activate the complement requires immunoglobulin M (IgM) or immunoglobulin G (IgG).

Dealing with physical barriers

Acidic pH

Bacteria belonging to the genus *Salmonella* usually invade their hosts through oral ingestion of contaminated food or water. Lysozyme found in saliva is not a significant barrier for these pathogens since, like many other Gram-negative bacteria, they are capable of resisting it thanks to their outer membrane shield around the PG layer (Van Kesteren *et al.*, 1942; Masschalck and Michiels, 2003). Therefore, the first real challenge that these bacteria have to face upon ingestion is the acidic pH of the stomach of the host (Ramos-Morales, 2012b) due to gastric secretions and hydrochloric acid. The pH value of the stomach is variable: a pH of 1.3 was observed in healthy humans in the fasted state and of 4.9 after meal ingestion (Russell *et al.*, 1993). Values of 3.9 and 4.0 were measured in rats and mice, respectively, in the fasted state (McConnell *et al.*, 2008). This low gastric pH can quickly kill enteric bacteria, but different bacterial species have developed mechanisms to survive during passage through the stomach. When compared with *E. coli*, that resists extreme acid stress and can survive at pH 2 for hours, *S. enterica* is much less acid resistant (Koutsoumanis and Sofos, 2004; Lin *et al.*, 1995). However, *S.* Typhimurium possesses at least three mechanisms of acid tolerance response (ATR) (Audia *et al.*, 2001): (i) the exponential phase ATR is induced at pH 4.5–5.8 and allows subsequent survival at pH 3; (ii) the RpoS dependent stationary phase ATR is pH independent and is part of a general stress response; (iii) the RpoS independent stationary phase ATR is induced by exposure to pH 4.5 and provides longer tolerance to pH 3 than the RpoS dependent system. These tolerance systems require acid shock proteins whose synthesis is regulated by RpoS, the iron regulatory protein Fur or the two-component systems PhoQ/PhoP or EnvZ/OmpR (Audia *et al.*, 2001; Foster, 1993). Small RNAs (sRNAs) have recently been identified as major regulators of stress response networks. For instance, the sRNA DsrA influences the ATR and virulence in *S.* Typhimurium (Ryan *et al.*, 2016).

Bacteria that are able to survive the passage through the stomach must face new challenges in the intestine before establishing an intracellular niche. This includes mucus, bile, antimicrobial peptides and the resident microbiota (Fig. 6.1).

Mucus

Epithelial cells in the human gut are covered with mucous secretions that protect the epithelium from the environment. The mucus barrier is composed of at least two glycan layers, which in turn incorporate multiple layers of glycoproteins (mucin) and complex oligosaccharides (glycocalyx) that protect cells from the local environment and infection. Microbes are able to use the host glycan as nutrients that regulate bacterial infection and virulence (Marcobal *et al.*, 2013). Intestinal intracellular pathogens have to penetrate the mucin layers and subsequently gain access to the cell membrane. In order to achieve this, *Salmonella* uses a set of glycosyl hydrolases (GHs) that are able to degrade glycans. Recently, Arabyan *et al.* (2016) investigated the specific enzymes used by *Salmonella* to degrade the glycan. They

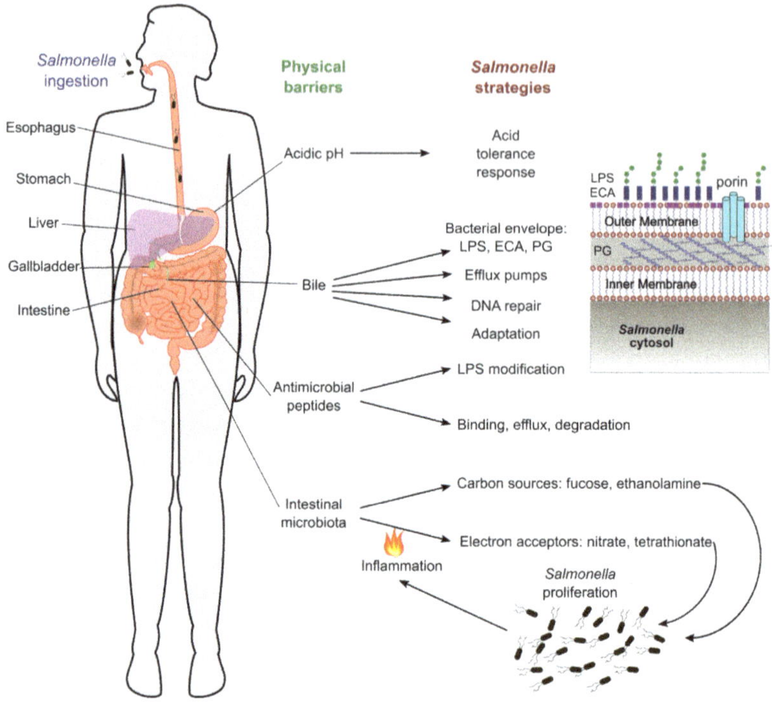

Figure 6.1 Schematic representation of *Salmonella* strategies to circumvent host physical barriers. After ingestion of contaminated food or water, *Salmonella* crosses the stomach where it is able to resist to acidic pH activating an acid tolerance response. Then *Salmonella* reaches the intestine where it has to face different chemical challenges, like the bile produced in the liver or antimicrobial peptides secreted by Paneth cells in the gut. In addition to that, *Salmonella* has to compete with the resident intestinal microbiota for essential nutrients. Interestingly, *Salmonella* triggers an inflammation response in the gut that offers different nutrients or adhesion receptor sites that can be exploited only by pathogens.

identified two specific glycan-degrading enzymes in *Salmonella*, NanH and MalS, as new virulence factors. These enzymes are expressed during *in vitro* infection of human colonic epithelial cells (Caco-2) and degrade the glycocalyx layer produced by these cells. The authors also analysed the host response to glycan degradation and they observed induced expression of host genes that hydrolyse mannose, fucose, and N-acetylneuraminic acid (Neu5Ac) from the glycan; as well as sialyltransferases, suggesting host glycan remodelling during infection (Arabyan *et al.*, 2016).

Bile

Bile is a secretory and excretory fluid produced in the liver whose composition varies depending on the individual's diet (Coleman, 1987). Secretory functions include: delivery to the intestinal tract of bile salts to aid fat digestion and absorption/secretion of polymeric IgA to prevent infection in the biliary and upper intestinal tracts. In addition, bile is the excretory vehicle of liver-derived metabolites of potentially toxic materials. Bile contains inorganic ions, bile salts, lipids (phosphatidylcholine, cholesterol) and proteins (plasma proteins, liver-specific proteins, polymeric IgA). It is a bactericidal agent that disrupts bacterial cell membranes due to the detergent activity of bile acids (Begley *et al.*, 2005), causes DNA damage (Prieto *et al.*, 2004, 2006) and alters the conformation of proteins. *Salmonella*, however, resists bile through different mechanisms. First, the bacterial cell envelope acts as a barrier to membrane-active agents like bile salts. Two components of the outer membrane are important for bile resistance: the LPS (Crawford *et al.*, 2012; Gunn, 2000) and the enterobacterial common antigen (ECA) (Ramos-Morales *et al.*, 2003). Recently, it has been shown that exposure to the bile salt sodium deoxycholate is associated with changes in the structure of the second layer of the cell envelope, the PG, and that PG remodelling contributes to bile resistance (Hernández *et al.*, 2015). Once bile enters the cell, additional mechanisms are activated. Efflux pumps mediate the export of bile salts from the bacterial cytoplasm to the outside medium and are responsible for resistance to many different toxic compounds (Nishino *et al.*, 2006). AcrAB-TolC is the best characterized efflux system in enteric bacteria (Ma *et al.*, 1993, 1995; Nikaido *et al.*, 2008; Zgurskaya and Nikaido, 1999). AcrB is an inner-membrane protein, TolC is the outer membrane component and AcrA is located in the periplasm. The genes encoding these proteins are regulated by the transcriptional regulator RamA, which is synthesized in response to bile (Baucheron *et al.*, 2014). In addition, bile-induced DNA damages can be repaired by Dam-directed mismatch repair and by base excision repair, and the impairment of DNA replication rescued by SOS-dependent translation DNA replication and RecBCD-dependent recombinational repair (Prieto *et al.*, 2006). Finally, *Salmonella* is also able to adapt to grow in extremely high bile concentrations if previously exposed to sublethal doses. This is a transitory state that involves multiple changes in gene expression (Prouty *et al.*, 2004). The RpoS-dependent general stress response plays a crucial role in the adaptation of *Salmonella* to bile (Hernández *et al.*, 2012).

Antimicrobial peptides

Different families of antimicrobial peptides protect the intestine. Defensins and cathelicidins are of special interest in the context of gastrointestinal infections (Wehkamp *et al.*, 2007). They are produced in the intestinal epithelium either constitutively or in an inducible manner. For instance, Paneth cells of the small intestine express the most abundant, constitutively expressed defensins. Human cathelicidin is expressed by epithelial cells in the stomach, small bowel and colon. In this context, butyrate, a short-chain fatty acid produced in the colon by bacterial fermentation of dietary fibres, has been shown to induce expression of the human cathelicidin gene. *Salmonella* is able to sense sublethal concentrations of antimicrobial peptides and induce several mechanisms of resistance that involve modification of the LPS or sequestering, efflux and proteolytic degradation of antimicrobial peptides (Matamouros and Miller, 2015). Regulatory proteins involved in this process include the alternative sigma factor RpoE and the two-component or phosphorelay systems PhoQ/PhoP, PmrA/PmrB and Rcs. PhoQ is an inner-membrane histidine kinase that senses antimicrobial peptides (Bader *et al.*, 2003, 2005) and phosphorylates the response regulator PhoP. Activation of PhoP prevents dephosphorylation of PmrA (Kato and Groisman, 2004). This protein directly controls expression of genes involved in LPS modifications necessary for resistance to antimicrobial peptides. The Rcs system is a phosphotransfer cascade that responds to antimicrobial peptides through RcsF (Farris *et al.*, 2010). This system controls the synthesis of colanic acid capsule and the transcription of other genes necessary for resistance (Detweiler *et al.*, 2003). Modifications of the bacterial surface that contribute to antimicrobial peptide resistance include:

1. Regulation of O-antigen length by the PmrA/PmrB system (Farizano *et al.*, 2012; Pescaretti *et al.*, 2011).
2. Modification of the anionic phosphate groups in the core and lipid A regions of the LPS by addition of aminoarabinose and phosphoethanolamine. PmrA is involved in the regulation of these modifications as well as in the inhibition of the activity of LpxT, which is responsible for increasing the negative charge of LPS by adding an additional phosphate. Together these changes lead to a reduction of the negative charge of the outer membrane and of the electrostatic interactions with antimicrobial peptides (Gunn *et al.*, 1998; Herrera *et al.*, 2010; Jones *et al.*, 2008a; Lee *et al.*, 2004; Zhou *et al.*, 2001b).
3. Decrease in the fluidity of the outer membrane by the incorporation, catalysed by PagP, of an additional palmitate to lipid A and to phosphatidylglycerol (Bishop *et al.*, 2000; Dalebroux *et al.*, 2014; Guo *et al.*, 1998).

Additional resistance mechanisms involve binding, efflux or degradation of antimicrobial peptides. The two PhoP-regulated proteins Mig-14 and VirK are suggested to bind antimicrobial peptides to prevent their penetration into *Salmonella* (Brodsky *et al.*, 2005). The *sapABCDF* operon encodes an ABC transporter system involved in resistance to antimicrobial peptides (Parra-Lopez *et al.*, 1993). It has been proposed that the SapABCDF transporter system functions as a complex with SapG and SapJ to mediate both peptide and K^+ transport (Parra-Lopez *et al.*, 1994). Finally, PgtE is a post-transcriptionally regulated component of the PhoQ/PhoP regulon with protease activity against alpha-helical cationic antimicrobial peptides (Guina *et al.*, 2000).

Intestinal microbiota

Before invading appropriate host cells and becoming an intracellular pathogen, *Salmonella* has to compete with the resident intestinal microbiota. The lumen of the human gastrointestinal tract is rich in nutrients and provides an appropriate niche for a diverse community of bacteria (Bäckhed *et al.*, 2005). They increase in number from the stomach to the colon, where they reach concentrations around 10^{12} per gram of luminal content. Most colonic bacteria are members of the Firmicutes and Bacteroidetes phyla, whereas Proteobacteria, Actinobacteria, Fusobacteria and Verrucomicrobia are less abundant (Eckburg *et al.*, 2005). This microbiota helps to prevent infection by providing a physical barrier for the attachment of bacterial pathogens to the mucosal surfaces and competing with pathogens for essential nutrients (van der Waaij *et al.*, 1971). *Salmonella*, however, can exploit the host innate immune response and its own metabolic capacities to overcome, at least temporarily, this barrier (Khan, 2014). First, the initial growth of *Salmonella* in the gut is powered by Hyb hydrogenases that enable the bacteria to use hydrogen produced by the resident microbiota as a central intermediate of metabolism (Maier *et al.*, 2013). Using a mouse colitis model, it has been shown that inflammatory host responses triggered by *S.* Typhimurium virulence factors (T3SS effectors) shift the competition in favour of the pathogen (Stecher *et al.*, 2007). The mechanisms leading to this result are not completely understood and different explanations have been proposed: (i) inflammation involves release of antibacterial factors that may kill or delay the growth of specific members of the microbiota that would normally inhibit *Salmonella* growth; and (ii) the inflamed intestine could offer different nutrients or adhesion receptor sites that can be exploited only by pathogens. In particular, reactive oxygen species generated by neutrophils during inflammation can react with endogenous thiosulfate to form tetrathionate. The *ttrRSBCA* locus, located in SPI2, confers to *S.* Typhimurium the ability to use tetrathionate as a terminal electron acceptor in anaerobic respiration (Winter *et al.*, 2010a). This confers a growth advantage because, unlike most bacteria in the gut, *S.* Typhimurium can use ethanolamine as a carbon source in the presence of tetrathionate as a respiratory electron acceptor (Thiennimitr *et al.*, 2011). It has also been suggested that the ability of *S.* Typhimurium to degrade fucose and anaerobically degrade 1,2-propanediol, probably using tetrathionate as a terminal electron acceptor, provides an advantage to this pathogen during competition with the microbiota (Staib and Fuchs, 2015). In addition, some *S.* Typhimurium isolates carry the gene *sopE* in a prophage. The product of this gene increases the severity of intestinal inflammation and drives the host to generate an additional terminal electron acceptor, nitrate, that suppresses genes responsible for utilization of energetically inferior electron acceptors such as tetrathionate (Lopez *et al.*, 2012). Stecher *et al.* (2012) described in an interesting study the discovery of a mechanism driving efficient conjugation between Enterobacteriaceae (*S.* Typhimurium and *E. coli*) in the host's intestine. Using a mouse colitis model, the authors observed parallel blooms of pathogenic *S.* Typhimurium and resident commensal *E. coli*. This phenomenon, probably due to dysbiosis of the anaerobic microbiota caused by *Salmonella* infection, led *E. coli* to account for > 80% of the total intestinal bacteria. In this scenario, the authors observed conjugative horizontal gene transfer efficiencies of 100%. This suggests that bacterial blooms occur after *Salmonella* infection fuels the reassortment of genetic material between different Enterobacteriaceae. This further implies that infected patients might enhance the spread of plasmid-encoded fitness, virulence- and antibiotic resistance determinants (Stecher *et al.*, 2012).

Evasion of the innate immune system

Salmonellae that are able to deal with the aforementioned physical barriers can invade the gut epithelium, preferentially in the distal ileum, using several routes. They usually invade microfold cells (M cells) that overlie Peyer's patches, but they can also induce their own uptake by epithelial cells or be engulfed by $CD18^+$ phagocytes, probably monocytes or dendritic cells (Watson and Holden, 2010). In a systemic infection, *Salmonella* can disseminate to internal organs [mesenteric lymph nodes (MLNs), liver and spleen]. Access to other cell types, like macrophages, could be a consequence of cell death induction (Fink and Cookson, 2007) and subsequent phagocytosis. Both intracellularly and during the transient extracellular passage from one host cell to another, *Salmonella* has to face and resist components of the innate immune system.

Salmonella finds cationic antimicrobial peptides not only in the intestinal lumen but also inside macrophages. Resistance mechanisms against this innate defence were described in the previous section. Interestingly, the PhoQ/PhoP two component system, that sense these antimicrobial peptides, are also additively activated by intracellular signals (acidic pH and divalent cation limitation). However the presence of antimicrobial peptides could be the most relevant signal for systemic virulence (Hicks *et al.*, 2015).

Manipulation of inflammatory pathways

Relevant PRRs for the recognition of *Salmonella* PAMPs are members of the TLR family, associated to host membranes, and members of the NLR family, that recognize the presence of PAMPs in the cytosol. This recognition leads to the activation of the NF-κB pathway (O'Dea and Hoffmann, 2010). Specific *Salmonella* virulence factors that enhance NF-κB activity are the T3SS1 effector SopE, which is recognized by NOD1 (Keestra *et al.*, 2013), and the T3SS2 effector SrfA, that promotes NF-κB activation by binding to TOLLIP, an inhibitor of IRAK1 (Lei *et al.*, 2016). This proinflammatory pathway can also be induced by the T3SS needle proteins PrgI and SsaG through TLR2 and TLR4 (Jessen *et al.*, 2014). In addition, SopE acts as a GTP exchange factor (GEF) for Rho family GTPases Cdc42 and Rac1, leading to activation of the c-Jun N-terminal kinase (JNK) pathway. Three different *Salmonella* proteins have been involved in inflammasome activation: (i) flagellin, an inducer of the NLRC4 inflammasome injected through T3SS1; (ii) PrgJ, a component of the basal body inner rod of T3SS1, also detected by NLRC4 (Miao *et al.*, 2010a); (iii) SopE contributes to inflammasome activation through its GEF activity, although the specific inflammasome sensor has not been determined (Hoffmann *et al.*, 2010; Müller *et al.*, 2009). NLRP3 has been suggested to have also an important role in innate immune defence against *S.* Typhimurium, since mice lacking both NLRP3 and NLRC4 are significantly more susceptible to infection (Broz *et al.*, 2010). It has been recently shown that NLRP3 associates with NLRC4 in macrophages infected with *S.* Typhimurium or transfected with flagellin, revealing an unexpected overlap between two distinct inflammasomes (Qu *et al.*, 2016). *S.* Typhimurium and other Gram-negative bacterial pathogens could induce a non-canonical inflammasome involving cytosolic recognition of LPS and activation of caspase-11 in mice and caspase-4 in humans (Broz and Monack, 2013a; Broz *et al.*, 2012; Casson *et al.*, 2015; Storek and Monack, 2015). This activation has been observed in the context of infections with mutants that aberrantly enter the cytosol (Aachoui *et al.*, 2013) after lysis of the SCV dependent on IFN-induced GTPases (Meunier *et al.*, 2014). However, wild-type *S.* Typhimurium can also lyse their nascent vacuole following invasion of tissue culture epithelial cells and enter the cytosol where it can eventually hyper-replicate (Birmingham *et al.*, 2006; Knodler *et al.*, 2014). Using a semi-quantitative single-cell analysis, Malik-Kale *et al.*,showed that although cytosolic hyper-replication occurs in less than 20% of infected epithelial cells, it accounts for the majority of net intracellular replication (Malik-Kale *et al.*, 2012). Interestingly, the transcriptional profiling of both intracellular bacterial subpopulations are different and it is known that cytosolic *Salmonella* are induced for T3SS1 and flagellated, whereas vacuolar bacteria are T3SS2-induced (Knodler *et al.*, 2010). Eventually, cytosolic hyper-replication leads to epithelial cell death via pyroptosis. This results in cell lysis, proinflammatory cytokine release and escape of the cytosolic bacteria into the extracellular space, providing a potential mechanism of dissemination (Knodler, 2015).

Anti-inflammatory response to *Salmonella* infection

Although an initial inflammatory response is beneficial for *Salmonella* to outcompete the microbiota, as the intracellular infection progresses, the pathogen still has mechanisms to inhibit inflammatory pathways (Fig. 6.2). Down-regulation of flagellin expression in systemic sites is a way to prevent NLRC4 inflammasome activation (Cummings *et al.*, 2006; Miao *et al.*, 2010b; Winter *et al.*, 2010b). Control of bacterial production of the tricarboxylic acid (TCA) cycle metabolite citrate may contribute to evade NLRP3 inflammasome

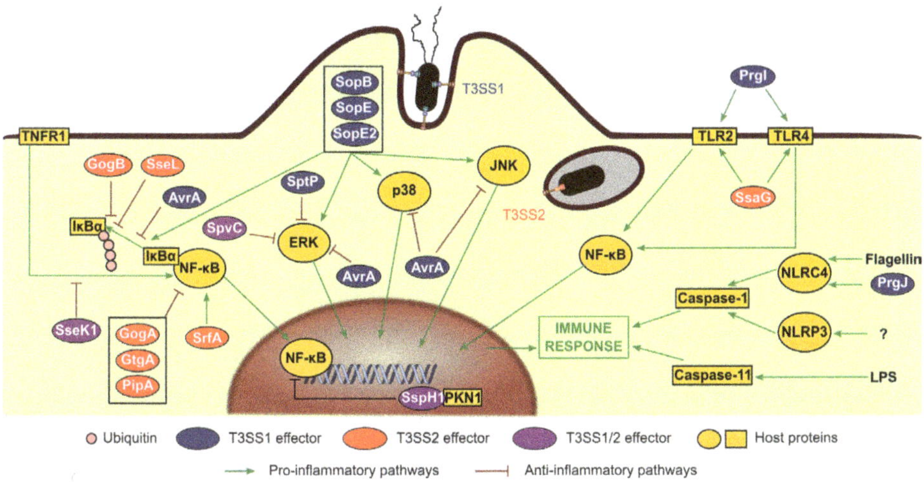

Figure 6.2 Schematic illustration of mechanisms of *Salmonella* to manipulate the host immune system by injecting effector proteins through two T3SS. *Salmonella* triggers its own internalization into the host cell in a T3SS1-dependent manner. Following internalization, *Salmonella* establishes its intracellular niche in a modified phagosome, called *Salmonella*-containing vacuole (SCV), where the pathogen manages to survive and proliferate. Biogenesis and maturation of the SCV depends on the activity of T3SS 1 and 2 effectors that manipulate the normal endocytic pathway to avoid the fusion and degradation by a lysosome. The actions of the T3SS proteins effectors on the inflammatory pathways to evade the host immune system are schematically represented.

activation in macrophages (Wynosky-Dolfi *et al.*, 2014). In addition, many T3SS effectors have an anti-inflammatory role. Both the T3SS1 effector AvrA and the T3SS2 effector SseL deubiquitinate IκBα and inhibit its degradation (Le Negrate *et al.*, 2008; Ye *et al.*, 2007). Binding of IκBα to NF-κB prevents its translocation to the nucleus. AvrA also inhibits JNK-induced signalling (Du and Galán, 2009; Jones *et al.*, 2008b; Wu *et al.*, 2012). GogA, GtgA and PipA are members of the same family of proteases that redundantly cleave specific members of the NF-κB family, including RelA (p65) and RelB (Sun *et al.*, 2016). SspH1 localizes to the mammalian nucleus and inhibits NF-κB-dependent gene expression through interaction with protein kinase 1 (PKN1) (Haraga *et al.*, 2006). The effector SptP consists of two domains. The N-terminal domain acts as a GAP for Cdc42 and Rac1 that mediates reversion of the effects of SopE on the actin cytoskeleton (Fu and Galán, 1999). The C-terminal domain possesses tyrosine phosphatase activity, which is (Kaniga *et al.*, 1996) involved in reversing the initial activation of the MAPK ERK, triggered by *Salmonella* infection (Murli *et al.*, 2001). SptP inhibits this MAPK pathway through inhibition of Raf1 activation in a process that involves both SptP activities (Lin *et al.*, 2003). This down-modulation of ERK activation could explain the contribution of SptP, together with SspH1, towards the inhibition of IL-8 production after invasion of intestinal epithelial cells (Haraga and Miller, 2003). Another effector involved in reduction of the production of inflammatory cytokines through the inactivation of ERK is SpvC. This effector is encoded in a virulence plasmid present in some *S. enterica* serovars and can be secreted by T3SS1 and T3SS2. SpvC inactivates ERK by irreversibly removing phosphate from a threonine residue in the conserved activation loop motif TXY using its phophothreoninelyase activity

(Mazurkiewicz *et al.*, 2008). Interestingly, during the interaction of *Salmonella* with the plant model *Arabidopsis*, SpvC interacts with and dephosphorylates activated MAPK6 to inhibit defence signalling (Neumann *et al.*, 2014). This suggests that some *Salmonella* effector proteins could have a conserved function during proliferation in different hosts. GogB is another anti-inflammatory effector that limits NF-κB activation by targeting the host SCF E3 ubiquitin ligase and inhibiting IκBα degradation (Pilar *et al.*, 2012). SseK1, an effector protein that can be secreted by T3SS1 and T3SS2 (Baisón-Olmo *et al.*, 2015), has a N-acetylglucosamine transferase activity that modifies the TNF-α receptor TNFR1 and its adaptor TRADD, preventing TNF-α mediated activation of NF-κB (Li *et al.*, 2013).

An important mediator of the anti-inflammatory effects of *Salmonella* is haem-oxygenase 1 (HO-1). HO-1 is an anti-stress regulatory molecule. It catabolyses the degradation of haem into biliverdin with the production of free iron and CO. Different studies have shown that HO-1 can be induced by a wide variety of bacteria. This is likely due to the increase in oxidative stress and inflammation (reviewed in Blancou *et al.*, 2011). Onyiah *et al.* (2013). using a colitis mouse model, showed that the enteric microbiota induces expression of HO-1 in mice and zebrafish. HO-1-derived CO enhances macrophage bactericidal activity and bacterial clearance *in vivo.* Similarly, others works have shown that an absence of HO-1 in humans and mice leads to chronic inflammation (Chung *et al.*, 2008; Yachie *et al.*, 1999), although the exact mechanisms remains unclear. One hypothesis emerged from the experiments performed by Choi and Lee, who showed a positive feedback of the system in which CO stimulates the synthesis of the anti-inflammatory cytokine interleukin 10 (IL-10) by macrophages (Otterbein *et al.*, 2000). IL-10 induces the expression of HO-1 potentially leading to a self-amplification of the anti-inflammatory effect (Lee and Chau, 2002). Others propose that HO-1 controls early innate immune response of *S.* Typhimurium-infected macrophages. In this study, a reduced survival of intracellular *Salmonella* in cells with low HO-1 expression was observed, probably due to the limited iron availability and to the activation of pro-inflammatory pathways that occurs in the absence of this enzyme (Mitterstiller *et al.*, 2016). On the other hand, *Salmonella*-infected macrophages with strongly up-regulated HO-1, provide a survival niche for *Salmonella* (Nix *et al.*, 2007). Thus, the induction of HO-1 expression has been suggested to be a strategy of the pathogen, given its harmful effects in the early control of *Salmonella* infection.

Another example of the immunosuppressive role of HO-1 was described by Riquelme *et al.* (2015), who suggest HO-1 prevents T-cell-mediated inflammatory disease by producing CO and impairing DC immunogenicity. CO has been shown to impair the mitochondrial function in DCs by reducing both the mitochondrial membrane potential and ATP production. This gas is able to impair the endo-lysosomal antigen trafficking route, reducing the ability of DCs to prime naive T cells. The role of HO-1 in immunity has been also studied in DCs, where CO is the molecule with the most prominent immunosuppressive capacity over their function (Rémy *et al.*, 2009; Tardif *et al.*, 2013). Contrary to this, other authors, like Zaki *et al.*, suggest that HO-1 contributes to improved control of *Salmonella* replication within macrophages by exerting a cytoprotective effect (Zaki *et al.*, 2009). Further studies are necessary to clarify the dual role of HO-1 in infections.

Autophagy

The relationship of *Salmonella* with autophagy is also of interest. During *S.* Typhimurium infection, T3SS1-mediated damage of the SCV allows ubiquitin and galectin-8 to target bacteria inside the affected SCVs or free in the cytoplasm (Birmingham and Brumell, 2006; Birmingham *et al.*, 2006; Thurston *et al.*, 2012). These signals recruit essential anti-bacterial autophagy components including several cargo receptors, the kinase TBK1 and WIPI2, that restrict *Salmonella* proliferation (Radtke *et al.*, 2007; Thurston *et al.*, 2016, 2009, 2012). Autophagy has also a role in sealing damaged endosomal membranes, allowing T3SS2 induction inside the SCV (Kreibich *et al.*, 2015; Owen and Casanova, 2015). In epithelial cells, *Salmonella* induces the formation of ubiquitinated aggregates near the SCV in a T3SS2-dependent manner. These aggregates are recognized by the autophagy machinery. The T3SS2 effector SseL is a deubiquitinase whose activity lowers autophagic flux and favours intracellular *Salmonella* replication (Mesquita *et al.*, 2012). In a recent study that used a fibroblast-infected model, López-Montero *et al.* (2016) showed that *S.* Typhimurium develops membranous aggregates connected to the phagosome where the bacteria controls its progeny and establishes an intracellular persistent infection. In addition, T3SS2 mediates active suppression of autophagic signalling in macrophages through recruitment of focal adhesion kinase (FAK) to SCVs and activation of the Akt-mTORC1 signalling pathway (Owen *et al.*, 2014). The inhibition of autophagy by *Salmonella* also prevents the induction of a protective cytokine response mediated by interferon beta (IFN-β) (Owen *et al.*, 2016). Two recent studies show that SpvB, a T3SS2 effector with ADP-ribosyltransferase activity that prevents polymerization of G-actin into F-actin filaments (Hochmann *et al.*, 2006; Lesnick *et al.*, 2001; Margarit *et al.*, 2006; Tezcan-Merdol *et al.*, 2001), can inhibit autophagosome formation and increase inflammatory injury in zebrafish intestine (Li *et al.*, 2016) mammalian cells and mice (Chu *et al.*, 2016). This inhibition occurs at the early stage of autophagy via depolymerization of actin filaments.

The complement system

Finally, mechanisms to evade or interfere with the complement system have evolved in *Salmonella*. The human pathogen *S.* Typhi expresses the Vi capsular polysaccharide during transit through the intestine (Tran *et al.*, 2010). This polysaccharide has anti-inflammatory properties, (Sharma and Qadri, 2004) and also inhibits complement deposition (Looney and Steigbigel, 1986; Wilson *et al.*, 2011). This, thanks to a lack of hydroxyl groups available for ester formation with C3b (Heyns and Kiessling, 1967). Evolution of these immune evasive properties of *S.* Typhi led to inactivation of the *fepE* gene, encoding a regulator of very-long O-antigen chains (Crawford *et al.*, 2013). The surface protease PgtE of *S.* Typhimurium affects complement activity by cleaving C3b, C4b, C5, B and H complement components (Ramu *et al.*, 2007; Riva *et al.*, 2015). Although factors B and H have opposite effects on complement activation, the overall effect of PgtE favours protection of *Salmonella* against the host immune system, since less C3-derived fragments accumulated on *Salmonella* and the association with neutrophils was reduced (Riva *et al.*, 2015).

Mechanisms to escape adaptive immunity

In addition to mechanisms developed by *Salmonella* to escape from innate immunity, this intracellular pathogen has acquired different mechanisms to evade different adaptive immunity mechanisms.

Interactions of *Salmonella enterica* with dendritic cells

Innate and adaptive immunities are linked by DCs. Therefore, the activation of an effective adaptive immune response relies on efficient DCs capable of priming of naive T cells. This is one of the most relevant immune evasion strategies for *Salmonella*. This pathogen has developed molecular mechanisms to prevent the presentation of bacterial antigens that prime naive T cells and therefore avoid the initiation of the adaptive immunity.

In general, virulent strains of *Salmonella* are capable of modulating phagocytosis by DCs, avoiding lysosomal degradation and preventing antigen presentation to T cells. Here, we summarize the mechanisms developed by *Salmonella* to escape from acquired immunity. It is well known that *Salmonella* is able to induce phagocytosis on epithelial cells by active translocation of T3SS1 effectors which modulate the actin cytoskeleton. This results in internalization of *Salmonella* into the epithelial cell (Scherer *et al.*, 2000; Steele-Mortimer *et al.*, 2002; Zhou *et al.*, 2001a). In contrast, the uptake of *Salmonella* by DCs is tightly regulated by effector proteins of T3SS1 (Bueno *et al.*, 2010), being able to control the amount of bacteria that enter into the cells, avoiding then a massive immune response that would restrict *Salmonella* replication and dissemination. Additionally, Riquelme *et al.* showed that IgG-opsonized *Salmonella* are recognized by different receptors (FcγRIII receptors and others that have not been identified) expressed on DCs surfaces. These immune complexes are internalized and degraded by the lysosomal route, restoring its capacity to present *Salmonella* antigens to T cells and initiate an adaptive immunity response. However, the exact molecular mechanism with which IgG interferes with the secretion of *Salmonella* virulent effectors or impair its capacity to evade capture in DCs remains unclear (Riquelme *et al.*, 2012).

After internalization, establishing a systemic infection depends on the pathogen's capacity to survive inside the host cells and to evade the immune response. In that sense, there are many studies on *Salmonella* virulence proteins that contribute to intracellular survival in DCs and dissemination in the host (Albaghdadi *et al.*, 2009). One of the most relevant strategies developed by *Salmonella* to survive and replicate intracellularly is the disruption of the host endocytic trafficking machinery to avoid its lysosomal degradation within the SCV. The ability to survive in DCs is dependent on SPI2 effectors (Albaghdadi *et al.*, 2009; Halici *et al.*, 2008; Jantsch *et al.*, 2003), which are injected to DCs cytosol from the SCV through a T3SS2. In fact, T3SS2 mutants show a reduced capacity to survive inside DCs and are attenuated in mice (Tobar *et al.*, 2006). However, there are several examples of T3SS1 effectors that participate in avoiding lysosomal fusion. Among these SPI1 effectors, SopB and SopE were found to play a role in SCV maturation (Hernandez *et al.*, 2004; Mukherjee *et al.*, 2001). SopB is a phosphatase that mediates phosphatidylinositol 3-phosphate production in the SCV, through Rab5 recruitment to the SCV and its effector Vps34 (Mallo *et al.*, 2008). Another example is SopE, which acts as a Rab5-specific exchange factor and mediates the recruitment of Rab5 in the GTP form in the SCV (Mukherjee *et al.*, 2001). There are also examples of T3SS2 effectors that impair trafficking of endocytic cargo to lysosomes. The effector SopD2 directly binds and inhibits the host GTPase Rab7, a regulatory switch central to endocytic trafficking and phagosome–lysosome fusion. Consequently, this limits Rab7 interaction with its dynein- and kinesin-binding effectors RILP and FYCO1. This in turn disrupts the regulation of microtubule motors (D'Costa *et al.*, 2015). SifA is another SPI2 effector required for bacterial survival inside DCs that alters lysosomal function

(Beuzón *et al.*, 2000; Boucrot *et al.*, 2003; Petrovska *et al.*, 2004). McGourty *et al.* showed that SifA prevents the delivery of hydrolytic enzymes to the SCV by inhibiting Rab9-dependent retrograde trafficking of mannose-6-phosphate receptors (MPRs). This requires binding of SifA to its host cell target SKIP. Translocated SifA forms a stable complex with SKIP and Rab9 in infected cells. Sequestration of Rab9 by SifA-SKIP accounts for the effect of SifA on MPR transport and lysosome function (McGourty *et al.*, 2012). There are other effectors of the T3SS2 that are involved in DC intracellular survival of *Salmonella* like SseJ, SseF, SspH2 and PipB2. Strains lacking these effectors show reduced survival inside DCs but the specific role of each of these virulence proteins remains unclear (Halici *et al.*, 2008).

One of the suggested strategies used by *Salmonella* to avoid antigen presentation is inhibition of lysosomal degradation (Tobar *et al.*, 2006, 2004). There are studies that suggest that the PhoQ/PhoP regulatory system is involved in this escape mechanism. This because *Salmonella* strains with mutations in the PhoQ/PhoP system are not able to escape from lysosomal degradation and therefore do not interfere with antigen processing and presentation (Niedergang *et al.*, 2000). Other studies report that *Salmonella* regulates the expression of MHC class II antigens by polyubiquitination of HLA-DR. (Lapaque *et al.*, 2009). These mechanisms rely on virulence factors encoded in SPI2. T3SS2 effectors like SifA, SlrP, SspH2, PipB2 and SopD have been shown to participate in this evasion process since they are required for the inhibition of MHC-II-dependent antigen presentation in DCs (Halici *et al.*, 2008). The ability of *Salmonella* to prevent antigen presentation by DCs seems to depend exclusively on SPI2-related proteins, since a mutant strain lacking the T3SS1 is still able to avoid antigen presentation by DCs (Bueno *et al.*, 2010). Importantly, this feature seems to be host specific and restricted to serovar Typhimurium, since other serovars like *S. enteritidis* and *S.* Typhi are not able to interfere with this function in murine DCs (Bueno *et al.*, 2008).

S. Typhimurium can interfere with the host's immune response during oral infection of mice through selective killing of $CD8\alpha^+$ DCs in a process that is dependent on MyD88 and TNFR1 (Sundquist and Wick, 2009). It has been recently shown that the pR_{ST98} plasmid of *S.* Typhi may also influence maturation, survival and cytokine production of DCs, preventing activation of T-cell-mediated immunity against antigens derived from this pathogen (Wei *et al.*, 2015).

Another strategy used by virulent *Salmonella* is based on down-regulation of flagellin, a target of the innate and adaptive immune responses during infection. This strategy consists in preventing T-cell activation by the active reduction of the availability of bacterial antigens for presentation to T cells (Alaniz *et al.*, 2006). Some data suggest that flagellin is differentially expressed by *Salmonella* populations infecting Peyer's patches (Cummings *et al.*, 2006). Furthermore, intracellular *Salmonella* can make flagellin unavailable for antigen processing by DCs (Alaniz *et al.*, 2006). Similar studies showed that during low-dose *Salmonella* infections, the microbe evades activation of flagellin-specific CD4 T cells (Srinivasan *et al.*, 2004).

Direct interactions of *Salmonella enterica* with T cells

Examples of *S.* Typhimurium directly blocking T-cell proliferation are prevalent in the literature. This inhibition was observed in both $CD4^+$ and $CD8^+$ T cells and was due to a down-regulation of TCR expression β-chain, which interferes with the first step in T-cell clonal expansion (van der Velden *et al.*, 2005, 2008). The authors of the aforementioned study also identified the enzyme L-asparaginase II as responsible for this inhibitory effect.

The activity of this enzyme causes depletion of exogenous L-asparagine, leading to down-regulation of TCR-β expression, suppression of T-cell blastogenesis, blockade of cytokine production, and, ultimately, inhibition of T-cell proliferation (Kullas *et al.*, 2012). In agreement with this, the cytolytic capacities of $CD8^+$ T cells have been shown to be initiated by TCR interactions (Lewinsohn *et al.*, 2011).

The phagosomal lifestyle of *S.* Typhimurium seems to be responsible for delayed MHC-I dependent antigen presentation and delayed expansion and contraction of the $CD8^+$ T-cell response. Interestingly, even memory $CD8^+$ T cells failed to undergo rapid expansion in response to infection with *S.* Typhimurium expressing the model antigen OVA (Luu *et al.*, 2006).

Additionally, *Salmonella* is able to induce apoptosis of Ag-specific CD4 T cells (Srinivasan *et al.*, 2009). This is driven by SPI-2 virulence genes. *In vivo* studies by Ertelt *et al.* suggested that *Salmonella* undergoes, in a SPI-2 dependent manner, a selective culling of activated $CD4^+$ T-cell subsets, which re-shapes the repertoire of antigen-specific T cells that persist later after infection (Ertelt *et al.*, 2011).

Non-cognate activation of T cells

In addition to conventional activation by cognate peptide–MHC complexes presented by DCs, T cells can be indirectly stimulated (bystander stimulation) by *Salmonella*. This alternative pathway involves the well-characterized TLRs and NLRs, as well as other less known PRRs. It results in production of inflammatory cytokines that prime T cells at the site of infection (Broz and Monack, 2013b). Indeed, *Salmonella* has been proposed as a particularly appropriate model pathogen for study of non-cognate CD4 T-cell responses based on (i) strength of the Th1 response during infection, (ii) the requirement for CD4 T cells in bacterial clearance and (iii) the well-characterized inflammatory response to conserved molecular patterns induced by *Salmonella* infection (O'Donnell and McSorley, 2014).

B lymphocytes' role

B lymphocytes have critical roles as positive and negative regulators of immunity. For instance, B cells have important contributions in acquired immunity against *Salmonella* infections, being essential for immunity during secondary challenge (McSorley and Jenkins, 2000). Several studies describe antibody dependent (Mastroeni *et al.*, 1993; McSorley and Jenkins, 2000; Mittrücker *et al.*, 2000) and antibody independent *Salmonella* specific B cells responses (Barr *et al.*, 2010; Mastroeni *et al.*, 2000; Ugrinovic *et al.*, 2003). These studies suggest that B cells can work as APCs and have a key role in the production of inflammatory cytokines during *Salmonella* infection. The immunosuppressive role of B lymphocytes, classically associated with B cell-derived IL-10, is related to the regulation of autoimmune disease. This IL-10 also increase susceptibility to pathogens (Fillatreau, 2011; Fillatreau *et al.*, 2008). In a recent study, Shen *et al.* described interleukin 35 (IL-35)-producing B cells as novel key players in the negative regulation of immunity (Shen *et al.*, 2014). The authors showed that mice whose B cells did not express IL-35, displayed improved resistance to *S.* Typhimurium infection. The increased immunity found in these mice was associated to higher activation of macrophages and T cells, as well as an increased stimulatory function of B cells as APCs. The data presented in this work demonstrate that B cells can inhibit antimicrobial immunity through production of IL-35 (Shen *et al.*, 2014).

One of the strategies developed by *Salmonella* to evade the action of B cells is to delay the formation of the germinal centre, which is necessary to produce high-affinity antibodies against a specific pathogen (Cunningham *et al.*, 2007). It has been shown that during *Salmonella* infection germinal centre formation and affinity maturation are delayed until the second month of infection. This differs greatly with conventional 7 day functional germinal centre formation. Furthermore, high-affinity antibodies start to be produced during the second week after immunization.

Persistence

One important aspect of typhoid *Salmonella* is its capacity to establish chronic infections. After adequate treatment the majority of patients who present acute typhoid fever recover; however, a significant percentage of patients (1–6%) develop a chronic infection (Levine *et al.*, 1982; Monack *et al.*, 2004a). These patients become asymptomatic chronic carriers and sporadically shed bacteria in their stool, allowing the pathogen to close its life cycle and to be transmitted to new hosts (Monack *et al.*, 2004b). Despite the importance of *Salmonella* chronic infections, relatively little is known about the host immune response or virulence mechanisms that characterize long-term systemic infections. Since the majority of chronically infected patients are asymptomatic, the identification of potential targets is challenging (Shpargel *et al.*, 1985; Sinnott and Teall, 1987). However, it is interesting to note that gut persistent infections are *Salmonella* pathogenicity island 1 (SPI-1) and *Salmonella* pathogenicity island 2 (SPI-2) dependent (Lam *et al.*, 2014). One study suggested that the human carrier state may be associated to an ineffective immune response (Thompson *et al.*, 2009).

In this section we describe some recent advances with respect to *S.* Typhi and *S.* Typhimurium persistent host colonization.

Biofilm formation

Biofilm development is an important component of bacterial survival and a common source of persistent infections (Costerton *et al.*, 1999). *Salmonella* are able to form biofilm in both biotic and abiotic surfaces. Moreover, there is a clear correlation between chronic infection and bacterial biofilm formation on the surface of gallstones (Gonzalez-Escobedo *et al.*, 2011; Prouty *et al.*, 2002). Epidemiological studies have revealed a strong association between chronic carriers of *S.* Typhi and gallstones (Schiøler *et al.*, 1983). Although, 90% of chronically affected individuals have gallstones (Karaki and Matsubara, 1984) or other biliary track pathologies, this is not required to develop a carrier state (Monack *et al.*, 2004b). *Salmonella* biofilms have been found on gallstones in both *S.* Typhi patients and a mouse model of chronic *S.* Typhimurium infection (Crawford *et al.*, 2010), suggesting both serovars share this biofilm strategy to survive in their host. Interestingly, bile has been shown to enhance *S.* Typhi biofilm formation (Gonzalez-Escobedo *et al.*, 2011). Consistently, *S.* Typhi exhibits specific binding to cholesterol-coated surfaces, like gallstones (Crawford *et al.*, 2008). FliC and OmpC are two proteins that were described to have a key role on initial adhesion to cholesterol-coated surfaces (Crawford *et al.*, 2010). A hallmark of biofilm formation is the self-production of an extracellular matrix that ensures the integrity of the biofilm. The biofilm matrix is composed of EPS (extracellular polymeric substances) and water. Some components of biofilm EPS like cellulose, colonic acid and O antigen capsule are crucial for *S.* Typhi persistence and biofilm development (Crawford *et al.*, 2008; Prouty and Gunn, 2003; Prouty *et al.*, 2002). Some studies show that bile present in the gallbladder induces the production of O-antigen, which facilitates *S.* Typhi biofilm formation on human gallstones (Crawford *et al.*, 2008; Hall-Stoodley and Stoodley, 2009).

Persistence in immune cells

The study of the molecular mechanisms underlying persistent infections is limited by the lack of appropriate animal models that resist challenges using virulent strains of *Salmonella*. A suitable mice model was established in 2004 to study *S.* Typhimurium persistent infections (Monack *et al.*, 2004b). In this model virulent *S.* Typhimurium strains cause long-term chronic infections. The persistent bacteria can then be monitored within macrophages in the MSNs up to 1 year post inoculation (Monack *et al.*, 2004b). In addition to that, *in vitro* assays showed that hemophagocytic macrophages might represent a survival niche for *Salmonella* (Nix *et al.*, 2007). Another study showed that persistent bacteria are sequestered within macrophages in systemic tissues, where they do not replicate and seem to be in a dormant-like state. This supports the idea of macrophages as a reservoir for persistent bacteria (Helaine *et al.*, 2010). Another study published recently, showed that *S.* Typhimurium preferentially associates with anti-inflammatory/M2 macrophages at later stages of infection in mice (Eisele *et al.*, 2013). This subset of macrophages does not express many of the defence mechanisms needed to eliminate invading microbes. Moreover, there are pathogens that have developed strategies to induce polarization of cells towards the M2 phenotype as a virulence mechanism (Jensen *et al.*, 2011).

In addition to macrophages, it has been shown that *Salmonella* remains for a long period of time within plasma cells, bone marrow B cell precursors, and all B cell subsets from the spleen (Castro-Eguiluz *et al.*, 2009; López-Medina *et al.*, 2015a). The ability of *Salmonella* to persist within B cells suggests that it developed a $CD8^+$ T-cell response evasion mechanism. In fact, *Salmonella* infection results in the enhanced expression of PD-L1 in B cells (López-Medina *et al.*, 2015a). PD-L1 is a member of the programmed death-1 (PD-1)/programmed death-ligands (PD-Ls) pathway that can turn off or reduce TCR signalling by recruiting phosphatases like SHP-1 and SHP-2 (Chemnitz *et al.*, 2004; Freeman *et al.*, 2000; Sheppard *et al.*, 2004). The activation of the PD-1 : PD-Ls axis leads to impairment of CD8 T cells during *Salmonella* infection (López-Medina *et al.*, 2015b).

T3SS effectors and persistence

In order to shed light on the molecular mechanisms involved in persistence, Monack's group performed a genome-wide screen to identify *Salmonella* genes required for persistent infection of the mice (Lawley *et al.*, 2006). The screen revealed that SPI1 was necessary to maintain a persistent infection for at least 1 month post challenge. The authors confirmed that SPI1 effectors SipB, SipC, and SipD, which have a role on invasion or translocation of other effectors, contribute to establish a persistent infection. Similarly, another study showed that SPI1 is required for persistently infected mice to transmit *S.* Typhimurium to naive cage-mates (Lawley *et al.*, 2008). Data from the genome screen mentioned above also showed that the SPI2 effector SseI was required for maintaining a long-term systemic infection. A follow-up study showed that SseI modulates macrophages and DCs by directly binding to IQGAP1, an important regulator of host cell migration (McLaughlin *et al.*, 2009). So, SseI blocks somehow the migration of macrophages and DCs interfering with the host immune system to clear systemic bacteria.

Other elements that have been shown to contribute to persistent *Salmonella* infection are those that protect against host-derived antimicrobial peptides. Examples of these factors are PgtE, a modifier of the bacterial outer surface membrane (Lawley *et al.*, 2006), and Mig-14, which provides resistance against anti-microbial peptides like VirK, RcsC, and YdeI (Erickson and Detweiler, 2006), both of them regulated by the PhoP/PhoQ system (Bader *et al.*, 2005).

Persistence and metabolism

Recently, a pioneering study described how bacteria modulate the host cell metabolism to create an appropriate environment for long-term infection (Eisele *et al.*, 2013). Firstly, the authors observed that PPARδ, which is a transcriptional factor involved in fatty acid metabolism, is up-regulated in *Salmonella* infected M2 macrophages. This situation produces a shift in the metabolic state of the cell that leads to an increase in the level of glucose available to bacteria. Moreover, the authors show that pharmacological activation of PPARδ increases glucose availability and enhances bacterial replication in macrophages, while *Salmonella* fails to persist in macrophage lines lacking PPARδ. All together, these data, support the idea that *Salmonella* has evolved to survive long periods of time in M2 macrophages by harnessing the unique metabolism of these cells (Eisele *et al.*, 2013).

Another relevant topic concerns genetic adaptation of bacteria during a persistent infection. A recent study describes the acquisition of adaptive mutations that facilitate persistence and survival in the host over the course of an infection (Søndberg and Jelsbak, 2016). In this work, the authors used a model of mice infected chronically with *S.* Typhimurium. They observed that bacteria acquire distinct single nucleotide polymorphisms (SNPs) in known regulators of metabolic and virulence genes. One such difference, the *kdgR*-SNP was confirmed to confer selective advantage during chronic infections and constitutes a true patho-adaptive mutation. Thus, the results provide evidence for rapid genetic adaptation to the host of *S.* Typhimurium during persistent infection (Søndberg and Jelsbak, 2016).

A non-mutational way for *Salmonella* to cause persistent infections even after antibiotic exposure is through the production of persister cells. These are rare cells, produced by all bacterial populations, that transiently become multidrug tolerant. It is thought this could be due to a state of low metabolic activity that leads to dormant or slow-growing cells (Bigger, 1944; Helaine and Kugelberg, 2014; Maisonneuve and Gerdes, 2014). *Salmonella* persisters are part of a non-replicating population formed immediately after uptake by macrophages and are induced by vacuolar acidification and nutritional deprivation (Helaine *et al.*, 2014). Furthermore, the majority of the 14 toxin–antitoxin (TA) modules contributed to intracellular persister formation. A more comprehensive analysis of *S.* Typhimurium TA modules identified multiple toxins with anti-proliferative activity A selected group of the corresponding TA modules were important for survival of intracellular bacteria inside fibroblasts (Lobato-Márquez *et al.*, 2015). One of these toxins, TacT, is an acetyltransferase that blocks the primary amine group of amino acids on charged tRNA molecules, thereby inhibiting translation and promoting persister formation (Cheverton *et al.*, 2016).

Transmission

During systemic salmonellosis, the life cycle completed when the pathogenic bacteria leave the former host to infect a naive one. This highlights the main issue with persistent infections: the constant shedding of virulent bacteria on faeces allows for propagation and maintenance within a population. In a chronic infection scenario, *Salmonella* can persist in MLNs, bone marrow, and gallbladder, and periodically discharge *Salmonella* from the gallbladder to the small intestine in bile. The molecular mechanisms that control transmission of bacteria from host reservoirs are poorly understood. However, an interesting study proposed that host transmission of *S.* Typhimurium is controlled by virulence factors and indigenous intestinal microbiota. The authors described a model in which persistently infected mice provide a natural mode of *S.* Typhimurium transmission. In this context, a subgroup of mice (30%), called 'supershedders' was found to shed high levels of *S.* Typhimurium in their stools, leading to rapid transmission of the infection to naive hosts (Lawley *et al.*, 2008). The authors showed that the development of the supershedder phenotype depends on SPI1 and SPI2. Moreover, treatment with antibiotics induced the supershedder state in mice, suggesting that the indigenous intestinal microbiota has a role in controlling this phenotype.

Concluding remarks

We have described recent advances concerning the strategies used by *Salmonella* to evade both innate and adaptive immune responses. Most of these evasion mechanisms are directly related to T3SS1 and 2 and to the different effectors injected by them (Table 6.1). In the last two decades more than 40 effectors have been identified, and the biochemical activity and host targets for some of them have been successfully characterized. However, the precise role of many of them and their contribution to the relationship between *Salmonella* and its hosts at the cellular and systemic level remain unknown. A better understanding of the specific functions of these effector proteins will allow us to unravel the complex network of action of these molecules and to decipher the underlying mechanisms that enable *Salmonella* to evade the host immune system.

For instance, we have to define the role of T3SS effectors in establishing different intracellular lifestyles and decipher which signals enable *Salmonella* to survive intracellularly versus proliferate and escape, depending on the cellular type.

Another challenge is to improve our understanding of *Salmonella* control of the fusion of SCV to the lysosome. The identification of new and appropriate SCV markers and the development of new imaging technologies will improve the detection of intracellular *Salmonella* and ultimately help us to answer these questions.

One of the most significant challenges concerns host specificity achieved by different *Salmonella* serovars. There is an obvious shortage of appropriate models. For example, there is no suitable animal model for the typhoid human pathogen *S.* Typhi. Therefore, *in vivo* studies for this serovar are deficient. Researchers tend to use the *S.* Typhimurium-infected mouse model to study systemic typhoid-like disease. However, extrapolation of results obtained in *S.* Typhimurium to *S.* Typhi is not always possible. The development of new animal and cellular models are slowly helping us surmount this lack of appropriate tools. One example is the recent work of Fresnay *et al.*, who established a controlled human infection model with wt *S.* Typhi using cutting-edge multichromatic flow cytometry to analyse the pre-challenge immunological status and its correlation with the subsequent clinical outcome (Fresnay *et al.*, 2016). Specifically, the authors investigated the relationship between *S.* Typhi-specific $CD8^+$ T-cell responses before exposure to wt *S.* Typhi and clinical outcome, i.e. whether the participants who were challenged developed disease or not. They observed higher baseline levels of multi functional *S.* Typhi-specific $CD8^+$ T in patients who did not

Table 6.1 Role of T3SS effectors in *Salmonella* evasion of the immune system

Effector	T3SS	Interaction with the immune system	References
AvrA	1	Inhibits NF-κB and JNK pro-inflammatory pathways	Collier-Hyams *et al*. (2002), Jones *et al*. (2008b)
CigR	2	Involved in biofilm formation	Yin *et al*. (2016)
GogA		Inhibits inflammation by targeting NF-κB signalling	Sun *et al*. (2016)
GogB	2	Limits inflammation preventing NF-κB	Pilar *et al*. (2012)
GtgA	2	Inhibits inflammation by targeting NF-κB signalling	Sun *et al*. (2016)
GtgE	1/2	Counters a Rab32-dependent host defence pathway	Spanò *et al*. (2016)
PipA		Inhibits inflammation by targeting NF-κB signalling	Sun *et al*. (2016)
PipB	2	Role in iNOS production in chicken oviduct epithelial cells	Li *et al*. (2009)
PrgJ	?	Activates caspase-1 through NLRC4	Miao *et al*. (2010)
PipB2	1/2	Inhibits dendritic cell migration	McLaughlin *et al*. (2014)
SifA	2	Inhibits dendritic cell migration	McLaughlin *et al*. (2014)
SipA/ SspA	1	Elicits accumulation of PERP to the apical surface of colonic epithelial cells that leads to inflammatory responses	Hallstrom *et al*. (2015)
SipB	1	Contributes to activation and release of IL-18	Dreher *et al*. (2002)
SipC/ SspC	1	Elicits accumulation of PERP to the apical surface of colonic epithelial cells that leads to inflammatory responses	Hallstrom and McCormick (2016)
SipD	1	Triggers cell death in macrophages	Arizmendi *et al*. (2016)
SlrP	1/2	Inhibits dendritic cell migration	McLaughlin *et al*. (2014)
SopA	1	Stimulates inflammation targeting TRIM56 and TRIM65	Kamanova *et al*. (2016)
SopB	1	Inhibits production of mitochondrial superoxide ROS Activates MAPK and NF-κB signalling through stimulation of Rho-family GTPases	Bruno *et al*. (2009), Ruan *et al*. (2016)
SopD	1/2?	Contributes to PMN migration and fluid secretion in bovine intestine	Zhang *et al*. (2002)
SopD2	2	Counters a Rab32-dependent host defence pathway	Spanò *et al*. (2016)
SopE	1	Induces caspase-1 dependent pro-inflammatory responses. Activates MAPK and NF-κB signalling through stimulation of Rho-family GTPases	Bruno *et al*. (2009), Hoffmann *et al*. (2010), Müller *et al*. (2009)
SopE2	1	Activates MAPK and NF-κB signalling through stimulation of Rho-family GTPases	Bruno *et al*. (2009)
SpiC/ SsaB	2	Inhibits dendritic cell migration	McLaughlin *et al*. (2014)
SptP	1	Inhibits p38 MAPK signalling. Suppresses degranulation of local mast cells	Choi *et al*. (2013), Tenor *et al*. (2004)

Table 6.1 Continued

Effector	T3SS	Interaction with the immune system	References
SpvB	2	Contributes to gut inflammation through a T3SS2-dependent pathway	Käppeli *et al.* (2011)
SpvC	1/2	Anti-inflammatory in mice. Interferes with immunity in plants	Haneda *et al.* (2012); Neumann *et al.* (2014)
SpvD	1/2	Inhibits nuclear transport of NF-κB p65	Rolhion *et al.* (2016)
SseF	2	Inhibits dendritic cell migration	McLaughlin *et al.* (2014)
SseI/ SrfH	2	Modulates cell migration of macrophages and dendritic cells	McLaughlin *et al.* (2009), Worley *et al.* (2006)
SseK1	1/2	Interferes with NF-κB activation	Li *et al.* (2013)
SseK3	2	Interferes with NF-κB activation	Yang *et al.* (2015)
SseL	2	Inhibits autophagy	Mesquita *et al.* (2012)
SspH1	1/2	Inhibits NF-κB-dependent gene expression	Haraga and Miller (2003), Haraga *et al.* (2006)
SspH2	2	Enhance SG1-dependent NLR-mediated immunity	Bhavsar *et al.* (2013), McLaughlin *et al.* (2014)
		Inhibits dendritic cell migration	
SteA	1/2	Represses genes involved in immune responses	Cardenal-Muñoz *et al.* (2014)
SteB		Biofilm formation	Dong *et al.* (2008), McGhie *et al.* (2009)

develop the disease. Their observations also indicate that *S.* Typhi-specific $CD8^+$ T migrated not only to mucosal sites, but also to secondary lymphoid tissues. These localized events, delay disease onset. They also suggested that co-production of MIP-1β with other cytokines is a key component in protection against *S.* Typhi (Fresnay *et al.*, 2016). This type of studies is key to identify an early selection of novel vaccine candidates for further evaluation in clinical trials. These models are also necessary to understand the role of certain T3SS effectors that may only be relevant in certain cell types or animal species.

Advances in techniques for single cell analysis allow us to study bacterial subpopulations that have different expression profiles of effectors and other virulence factors.

Finally, a more precise understanding of the interactions between *Salmonella* and the host immune system could help us design new approaches to modify the progression of *Salmonella* infections, strengthening key components of the immune response.

Acknowledgements

We thank Sebastián Aguilar for useful discussions and critical reading of the manuscript. Work in the laboratory of FR-M is supported by Ministerio de Economía y Competitividad of Spain and the European Regional Development Fund (grants SAF2013-46229-R and SAF2016-75365-R).

References

Aachoui, Y., Leaf, I.A., Hagar, J.A., Fontana, M.F., Campos, C.G., Zak, D.E., Tan, M.H., Cotter, P.A., Vance, R.E., Aderem, A., *et al.* (2013). Caspase-11 protects against bacteria that escape the vacuole. Science *339*, 975–978. https://doi.org/10.1126/science.1230751.

Alaniz, R.C., Cummings, L.A., Bergman, M.A., Rassoulian-Barrett, S.L., and Cookson, B.T. (2006). *Salmonella* Typhimurium coordinately regulates FliC location and reduces dendritic cell activation and antigen presentation to CD4+ T cells. J. Immunol. *177*, 3983–3993.

Albaghdadi, H., Robinson, N., Finlay, B., Krishnan, L., and Sad, S. (2009). Selectively reduced intracellular proliferation of *Salmonella enterica* serovar Typhimurium within APCs limits antigen presentation and development of a rapid CD8 T cell response. J. Immunol. *183*, 3778–3787.

Arabyan, N., Park, D., Foutouhi, S., Weis, A.M., Huang, B.C., Williams, C.C., Desai, P., Shah, J., Jeannotte, R., Kong, N., *et al.* (2016). *Salmonella* degrades the host glycocalyx leading to altered infection and glycan remodeling. Sci. Rep. *6*, 29525. https://doi.org/10.1038/srep29525.

Arizmendi, O., Picking, W.D., and Picking, W.L. (2016). Macrophage apoptosis triggered by IpaD from *Shigella flexneri*. Infect. Immun. *84*, 1857–1865. https://doi.org/10.1128/IAI.01483-15.

Audia, J.P., Webb, C.C., and Foster, J.W. (2001). Breaking through the acid barrier: an orchestrated response to proton stress by enteric bacteria. Int. J. Med. Microbiol. *291*, 97–106.

Bäckhed, F., Ley, R.E., Sonnenburg, J.L., Peterson, D.A., and Gordon, J.I. (2005). Host-bacterial mutualism in the human intestine. Science *307*, 1915–1920.

Bader, M.W., Navarre, W.W., Shiau, W., Nikaido, H., Frye, J.G., McClelland, M., Fang, F.C., and Miller, S.I. (2003). Regulation of *Salmonella* Typhimurium virulence gene expression by cationic antimicrobial peptides. Mol. Microbiol. *50*, 219–230.

Bader, M.W., Sanowar, S., Daley, M.E., Schneider, A.R., Cho, U., Xu, W., Klevit, R.E., Le Moual, H., and Miller, S.I. (2005). Recognition of antimicrobial peptides by a bacterial sensor kinase. Cell *122*, 461–472.

Baisón-Olmo, F., Galindo-Moreno, M., and Ramos-Morales, F. (2015). Host cell type-dependent translocation and PhoP-mediated positive regulation of the effector SseK1 of *Salmonella enterica*. Front. Microbiol. *6*, 396.

Barr, T.A., Brown, S., Mastroeni, P., and Gray, D. (2010). TLR and B cell receptor signals to B cells differentially program primary and memory Th1 responses to *Salmonella enterica*. J. Immunol. *185*, 2783–2789. https://doi.org/10.4049/jimmunol.1001431.

Baucheron, S., Nishino, K., Monchaux, I., Canepa, S., Maurel, M.-C., Coste, F., Roussel, A., Cloeckaert, A., and Giraud, E. (2014). Bile-mediated activation of the *acrAB* and *tolC* multidrug efflux genes occurs mainly through transcriptional derepression of *ramA* in *Salmonella enterica* serovar Typhimurium. J. Antimicrob. Chemother. *69*, 2400–2406.

Begley, M., Gahan, C.G., and Hill, C. (2005). The interaction between bacteria and bile. FEMS Microbiol. Rev. *29*, 625–651.

Beuzón, C.R., Méresse, S., Unsworth, K.E., Ruíz-Albert, J., Garvis, S., Waterman, S.R., Ryder, T.A., Boucrot, E., and Holden, D.W. (2000). *Salmonella* maintains the integrity of its intracellular vacuole through the action of SifA. EMBO J. *19*, 3235–3249. https://doi.org/10.1093/emboj/19.13.3235.

Bhavsar, A.P., Brown, N.F., Stoepel, J., Wiermer, M., Martin, D.D., Hsu, K.J., Imami, K., Ross, C.J., Hayden, M.R., Foster, L.J., *et al.* (2013). The *Salmonella* type III effector SspH2 specifically exploits the NLR co-chaperone activity of SGT1 to subvert immunity. PLOS Pathog. *9*, e1003518. https://doi.org/10.1371/journal.ppat.1003518.

Bigger, J. (1944). Treatment of staphylococcal infections with penicillin by intermittent sterilisation. Lancet *244*, 497–500.

Birmingham, C.L., and Brumell, J.H. (2006). Autophagy recognizes intracellular *Salmonella enterica* serovar Typhimurium in damaged vacuoles. Autophagy *2*, 156–158.

Birmingham, C.L., Smith, A.C., Bakowski, M.A., Yoshimori, T., and Brumell, J.H. (2006). Autophagy controls *Salmonella* infection in response to damage to the *Salmonella*-containing vacuole. J. Biol. Chem. *281*, 11374–11383.

Bishop, R.E., Gibbons, H.S., Guina, T., Trent, M.S., Miller, S.I., and Raetz, C.R. (2000). Transfer of palmitate from phospholipids to lipid A in outer membranes of gram-negative bacteria. EMBO J. *19*, 5071–5080. https://doi.org/10.1093/emboj/19.19.5071.

Blancou, P., Tardif, V., Simon, T., Rémy, S., Carreño, L., Kalergis, A., and Anegon, I. (2011). Immunoregulatory properties of heme oxygenase-1. Methods Mol. Biol. *677*, 247–268. https://doi.org/10.1007/978-1-60761-869-0_18.

Boucrot, E., Beuzón, C.R., Holden, D.W., Gorvel, J.P., and Méresse, S. (2003). *Salmonella* Typhimurium SifA effector protein requires its membrane-anchoring C-terminal hexapeptide for its biological function. J. Biol. Chem. *278*, 14196–14202.

Brodsky, I.E., Ghori, N., Falkow, S., and Monack, D. (2005). Mig-14 is an inner membrane-associated protein that promotes *Salmonella* Typhimurium resistance to CRAMP, survival within activated macrophages and persistent infection. Mol. Microbiol. *55*, 954–972.

Broz, P., and Monack, D.M. (2013a). Noncanonical inflammasomes: caspase-11 activation and effector mechanisms. PLOS Pathog. *9*, e1003144. https://doi.org/10.1371/journal.ppat.1003144.

Broz, P., and Monack, D.M. (2013b). Newly described pattern recognition receptors team up against intracellular pathogens. Nat. Rev. Immunol. *13*, 551–565. https://doi.org/10.1038/nri3479.

Broz, P., Newton, K., Lamkanfi, M., Mariathasan, S., Dixit, V.M., and Monack, D.M. (2010). Redundant roles for inflammasome receptors NLRP3 and NLRC4 in host defense against *Salmonella*. J. Exp. Med. *207*, 1745–1755. https://doi.org/10.1084/jem.20100257.

Broz, P., Ruby, T., Belhocine, K., Bouley, D.M., Kayagaki, N., Dixit, V.M., and Monack, D.M. (2012). Caspase-11 increases susceptibility to *Salmonella* infection in the absence of caspase-1. Nature *490*, 288–291. https://doi.org/10.1038/nature11419.

Bruno, V.M., Hannemann, S., Lara-Tejero, M., Flavell, R.A., Kleinstein, S.H., Galán, J.E., H., S., Bruno, V.M., and Hannemann, S. (2009). *Salmonella* Typhimurium type III secretion effectors stimulate innate immune responses in cultured epithelial cells. PLOS Pathog. *5*, e1000538.

Bueno, S.M., González, P.A., Carreño, L.J., Tobar, J.A., Mora, G.C., Pereda, C.J., Salazar-Onfray, F., and Kalergis, A.M. (2008). The capacity of *Salmonella* to survive inside dendritic cells and prevent antigen presentation to T cells is host specific. Immunology *124*, 522–533.

Bueno, S.M., Wozniak, A., Leiva, E.D., Riquelme, S.A., Carreño, L.J., Hardt, W.-D., Riedel, C.A., and Kalergis, A.M. (2010). *Salmonella* pathogenicity island 1 differentially modulates bacterial entry to dendritic and non-phagocytic cells. Immunology *130*, 273–287.

Cardenal-Muñoz, E., Gutiérrez, G., and Ramos-Morales, F. (2014). Global impact of *Salmonella* type III secretion effector SteA on host cells. Biochem. Biophys. Res. Commun. *449*, 419–424. https://doi.org/10.1016/j.bbrc.2014.05.056.

Casson, C.N., Yu, J., Reyes, V.M., Taschuk, F.O., Yadav, A., Copenhaver, A.M., Nguyen, H.T., Collman, R.G., and Shin, S. (2015). Human caspase-4 mediates noncanonical inflammasome activation against gram-negative bacterial pathogens. Proc. Natl. Acad. Sci. U. S. A. *112*, 6688–6693.

Castro-Eguiluz, D., Pelayo, R., Rosales-Garcia, V., Rosales-Reyes, R., Alpuche-Aranda, C., and Ortiz-Navarrete, V. (2009). B cell precursors are targets for *Salmonella* infection. Microb. Pathog. *47*, 52–56. https://doi.org/10.1016/j.micpath.2009.04.005.

Chemnitz, J.M., Parry, R.V., Nichols, K.E., June, C.H., and Riley, J.L. (2004). SHP-1 and SHP-2 associate with immunoreceptor tyrosine-based switch motif of programmed death 1 upon primary human T cell stimulation, but only receptor ligation prevents T cell activation. J. Immunol. *173*, 945–954.

Cheverton, A.M., Gollan, B., Przydacz, M., Wong, C.T., Mylona, A., Hare, S.A., and Helaine, S. (2016). A *Salmonella* toxin promotes persister formation through acetylation of tRNA. Mol. Cell *63*, 86–96. https://doi.org/10.1016/j.molcel.2016.05.002.

Choi, H.W., Brooking-Dixon, R., Neupane, S., Lee, C.-J., Miao, E.A., Staats, H.F., and Abraham, S.N. (2013). *Salmonella* Typhimurium impedes innate immunity with a mast-cell-suppressing protein tyrosine phosphatase, SptP. Immunity *39*, 1108–1120.

Chu, Y., Gao, S., Wang, T., Yan, J., Xu, G., Li, Y., Niu, H., Huang, R., and Wu, S. (2016). A novel contribution of *spvB* to pathogenesis of *Salmonella* Typhimurium by inhibiting autophagy in host cells. Oncotarget *7*, 8295–8309. https://doi.org/10.18632/oncotarget.6989.

Chung, S.W., Liu, X., Macias, A.A., Baron, R.M., and Perrella, M.A. (2008). Heme oxygenase-1-derived carbon monoxide enhances the host defense response to microbial sepsis in mice. J. Clin. Invest. *118*, 239–247. https://doi.org/10.1172/JCI32730.

Coleman, R. (1987). Biochemistry of bile secretion. Biochem. J. *244*, 249–261.

Collier-Hyams, L.S., Zeng, H., Sun, J., Tomlinson, A.D., Bao, Z.Q., Chen, H., Madara, J.L., Orth, K., and Neish, A.S. (2002). Cutting edge: *Salmonella* AvrA effector inhibits the key proinflammatory, anti-apoptotic NF-kappa B pathway. J. Immunol. *169*, 2846–2850.

Costerton, J.W., Stewart, P.S., and Greenberg, E.P. (1999). Bacterial biofilms: a common cause of persistent infections. Science *284*, 1318–1322.

Crawford, R.W., Gibson, D.L., Kay, W.W., and Gunn, J.S. (2008). Identification of a bile-Induced exopolysaccharide required for *Salmonella* biofilm formation on gallstone surfaces. Infect. Immun. *76*, 5341–5349.

Crawford, R.W., Rosales-Reyes, R., Ramírez-Aguilar, Mde. L., Chapa-Azuela, O., Alpuche-Aranda, C., and Gunn, J.S. (2010). Gallstones play a significant role in *Salmonella* spp. gallbladder colonization and carriage. Proc. Natl. Acad. Sci. U.S.A. *107*, 4353–4358. https://doi.org/10.1073/pnas.1000862107.

Crawford, R.W., Keestra, A.M., Winter, S.E., Xavier, M.N., Tsolis, R.M., Tolstikov, V., and Bäumler, A.J. (2012). Very long O-antigen chains enhance fitness during *Salmonella*-induced colitis by increasing bile resistance. PLOS Pathog. *8*, e1002918. https://doi.org/10.1371/journal.ppat.1002918.

Crawford, R.W., Wangdi, T., Spees, A.M., Xavier, M.N., Tsolis, R.M., and Bäumler, A.J. (2013). Loss of very-long O-antigen chains optimizes capsule-mediated immune evasion by *Salmonella enterica* serovar Typhi. MBio *4*, 1–8.

Cummings, L.A., Wilkerson, W.D., Bergsbaken, T., and Cookson, B.T. (2006). *In vivo, fliC* expression by *Salmonella enterica* serovar Typhimurium is heterogeneous, regulated by ClpX, and anatomically restricted. Mol. Microbiol. *61*, 795–809.

Cunningham, A.F., Gaspal, F., Serre, K., Mohr, E., Henderson, I.R., Scott-Tucker, A., Kenny, S.M., Khan, M., Toellner, K.-M., Lane, P.J.L., et al. (2007). *Salmonella* induces a switched antibody response without germinal centers that impedes the extracellular spread of infection. J. Immunol. *178*, 6200–6207.

D'Costa, V.M., Braun, V., Landekic, M., Shi, R., Proteau, A., McDonald, L., Cygler, M., Grinstein, S., and Brumell, J.H. (2015). *Salmonella* disrupts host endocytic trafficking by SopD2-mediated inhibition of Rab7. Cell Rep. *12*, 1508–1518.

Dalebroux, Z.D., Matamouros, S., Whittington, D., Bishop, R.E., and Miller, S.I. (2014). PhoPQ regulates acidic glycerophospholipid content of the *Salmonella* Typhimurium outer membrane. Proc. Natl. Acad. Sci. U. S. A. *111*, 1963–1968.

Delves, P.J., Martin, S.J., Burton, D.R., and Roitt, I.M. (2011). Roitt's essential immunology.

Detweiler, C.S., Monack, D.M., Brodsky, I.E., Mathew, H., and Falkow, S. (2003). *virK, somA* and *rcsC* are important for systemic *Salmonella enterica* serovar Typhimurium infection and cationic peptide resistance. Mol. Microbiol. *48*, 385–400.

Dong, H., Zhang, X., Pan, Z., Peng, D., and Liu, X. (2008). Identification of genes for biofilm formation in a *Salmonella enteritidis* strain by transposon mutagenesis. Wei Sheng Wu Xue Bao *48*, 869–873.

Dreher, D., Kok, M., Obregon, C., Kiama, S.G., Gehr, P., and Nicod, L.P. (2002). *Salmonella* virulence factor SipB induces activation and release of IL-18 in human dendritic cells. J. Leukoc. Biol. *72*, 743–751.

Du, F., and Galán, J.E. (2009). Selective inhibition of type III secretion activated signaling by the *Salmonella* effector AvrA. PLOS Pathog. *5*, e1000595. https://doi.org/10.1371/journal.ppat.1000595.

Eckburg, P.B., Bik, E.M., Bernstein, C.N., Purdom, E., Dethlefsen, L., Sargent, M., Gill, S.R., Nelson, K.E., and Relman, D.A. (2005). Diversity of the human intestinal microbial flora. Science *308*, 1635–1638.

Eisele, N.A., Ruby, T., Jacobson, A., Manzanillo, P.S., Cox, J.S., Lam, L., Mukundan, L., Chawla, A., and Monack, D.M. (2013). *Salmonella* require the fatty acid regulator PPARδ for the establishment of a metabolic environment essential for long-term persistence. Cell Host Microbe *14*, 171–182. https://doi.org/10.1016/j.chom.2013.07.010.

Erickson, K.D., and Detweiler, C.S. (2006). The Rcs phosphorelay system is specific to enteric pathogens/commensals and activates *ydeI*, a gene important for persistent *Salmonella* infection of mice. Mol. Microbiol. *62*, 883–894.

Ertelt, J.M., Johanns, T.M., Mysz, M.A., Nanton, M.R., Rowe, J.H., Aguilera, M.N., and Way, S.S. (2011). Selective culling of high avidity antigen-specific CD4+ T cells after virulent *Salmonella* infection. Immunology *134*, 487–497.

Farizano, J.V., Pescaretti, M. de las M., López, F.E., Hsu, F.-F., and Delgado, M.A. (2012). The PmrAB system-inducing conditions control both lipid A remodeling and O-antigen length distribution, influencing the *Salmonella* Typhimurium-host interactions. J. Biol. Chem. *287*, 38778–38789.

Farris, C., Sanowar, S., Bader, M.W., Pfuetzner, R., and Miller, S.I. (2010). Antimicrobial peptides activate the Rcs regulon through the outer membrane lipoprotein RcsF. J. Bacteriol. *192*, 4894–4903. https://doi.org/10.1128/JB.00505-10.

Fillatreau, S. (2011). Novel regulatory functions for Toll-like receptor-activated B cells during intracellular bacterial infection. Immunol. Rev. *240*, 52–71. https://doi.org/10.1111/j.1600-065X.2010.00991.x.

Fillatreau, S., Gray, D., and Anderton, S.M. (2008). Not always the bad guys: B cells as regulators of autoimmune pathology. Nat. Rev. Immunol. *8*, 391–397. https://doi.org/10.1038/nri2315.

Fink, S.L., and Cookson, B.T. (2007). Pyroptosis and host cell death responses during *Salmonella* infection. Cell. Microbiol. *9*, 2562–2570.

Foster, J.W. (1993). The acid tolerance response of *Salmonella* Typhimurium involves transient synthesis of key acid shock proteins. J. Bacteriol. *175*, 1981–1987.
Freeman, G.J., Long, A.J., Iwai, Y., Bourque, K., Chernova, T., Nishimura, H., Fitz, L.J., Malenkovich, N., Okazaki, T., Byrne, M.C., et al. (2000). Engagement of the PD-1 immunoinhibitory receptor by a novel B7 family member leads to negative regulation of lymphocyte activation. J. Exp. Med. *192*, 1027–1034.
Fresnay, S., McArthur, M.A., Magder, L., Darton, T.C., Jones, C., Waddington, C.S., Blohmke, C.J., Angus, B., Levine, M.M., Pollard, A.J., *et al.* (2016). *Salmonella* Typhi-specific multifunctional CD8+ T cells play a dominant role in protection from typhoid fever in humans. J. Transl. Med. *14*, 62. https://doi.org/10.1186/s12967-016-0819-7.
Fu, Y., and Galán, J.E. (1999). A *Salmonella* protein antagonizes Rac-1 and Cdc42 to mediate host-cell recovery after bacterial invasion. Nature *401*, 293–297. https://doi.org/10.1038/45829.
Galán, J.E., and Curtiss, R. (1989). Cloning and molecular characterization of genes whose products allow *Salmonella* Typhimurium to penetrate tissue culture cells. Proc. Natl. Acad. Sci. U. S.A. *86*, 6383–6387.
Garai, P., Gnanadhas, D.P., and Chakravortty, D. (2012). *Salmonella enterica* serovars Typhimurium and Typhi as model organisms: Revealing paradigm of host-pathogen interactions. Virulence *3*, 377–388.
Gerlach, R.G., and Hensel, M. (2007). *Salmonella* pathogenicity islands in host specificity, host pathogen-interactions and antibiotics resistance of *Salmonella enterica*. Berliner Und Münchener Tierärztliche Wochenschrift *120*, 317–327.
Gomes, L.C., and Dikic, I. (2014). Autophagy in antimicrobial immunity. Mol. Cell *54*, 224–233. https://doi.org/10.1016/j.molcel.2014.03.009.
Gonzalez-Escobedo, G., Marshall, J.M., and Gunn, J.S. (2011). Chronic and acute infection of the gall bladder by *Salmonella* Typhi: understanding the carrier state. Nat. Rev. Microbiol. *9*, 9–14. https://doi.org/10.1038/nrmicro2490.
Guina, T., Yi, E.C., Wang, H., Hackett, M., and Miller, S.I. (2000). A PhoP-regulated outer membrane protease of *Salmonella enterica* serovar Typhimurium promotes resistance to alpha-helical antimicrobial peptides. J. Bacteriol. *182*, 4077–4086.
Gunn, J.S. (2000). Mechanisms of bacterial resistance and response to bile. Microbes Infect. *2*, 907–913.
Gunn, J.S., Lim, K.B., Krueger, J., Kim, K., Guo, L., Hackett, M., and Miller, S.I. (1998). PmrA-PmrB-regulated genes necessary for 4-aminoarabinose lipid A modification and polymyxin resistance. Mol. Microbiol. *27*, 1171–1182.
Guo, L., Lim, K.B., Poduje, C.M., Daniel, M., Gunn, J.S., Hackett, M., and Miller, S.I. (1998). Lipid A acylation and bacterial resistance against vertebrate antimicrobial peptides. Cell *95*, 189–198.
Halici, S., Zenk, S.F., Jantsch, J., and Hensel, M. (2008). Functional analysis of the *Salmonella* pathogenicity island 2-mediated inhibition of antigen presentation in dendritic cells. Infect. Immun. *76*, 4924–4933.
Hall-Stoodley, L., and Stoodley, P. (2009). Evolving concepts in biofilm infections. Cell. Microbiol. *11*, 1034–1043. https://doi.org/10.1111/j.1462-5822.2009.01323.x.
Hallstrom, K.N., and McCormick, B.A. (2016). The type three secreted effector SipC regulates the trafficking of PERP during *Salmonella* infection. Gut Microbes *7*, 136–145. https://doi.org/10.1080/19490976.2015.1128626.
Hallstrom, K.N., Srikanth, C.V., Agbor, T.A., Dumont, C.M., Peters, K.N., Paraoan, L., Casanova, J.E., Boll, E.J., and McCormick, B.A. (2015). PERP, a host tetraspanning membrane protein, is required for *Salmonella*-induced inflammation. Cell. Microbiol. *17*, 843–859. https://doi.org/10.1111/cmi.12406.
Haneda, T., Ishii, Y., Shimizu, H., Ohshima, K., Iida, N., Danbara, H., and Okada, N. (2012). *Salmonella* type III effector SpvC, a phosphothreonine lyase, contributes to reduction in inflammatory response during intestinal phase of infection. Cell. Microbiol. *14*, 485–499. https://doi.org/10.1111/j.1462-5822.2011.01733.x.
Haraga, A., and Miller, S.I. (2003). A *Salmonella enterica* serovar Typhimurium translocated leucine-rich repeat effector protein inhibits NF-kappa B-dependent gene expression. Infect. Immun. *71*, 4052–4058.
Haraga, A., and Miller, S.I. (2006). A *Salmonella* type III secretion effector interacts with the mammalian serine/threonine protein kinase PKN1. Cell. Microbiol. *8*, 837–846.
Helaine, S., and Kugelberg, E. (2014). Bacterial persisters: formation, eradication, and experimental systems. Trends Microbiol. *22*, 417–424. https://doi.org/10.1016/j.tim.2014.03.008.
Helaine, S., Thompson, J.A., Watson, K.G., Liu, M., Boyle, C., and Holden, D.W. (2010). Dynamics of intracellular bacterial replication at the single cell level. Proc. Natl. Acad. Sci. U.S.A. *107*, 3746–3751. https://doi.org/10.1073/pnas.1000041107.
Helaine, S., Cheverton, A.M., Watson, K.G., Faure, L.M., Matthews, S.A., and Holden, D.W. (2014). Internalization of *Salmonella* by macrophages induces formation of nonreplicating persisters. Science *343*, 204–208. https://doi.org/10.1126/science.1244705.

Hernandez, L.D., Hueffer, K., Wenk, M.R., and Galán, J.E. (2004). *Salmonella* modulates vesicular traffic by altering phosphoinositide metabolism. Science *30*, 1805–1807.

Hernández, S.B., Cota, I., Ducret, A., Aussel, L., and Casadesús, J. (2012). Adaptation and preadaptation of *Salmonella enterica* to Bile. PLOS Genet. *8*, e1002459. https://doi.org/10.1371/journal.pgen.1002459.

Hernández, S.B., Cava, F., Pucciarelli, M.G., García-Del Portillo, F., de Pedro, M.A., and Casadesús, J. (2015). Bile-induced peptidoglycan remodelling in *Salmonella enterica*. Environ. Microbiol. *17*, 1081–1089. https://doi.org/10.1111/1462-2920.12491.

Herrera, C.M., Hankins, J.V., and Trent, M.S. (2010). Activation of PmrA inhibits LpxT-dependent phosphorylation of lipid A promoting resistance to antimicrobial peptides. Mol. Microbiol. *76*, 1444–1460. https://doi.org/10.1111/j.1365-2958.2010.07150.x.

Heyns, K., and Kiessling, G. (1967). Strukturaufklarung des Vi-antigens aus *Citrobacter freundii* (*E. coli*) 5396/3. Carbohydr. Res. *3*, 340–353.

Hicks, K.G., Delbecq, S.P., Sancho-Vaello, E., Blanc, M.P., Dove, K.K., Prost, L.R., Daley, M.E., Zeth, K., Klevit, R.E., and Miller, S.I. (2015). Acidic pH and divalent cation sensing by PhoQ are dispensable for systemic salmonellae virulence. Elife *4*, e06792. https://doi.org/10.7554/eLife.06792.

Hochmann, H., Pust, S., von Figura, G., Aktories, K., and Barth, H. (2006). *Salmonella enterica* SpvB ADP-ribosylates actin at position arginine-177-characterization of the catalytic domain within the SpvB protein and a comparison to binary clostridial actin-ADP-ribosylating toxins. Biochemistry *45*, 1271–1277.

Hoffmann, C., Galle, M., Dilling, S., Käppeli, R., Müller, A.J., Songhet, P., Beyaert, R., and Hardt, W.D. (2010). In macrophages, caspase-1 activation by SopE and the type III secretion system-1 of S. Typhimurium can proceed in the absence of flagellin. PLOS ONE *5*, e12477. https://doi.org/10.1371/journal.pone.0012477.

Jantsch, J., Cheminay, C., Chakravortty, D., Lindig, T., Hein, J., and Hensel, M. (2003). Intracellular activities of *Salmonella enterica* in murine dendritic cells. Cell. Microbiol. *5*, 933–945.

Jensen, K.D.C., Wang, Y., Wojno, E.D.T., Shastri, A.J., Hu, K., Cornel, L., Boedec, E., Ong, Y.C., Chien, Y.H., Hunter, C.A., et al. (2011). Toxoplasma polymorphic effectors determine macrophage polarization and intestinal inflammation. Cell Host Microbe *9*, 472–483.

Jessen, D.L., Osei-Owusu, P., Toosky, M., Roughead, W., Bradley, D.S., and Nilles, M.L. (2014). Type III secretion needle proteins induce cell signaling and cytokine secretion via Toll-like receptors. Infect. Immun. *82*, 2300–2309. https://doi.org/10.1128/IAI.01705-14.

Jones, J.W., Shaffer, S.A., Ernst, R.K., Goodlett, D.R., and Turecek, F. (2008). Determination of pyrophosphorylated forms of lipid A in Gram-negative bacteria using a multivaried mass spectrometric approach. Proc. Natl. Acad. Sci. U.S.A. *105*, 12742–12747. https://doi.org/10.1073/pnas.0800445105.

Jones, R.M., Wu, H., Wentworth, C., Luo, L., Collier-Hyams, L., and Neish, A.S. (2008). *Salmonella* AvrA coordinates suppression of host immune and apoptotic defenses via JNK pathway blockade. Cell Host Microbe *3*, 233–244. https://doi.org/10.1016/j.chom.2008.02.016.

Kamanova, J., Sun, H., Lara-Tejero, M., and Galán, J.E. (2016). The *Salmonella* effector protein SopA modulates innate immune responses by targeting TRIM E3 ligase family members. PLOS Pathog. *12*, e1005552. https://doi.org/10.1371/journal.ppat.1005552.

Kaniga, K., Uralil, J., Bliska, J.B., and Galán, J.E. (1996). A secreted protein tyrosine phosphatase with modular effector domains in the bacterial pathogen *Salmonella* Typhimurium. Mol. Microbiol. *21*, 633–641.

Käppeli, R., Kaiser, P., Stecher, B., and Hardt, W.D. (2011). Roles of *spvB* and *spvC* in S. Typhimurium colitis via the alternative pathway. Int. J. Med. Microbiol. *301*, 117–124. https://doi.org/10.1016/j.ijmm.2010.08.017.

Karaki, K., and Matsubara, Y. (1984). Surgical treatment of chronic biliary typhoid and paratyphoid carriers. Nihon Shokakibyo Gakkai Zasshi *81*, 2978–2985.

Kato, A., and Groisman, E.A. (2004). Connecting two-component regulatory systems by a protein that protects a response regulator from dephosphorylation by its cognate sensor. Genes Dev. *18*, 2302–2313. https://doi.org/10.1101/gad.1230804.

Keestra, A.M., Winter, M.G., Auburger, J.J., Frässle, S.P., Xavier, M.N., Winter, S.E., Kim, A., Poon, V., Ravesloot, M.M., Waldenmaier, J.F., *et al.* (2013). Manipulation of small Rho GTPases is a pathogen-induced process detected by NOD1. Nature *496*, 233–237. https://doi.org/10.1038/nature12025.

Kenneth, M. (2011). Innate Immunity : The First Lines of Defense. Janeway's Immunobiol. 37–73.

Khan, C.M. (2014). The dynamic interactions between *Salmonella* and the microbiota, within the challenging niche of the gastrointestinal tract. Int. Sch. Res. Notices *2014*, 846049. https://doi.org/10.1155/2014/846049.

Kim, Y.K., Shin, J.-S., Nahm, M.H., Janeway, C., Medzhitov, R., Lemaitre, B., Nicolas, E., Michaut, L., Reichhart, J., Hoffmann, J., et al. (2016). NOD-Like receptors in infection, immunity, and diseases. Yonsei Med. J. *57*, 5.

Knodler, L.A. (2015). *Salmonella enterica*: living a double life in epithelial cells. Curr. Opin. Microbiol. *23*, 23–31. https://doi.org/10.1016/j.mib.2014.10.010.

Knodler, L.A., Vallance, B.A., Celli, J., Winfree, S., Hansen, B., Montero, M., and Steele-Mortimer, O. (2010). Dissemination of invasive *Salmonella* via bacterial-induced extrusion of mucosal epithelia. Proc. Natl. Acad. Sci. U.S.A. *107*, 17733–17738. https://doi.org/10.1073/pnas.1006098107.

Knodler, L.A., Nair, V., Steele-Mortimer, O., Voth, D., Heinzen, R., Vergne, I., Chua, J., Singh, S., Deretic, V., Celli, J., et al. (2014). Quantitative assessment of cytosolic *Salmonella* in epithelial cells. PLOS ONE *9*, e84681.

Koutsoumanis, K.P., and Sofos, J.N. (2004). Comparative acid stress response of *Listeria monocytogenes*, *Escherichia coli* O157:H7 and *Salmonella* Typhimurium after habituation at different pH conditions. Lett. Appl. Microbiol. *38*, 321–326.

Kreibich, S., Emmenlauer, M., Fredlund, J., Rämö, P., Münz, C., Dehio, C., Enninga, J., and Hardt, W.D. (2015). Autophagy proteins promote repair of endosomal membranes damaged by the *Salmonella* Type Three Secretion System 1. Cell Host Microbe *18*, 527–537. https://doi.org/10.1016/j.chom.2015.10.015.

Kullas, A.L., McClelland, M., Yang, H.J., Tam, J.W., Torres, A., Porwollik, S., Mena, P., McPhee, J.B., Bogomolnaya, L., Andrews-Polymenis, H., *et al.* (2012). L-asparaginase II produced by *Salmonella* Typhimurium inhibits T cell responses and mediates virulence. Cell Host Microbe *12*, 791–798. https://doi.org/10.1016/j.chom.2012.10.018.

Lam, L.H., Monack, D.M., Parry, C., Hien, T., Dougan, G., White, N., Farrar, J., Levin, M., Black, R., Lanata, C., et al. (2014). Intraspecies competition for niches in the distal gut dictate transmission during persistent *Salmonella* infection. PLOS Pathog. *10*, e1004527.

Lapaque, N., Hutchinson, J.L., Jones, D.C., Méresse, S., Holden, D.W., Trowsdale, J., and Kelly, A.P. (2009). *Salmonella* regulates polyubiquitination and surface expression of MHC class II antigens. Proc. Natl. Acad. Sci. U.S.A. *106*, 14052–14057. https://doi.org/10.1073/pnas.0906735106.

Lawley, T.D., Chan, K., Thompson, L.J., Kim, C.C., Govoni, G.R., and Monack, D.M. (2006). Genome-wide screen for *Salmonella* genes required for long-term systemic infection of the mouse. PLOS Pathog. *2*, e11. https://doi.org/10.1371/journal.ppat.0020011.

Lawley, T.D., Bouley, D.M., Hoy, Y.E., Gerke, C., Relman, D.A., and Monack, D.M. (2008). Host transmission of *Salmonella enterica* serovar Typhimurium is controlled by virulence factors and indigenous intestinal microbiota. Infect. Immun. *76*, 403–416.

Le Negrate, G., Faustin, B., Welsh, K., Loeffler, M., Krajewska, M., Hasegawa, P., Mukherjee, S., Orth, K., Krajewski, S., Godzik, A., et al. (2008). *Salmonella* secreted factor L deubiquitinase of *Salmonella* Typhimurium inhibits NF-kappaB, suppresses IkappaBalpha ubiquitination and modulates innate immune responses. J. Immunol. *180*, 5045–5056.

Lee, T.S., and Chau, L.Y. (2002). Heme oxygenase-1 mediates the anti-inflammatory effect of interleukin-10 in mice. Nat. Med. *8*, 240–246. https://doi.org/10.1038/nm0302-240.

Lee, H., Hsu, F.-F., Turk, J., and Groisman, E.A. (2004). The PmrA-regulated *pmrC* gene mediates phosphoethanolamine modification of lipid A and polymyxin resistance in *Salmonella enterica*. J. Bacteriol. *186*, 4124–4133.

Lei, L., Wang, W., Xia, C., and Liu, F. (2016). *Salmonella* virulence factor SsrAB Regulated factor modulates inflammatory responses by enhancing the activation of NF-κB signaling pathway. J. Immunol. *196*, 792–802.

Lesnick, M.L., Reiner, N.E., Fierer, J., and Guiney, D.G. (2001). The *Salmonella spvB* virulence gene encodes an enzyme that ADP-ribosylates actin and destabilizes the cytoskeleton of eukaryotic cells. Mol. Microbiol. *39*, 1464–1470.

Levine, M.M., Black, R.E., and Lanata, C. (1982). Precise estimation of the numbers of chronic carriers of *Salmonella* Typhi in Santiago, Chile, an endemic area. J. Infect. Dis. *146*, 724–726.

Lewinsohn, D.A., Gold, M.C., and Lewinsohn, D.M. (2011). Views of immunology: effector T cells. Immunol. Rev. *240*, 25–39. https://doi.org/10.1111/j.1600-065X.2010.00997.x.

Li, S., Zhang, M.Z., Yan, L., Lillehoj, H., Pace, L.W., and Zhang, S. (2009). Induction of CXC chemokine messenger-RNA expression in chicken oviduct epithelial cells by *Salmonella enterica* serovar Enteritidis via the type three secretion system-1. Avian Dis. *53*, 396–404.

Li, S., Zhang, L., Yao, Q., Li, L., Dong, N., Rong, J., Gao, W., Ding, X., Sun, L., Chen, X., *et al.* (2013). Pathogen blocks host death receptor signalling by arginine GlcNAcylation of death domains. Nature *501*, 242–246. https://doi.org/10.1038/nature12436.

Li, Y., Wang, T., Gao, S., Xu, G., Niu, H., Huang, R., and Wu, S. (2016). *Salmonella* plasmid virulence gene *spvB* enhances bacterial virulence by inhibiting autophagy in a zebrafish infection model. Fish Shellfish Immunol. *49*, 252–259.

Liévin-Le Moal, V., and Servin, A.L. (2006). The front line of enteric host defense against unwelcome intrusion of harmful microorganisms: mucins, antimicrobial peptides, and microbiota. Clin. Microbiol. Rev. *19*, 315–337.

Lin, J., Lee, I.S., Frey, J., Slonczewski, J.L., and Foster, J.W. (1995). Comparative analysis of extreme acid survival in *Salmonella* Typhimurium, *Shigella flexneri*, and *Escherichia coli*. J. Bacteriol. *177*, 4097–4104.

Lin, S.L., Le, T.X., and Cowen, D.S. (2003). SptP, a *Salmonella* Typhimurium type III-secreted protein, inhibits the mitogen-activated protein kinase pathway by inhibiting Raf activation. Cell. Microbiol. *5*, 267–275.

Lobato-Márquez, D., Moreno-Córdoba, I., Figueroa, V., Díaz-Orejas, R., and García-del Portillo, F. (2015). Distinct type I and type II toxin-antitoxin modules control *Salmonella* lifestyle inside eukaryotic cells. Sci. Rep. *5*, 9374.

Looney, R.J., and Steigbigel, R.T. (1986). Role of the Vi antigen of *Salmonella* Typhi in resistance to host defense in vitro. J. Lab. Clin. Med. *108*, 506–516.

Lopez, C.A., Winter, S.E., Rivera-Chávez, F., Xavier, M.N., Poon, V., Nuccio, S.-P., Tsolis, R.M., and Bäumler, A.J. (2012). Phage-mediated acquisition of a type III secreted effector protein boosts growth of *Salmonella* by nitrate respiration. MBio 3.

López-Medina, M., Pérez-López, A., Alpuche-Aranda, C., and Ortiz-Navarrete, V. (2015a). *Salmonella* induces PD-L1 expression in B cells. Immunol. Lett. *167*, 131–140. https://doi.org/10.1016/j.imlet.2015.08.004.

López-Medina, M., Carrillo-Martín, I., Leyva-Rangel, J., Alpuche-Aranda, C., and Ortiz-Navarrete, V. (2015b). *Salmonella* impairs CD8 T cell response through PD-1: PD-L axis. Immunobiology *220*, 1369–1380. https://doi.org/10.1016/j.imbio.2015.07.005.

López-Montero, N., Ramos-Marquès, E., Risco, C., and García-Del Portillo, F. (2016). Intracellular *Salmonella* induces aggrephagy of host endomembranes in persistent infections. Autophagy *12*, 1–16.

Luu, R.A., Gurnani, K., Dudani, R., Kammara, R., van Faassen, H., Sirard, J.C., Krishnan, L., and Sad, S. (2006). Delayed expansion and contraction of CD8+ T cell response during infection with virulent *Salmonella* Typhimurium. J. Immunol. *177*, 1516–1525.

Ma, D., Cook, D.N., Alberti, M., Pon, N.G., Nikaido, H., and Hearst, J.E. (1993). Molecular cloning and characterization of *acrA* and *acrE* genes of *Escherichia coli*. J. Bacteriol. *175*, 6299–6313.

Ma, D., Cook, D.N., Alberti, M., Pon, N.G., Nikaido, H., and Hearst, J.E. (1995). Genes *acrA* and *acrB* encode a stress-induced efflux system of *Escherichia coli*. Mol. Microbiol. *16*, 45–55.

Maier, L., Vyas, R., Cordova, C.D., Lindsay, H., Schmidt, T.S., Brugiroux, S., Periaswamy, B., Bauer, R., Sturm, A., Schreiber, F., *et al.* (2013). Microbiota-derived hydrogen fuels *Salmonella* Typhimurium invasion of the gut ecosystem. Cell Host Microbe *14*, 641–651. https://doi.org/10.1016/j.chom.2013.11.002.

Maisonneuve, E., and Gerdes, K. (2014). Molecular mechanisms underlying bacterial persisters. Cell *157*, 539–548. https://doi.org/10.1016/j.cell.2014.02.050.

Malik-Kale, P., Winfree, S., Steele-Mortimer, O., Marlovits, T., Stebbins, C., Steele-Mortimer, O., Beuzon, C., Salcedo, S., Holden, D., Birmingham, C., et al. (2012). The bimodal lifestyle of intracellular *Salmonella* in epithelial cells: replication in the cytosol obscures defects in vacuolar replication. PLOS ONE 7, e38732.

Mallo, G.V., Espina, M., Smith, A.C., Terebiznik, M.R., Alemán, A., Finlay, B.B., Rameh, L.E., Grinstein, S., and Brumell, J.H. (2008). SopB promotes phosphatidylinositol 3-phosphate formation on *Salmonella* vacuoles by recruiting Rab5 and Vps34. J. Cell Biol. *182*, 741–752. https://doi.org/10.1083/jcb.200804131.

Marcobal, A., Southwick, A.M., Earle, K.A., and Sonnenburg, J.L. (2013). A refined palate: bacterial consumption of host glycans in the gut. Glycobiology 23, 1038–1046. https://doi.org/10.1093/glycob/cwt040.

Margarit, S.M., Davidson, W., Frego, L., and Stebbins, C.E. (2006). A steric antagonism of actin polymerization by a *Salmonella* virulence protein. Structure *14*, 1219–1229.

Masschalck, B., and Michiels, C.W. (2003). Antimicrobial properties of lysozyme in relation to foodborne vegetative bacteria. Crit. Rev. Microbiol. *29*, 191–214. https://doi.org/10.1080/713610448.

Mastroeni, P., Villarreal-Ramos, B., and Hormaeche, C.E. (1993). Adoptive transfer of immunity to oral challenge with virulent salmonellae in innately susceptible BALB/c mice requires both immune serum and T cells. Infect. Immun. *61*, 3981–3984.

Mastroeni, P., Simmons, C., Fowler, R., Hormaeche, C.E., and Dougan, G. (2000). Igh-6$^{-/-}$ (B-cell-deficient) mice fail to mount solid acquired resistance to oral challenge with virulent *Salmonella enterica* serovar Typhimurium and show impaired Th1 T-cell responses to *Salmonella* antigens. Infect. Immun. *68*, 46–53.

Matamouros, S., and Miller, S.I. (2015). *S.* Typhimurium strategies to resist killing by cationic antimicrobial peptides. Biochim. Biophys. Acta *1848*, 3021–3025.

Mazurkiewicz, P., Thomas, J., Thompson, J.A., Liu, M., Arbibe, L., Sansonetti, P., and Holden, D.W. (2008). SpvC is a *Salmonella* effector with phosphothreonine lyase activity on host mitogen-activated protein kinases. Mol. Microbiol. *67*, 1371–1383. https://doi.org/10.1111/j.1365-2958.2008.06134.x.

McConnell, E.L., Basit, A.W., and Murdan, S. (2008). Measurements of rat and mouse gastrointestinal pH, fluid and lymphoid tissue, and implications for in-vivo experiments. J. Pharm. Pharmacol. *60*, 63–70. https://doi.org/10.1211/jpp.60.1.0008.

McGhie, E.J., Brawn, L.C., Hume, P.J., Humphreys, D., and Koronakis, V. (2009). *Salmonella* takes control: effector-driven manipulation of the host. Curr. Opin. Microbiol. *12*, 117–124. https://doi.org/10.1016/j.mib.2008.12.001.

McGourty, K., Thurston, T.L., Matthews, S.A., Pinaud, L., Mota, L.J., and Holden, D.W. (2012). *Salmonella* Inhibits Retrograde Trafficking of Mannose-6-Phosphate Receptors and Lysosome Function. Science (80-.). *338*.

McLaughlin, L.M., Govoni, G.R., Gerke, C., Gopinath, S., Peng, K., Laidlaw, G., Chien, Y.H., Jeong, H.W., Li, Z., Brown, M.D., *et al.* (2009). The *Salmonella* SPI2 effector SseI mediates long-term systemic infection by modulating host cell migration. PLOS Pathog. *5*, e1000671. https://doi.org/10.1371/journal.ppat.1000671.

McLaughlin, L.M., Xu, H., Carden, S.E., Fisher, S., Reyes, M., Heilshorn, S.C., and Monack, D.M. (2014). A microfluidic-based genetic screen to identify microbial virulence factors that inhibit dendritic cell migration. Integr. Biol. (Camb). *6*, 438–449.

McSorley, S.J., and Jenkins, M.K. (2000). Antibody is required for protection against virulent but not attenuated *Salmonella enterica* serovar Typhimurium. Infect. Immun. *68*, 3344–3348.

Mesquita, F.S., Thomas, M., Sachse, M., Santos, A.J., Figueira, R., and Holden, D.W. (2012). The *Salmonella* deubiquitinase SseL inhibits selective autophagy of cytosolic aggregates. PLOS Pathog. *8*, e1002743. https://doi.org/10.1371/journal.ppat.1002743.

Meunier, E., Dick, M.S., Dreier, R.F., Schürmann, N., Kenzelmann Broz, D., Warming, S., Roose-Girma, M., Bumann, D., Kayagaki, N., Takeda, K., *et al.* (2014). Caspase-11 activation requires lysis of pathogen-containing vacuoles by IFN-induced GTPases. Nature *509*, 366–370. https://doi.org/10.1038/nature13157.

Miao, E.A., Mao, D.P., Yudkovsky, N., Bonneau, R., Lorang, C.G., Warren, S.E., Leaf, I.A., and Aderem, A. (2010a). Innate immune detection of the type III secretion apparatus through the NLRC4 inflammasome. Proc. Natl. Acad. Sci. U.S.A. *107*, 3076–3080. https://doi.org/10.1073/pnas.0913087107.

Miao, E.A., Leaf, I.A., Treuting, P.M., Mao, D.P., Dors, M., Sarkar, A., Warren, S.E., Wewers, M.D., and Aderem, A. (2010b). Caspase-1-induced pyroptosis is an innate immune effector mechanism against intracellular bacteria. Nat. Immunol. *11*, 1136–1142.

Mitterstiller, A.M., Haschka, D., Dichtl, S., Nairz, M., Demetz, E., Talasz, H., Soares, M.P., Einwallner, E., Esterbauer, H., Fang, F.C., *et al.* (2016). Heme oxygenase 1 controls early innate immune response of macrophages to *Salmonella* Typhimurium infection. Cell. Microbiol. *18*, 1374–1389. https://doi.org/10.1111/cmi.12578.

Mittrücker, H.W., Raupach, B., Köhler, A., and Kaufmann, S.H. (2000). Cutting edge: role of B lymphocytes in protective immunity against *Salmonella* Typhimurium infection. J. Immunol. *164*, 1648–1652.

Monack, D.M., Mueller, A., and Falkow, S. (2004a). Persistent bacterial infections: the interface of the pathogen and the host immune system. Nat. Rev. Microbiol. *2*, 747–765. https://doi.org/10.1038/nrmicro955.

Monack, D.M., Bouley, D.M., and Falkow, S. (2004b). *Salmonella* Typhimurium persists within macrophages in the mesenteric lymph nodes of chronically infected Nramp1+/+ mice and can be reactivated by IFNgamma neutralization. J. Exp. Med. *199*, 231–241.

Mukherjee, K., Parashuraman, S., Raje, M., and Mukhopadhyay, A. (2001). SopE acts as an Rab5-specific nucleotide exchange factor and recruits non-prenylated Rab5 on *Salmonella*-containing phagosomes to promote fusion with early endosomes. J. Biol. Chem. *276*, 23607–23615.

Müller, A.J., Hoffmann, C., Galle, M., Van Den Broeke, A., Heikenwalder, M., Falter, L., Misselwitz, B., Kremer, M., Beyaert, R., and Hardt, W.D. (2009). The *S.* Typhimurium effector SopE induces caspase-1 activation in stromal cells to initiate gut inflammation. Cell Host Microbe *6*, 125–136. https://doi.org/10.1016/j.chom.2009.07.007.

Murli, S., Watson, R.O., and Galán, J.E. (2001). Role of tyrosine kinases and the tyrosine phosphatase SptP in the interaction of *Salmonella* with host cells. Cell. Microbiol. *3*, 795–810.

Murshid, A., Borges, T.J., Lang, B.J., and Calderwood, S.K. (2016). The Scavenger Receptor SREC-I Cooperates with Toll-Like Receptors to Trigger Inflammatory Innate Immune Responses. Front. Immunol. *7*, 226. https://doi.org/10.3389/fimmu.2016.00226.

Neumann, C., Fraiture, M., Hernàndez-Reyes, C., Akum, F.N., Virlogeux-Payant, I., Chen, Y., Pateyron, S., Colcombet, J., Kogel, K.-H., Hirt, H., et al. (2014). The *Salmonella* effector protein SpvC, a phosphothreonine lyase is functional in plant cells. Front. Microbiol. *5*, 548.

Niedergang, F., Sirard, J.C., Blanc, C.T., and Kraehenbuhl, J.P. (2000). Entry and survival of *Salmonella* Typhimurium in dendritic cells and presentation of recombinant antigens do not require macrophage-specific virulence factors. Proc. Natl. Acad. Sci. U. S. A. *97*, 14650–14655.

Nikaido, E., Yamaguchi, A., and Nishino, K. (2008). AcrAB multidrug efflux pump regulation in *Salmonella enterica* serovar Typhimurium by RamA in response to environmental signals. J. Biol. Chem. *283*, 24245–24253. https://doi.org/10.1074/jbc.M804544200.

Nishino, K., Latifi, T., and Groisman, E.A. (2006). Virulence and drug resistance roles of multidrug efflux systems of *Salmonella enterica* serovar Typhimurium. Mol. Microbiol. *59*, 126–141.

Nix, R.N., Altschuler, S.E., Henson, P.M., and Detweiler, C.S. (2007). Hemophagocytic macrophages harbor *Salmonella enterica* during persistent infection. PLOS Pathog. *3*, e193.

O'Dea, E., and Hoffmann, A. (2010). The regulatory logic of the NF-kappaB signaling system. Cold Spring Harb. Perspect. Biol. *2*, a000216.

O'Donnell, H., and McSorley, S.J. (2014). *Salmonella* as a model for non-cognate Th1 cell stimulation. Front. Immunol. *5*, 1–13.

Ochman, H., Soncini, F.C., Solomon, F., and Groisman, E.A. (1996). Identification of a pathogenicity island required for *Salmonella* survival in host cells. Proc. Natl. Acad. Sci. U.S.A. *93*, 7800–7804.

Onyiah, J.C., Sheikh, S.Z., Maharshak, N., Steinbach, E.C., Russo, S.M., Kobayashi, T., Mackey, L.C., Hansen, J.J., Moeser, A.J., Rawls, J.F., et al. (2013). Carbon monoxide and heme oxygenase-1 prevent untestinal inflammation in mice by promoting bacterial clearance. Gastroenterology *144*, 789–798.

Otterbein, L.E., Bach, F.H., Alam, J., Soares, M., Tao Lu, H., Wysk, M., Davis, R.J., Flavell, R.A., and Choi, A.M. (2000). Carbon monoxide has anti-inflammatory effects involving the mitogen-activated protein kinase pathway. Nat. Med. *6*, 422–428. https://doi.org/10.1038/74680.

Owen, K.A., and Casanova, J.E. (2015). *Salmonella* manipulates autophagy to 'serve and protect'. Cell Host Microbe *18*, 517–519. https://doi.org/10.1016/j.chom.2015.10.020.

Owen, K.A., Meyer, C.B., Bouton, A.H., and Casanova, J.E. (2014). Activation of focal adhesion kinase by *Salmonella* suppresses autophagy via an Akt/mTOR signaling pathway and promotes bacterial survival in macrophages. PLOS Pathog. *10*, e1004159.

Owen, K.A., Anderson, C.J., and Casanova, J.E. (2016). *Salmonella* suppresses the TRIF-dependent type I interferon response in macrophages. MBio 7, e02051-15. https://doi.org/10.1128/mBio.02051-15.

Parra-Lopez, C., Baer, M.T., and Groisman, E.A. (1993). Molecular genetic analysis of a locus required for resistance to antimicrobial peptides in *Salmonella* Typhimurium. EMBO J. *12*, 4053–4062.

Parra-Lopez, C., Lin, R., Aspedon, A., and Groisman, E.A. (1994). A *Salmonella* protein that is required for resistance to antimicrobial peptides and transport of potassium. EMBO J. *13*, 3964–3972.

Pescaretti, M. de L.M., López, F.E., Morero, R.D., and Delgado, M.A. (2011). The PmrA/PmrB regulatory system controls the expression of the *wzzfepE* gene involved in the O-antigen synthesis of *Salmonella enterica* serovar Typhimurium. Microbiology *157*, 2515–2521.

Petrovska, L., Aspinall, R.J., Barber, L., Clare, S., Simmons, C.P., Stratford, R., Khan, S.A., Lemoine, N.R., Frankel, G., Holden, D.W., et al. (2004). *Salmonella enterica* serovar Typhimurium interaction with dendritic cells: Impact of the sifA gene. Cell. Microbiol. *6*, 1071–1084.

Pilar, A.V.C., Reid-Yu, S. a, Cooper, C. a, Mulder, D.T., and Coombes, B.K. (2012). GogB is an anti-inflammatory effector that limits tissue damage during *Salmonella* infection through interaction with human FBXO22 and Skp1. PLOS Pathog. *8*, e1002773.

Prieto, A.I., Ramos-Morales, F., and Casadesús, J. (2004). Bile-induced DNA damage in *Salmonella enterica*. Genetics *168*, 1787–1794.

Prieto, A.I., Ramos-Morales, F., and Casadesús, J. (2006). Repair of DNA damage induced by bile salts in *Salmonella enterica*. Genetics *174*, 575–584.

Prouty, A.M., and Gunn, J.S. (2003). Comparative analysis of *Salmonella enterica* serovar Typhimurium biofilm formation on gallstones and on glass. Infect Immun *71*, 7154–7158.

Prouty, A.M., Schwesinger, W.H., and Gunn, J.S. (2002). Biofilm formation and interaction with the surfaces of gallstones by *Salmonella* spp. Infect. Immun. *70*, 2640–2649.

Prouty, A.M., Brodsky, I.E., Manos, J., Belas, R., Falkow, S., and Gunn, J.S. (2004). Transcriptional regulation of *Salmonella enterica* serovar Typhimurium genes by bile. FEMS Immunol. Med. Microbiol. *41*, 177–185. https://doi.org/10.1016/j.femsim.2004.03.002.

Qu, Y., Misaghi, S., Newton, K., Maltzman, A., Izrael-Tomasevic, A., Arnott, D., and Dixit, V.M. (2016). NLRP3 recruitment by NLRC4 during *Salmonella* infection. J. Exp. Med. *213*, 877–885. https://doi.org/10.1084/jem.20132234.

Radtke, A.L., Delbridge, L.M., Balachandran, S., Barber, G.N., and O'Riordan, M.X. (2007). TBK1 protects vacuolar integrity during intracellular bacterial infection. PLOS Pathog. *3*, e29.

Ramos-Morales, F. (2012a). Impact of *Salmonella enterica* type III secretion system effectors on the eukaryotic host cell. ISRN Cell Biol. *2012*, 1–36.

Ramos-Morales, F. (2012b). Acidic pH: enemy or ally for enteric bacteria? Virulence *3*, 103–106. https://doi.org/10.4161/viru.19382.

Ramos-Morales, F., Prieto, A.I., Beuzón, C.R., Holden, D.W., and Casadesús, J. (2003). Role for *Salmonella enterica* enterobacterial common antigen in bile resistance and virulence. J. Bacteriol. *185*, 5328–5332.

Ramu, P., Tanskanen, R., Holmberg, M., Lähteenmäki, K., Korhonen, T.K., and Meri, S. (2007). The surface protease PgtE of *Salmonella enterica* affects complement activity by proteolytically cleaving C3b, C4b and C5. FEBS Lett. *581*, 1716–1720.

Rathinam, V.A., and Fitzgerald, K.A. (2016). Inflammasome complexes: emerging mechanisms and effector functions. Cell *165*, 792–800. https://doi.org/10.1016/j.cell.2016.03.046.

Rémy, S., Blancou, P., Tesson, L., Tardif, V., Brion, R., Royer, P.J., Motterlini, R., Foresti, R., Painchaut, M., Pogu, S., *et al*. (2009). Carbon monoxide inhibits TLR-induced dendritic cell immunogenicity. J. Immunol. *182*, 1877–1884. https://doi.org/10.4049/jimmunol.0802436.

Riquelme, S.A., Bueno, S.M., and Kalergis, A.M. (2012). IgG keeps virulent *Salmonella* from evading dendritic cell uptake. Immunology *136*, 291–305. https://doi.org/10.1111/j.1365-2567.2012.03578.x.

Riquelme, S.A., Pogu, J., Anegon, I., Bueno, S.M., and Kalergis, A.M. (2015). Carbon monoxide impairs mitochondria-dependent endosomal maturation and antigen presentation in dendritic cells. Eur. J. Immunol. *45*, 3269–3288.

Riva, R., Korhonen, T.K., and Meri, S. (2015). The outer membrane protease PgtE of *Salmonella enterica* interferes with the alternative complement pathway by cleaving factors B and H. Front. Microbiol. *6*, 1–9.

Rolhion, N., Furniss, R.C., Grabe, G., Ryan, A., Liu, M., Matthews, S.A., and Holden, D.W. (2016). Inhibition of nuclear transport of NF-κB p65 by the *Salmonella* type III secretion system effector SpvD. PLOS Pathog. *12*, e1005653. https://doi.org/10.1371/journal.ppat.1005653.

Ruan, H., Zhang, Z., Tian, L., Wang, S., Hu, S., and Qiao, J.J. (2016). The *Salmonella* effector SopB prevents ROS-induced apoptosis of epithelial cells by retarding TRAF6 recruitment to mitochondria. Biochem. Biophys. Res. Commun. *478*, 618–623. https://doi.org/10.1016/j.bbrc.2016.07.116.

Russell, T.L., Berardi, R.R., Barnett, J.L., Dermentzoglou, L.C., Jarvenpaa, K.M., Schmaltz, S.P., and Dressman, J.B. (1993). Upper gastrointestinal pH in seventy-nine healthy, elderly, North American men and women. Pharm. Res. *10*, 187–196.

Ryan, D., Ojha, U.K., Jaiswal, S., Padhi, C., and Suar, M. (2016). The small RNA DsrA influences the acid tolerance response and virulence of *Salmonella enterica* serovar Typhimurium. Front. Microbiol. *7*, 599. https://doi.org/10.3389/fmicb.2016.00599.

Scherer, C.A., Cooper, E., and Miller, S.I. (2000). The *Salmonella* type III secretion translocon protein SspC is inserted into the epithelial cell plasma membrane upon infection. Mol. Microbiol. 37, 1133–1145.

Schiøler, H., Christiansen, E.D., Høybye, G., Rasmussen, S.N., and Greibe, J. (1983). Biliary calculi in chronic *Salmonella* carriers and healthy controls: a controlled study. Scand. J. Infect. Dis. *15*, 17–19.

Sharma, A., and Qadri, A. (2004). Vi polysaccharide of *Salmonella* Typhi targets the prohibitin family of molecules in intestinal epithelial cells and suppresses early inflammatory responses. Proc. Natl. Acad. Sci. U.S.A. *101*, 17492–17497.

Sharma, D., and Kanneganti, T.D. (2016). The cell biology of inflammasomes: Mechanisms of inflammasome activation and regulation. J. Cell Biol. *213*, 617–629. https://doi.org/10.1083/jcb.201602089.

Shea, J.E., Hensel, M., Gleeson, C., and Holden, D.W. (1996). Identification of a virulence locus encoding a second type III secretion system in *Salmonella* Typhimurium. Proc. Natl. Acad. Sci. U.S.A. *93*, 2593–2597.

Shen, P., Roch, T., Lampropoulou, V., O'Connor, R.A., Stervbo, U., Hilgenberg, E., Ries, S., Dang, V.D., Jaimes, Y., Daridon, C., *et al.* (2014). IL-35-producing B cells are critical regulators of immunity during autoimmune and infectious diseases. Nature *507*, 366–370. https://doi.org/10.1038/nature12979.

Sheppard, K.A., Fitz, L.J., Lee, J.M., Benander, C., George, J.A., Wooters, J., Qiu, Y., Jussif, J.M., Carter, L.L., Wood, C.R., *et al.* (2004). PD-1 inhibits T-cell receptor induced phosphorylation of the ZAP70/CD3zeta signalosome and downstream signaling to PKCtheta. FEBS Lett. *574*, 37–41. https://doi.org/10.1016/j.febslet.2004.07.083.

Shpargel, J.S., Berardi, R.S., and Lenz, D. (1985). *Salmonella* Typhi carrier state 52 years after illness with typhoid fever: a case study. Am. J. Infect. Control *13*, 122–123.

Sinnott, C.R., and Teall, A.J. (1987). Persistent gallbladder carriage of *Salmonella* Typhi. Lancet *1*, 976.

Søndberg, E., and Jelsbak, L. (2016). *Salmonella* Typhimurium undergoes distinct genetic adaption during chronic infections of mice. BMC Microbiol. *16*, 30.

Spanò, S., Gao, X., Hannemann, S., Lara-Tejero, M., and Galán, J.E. (2016). A bacterial pathogen targets a host Rab-family GTPase defense pathway with a GAP. Cell Host Microbe *19*, 216–226. https://doi.org/10.1016/j.chom.2016.01.004.

Srinivasan, A., Foley, J., Ravindran, R., and McSorley, S.J. (2004). Low-dose *Salmonella* infection evades activation of flagellin-specific CD4 T cells. J. Immunol. *173*, 4091–4099.

Srinivasan, A., Nanton, M., Griffin, A., and McSorley, S.J. (2009). Culling of activated CD4 T cells during typhoid is driven by *Salmonella* virulence genes. J. Immunol. *182*, 7838–7845. https://doi.org/10.4049/jimmunol.0900382.

Staib, L., and Fuchs, T.M. (2015). Regulation of fucose and 1,2-propanediol utilization by *Salmonella enterica* serovar Typhimurium. Front. Microbiol. *6*, 1116. https://doi.org/10.3389/fmicb.2015.01116.

Stecher, B., Robbiani, R., Walker, A.W., Westendorf, A.M., Barthel, M., Kremer, M., Chaffron, S., Macpherson, A.J., Buer, J., Parkhill, J., et al. (2007). *Salmonella enterica* serovar Typhimurium exploits inflammation to compete with the intestinal microbiota. PLOS Biol. *5*, 2177–2189.

Stecher, B., Denzler, R., Maier, L., Bernet, F., Sanders, M.J., Pickard, D.J., Barthel, M., Westendorf, A.M., Krogfelt, K.A., Walker, A.W., *et al.* (2012). Gut inflammation can boost horizontal gene transfer between pathogenic and commensal Enterobacteriaceae. Proc. Natl. Acad. Sci. U.S.A. *109*, 1269–1274. https://doi.org/10.1073/pnas.1113246109.

Steele-Mortimer, O., Brumell, J.H., Knodler, L.A., Méresse, S., Lopez, A., and Finlay, B.B. (2002). The invasion-associated type III secretion system of *Salmonella enterica* serovar Typhimurium is necessary for intracellular proliferation and vacuole biogenesis in epithelial cells. Cell. Microbiol. *4*, 43–54.

Storek, K.M., and Monack, D.M. (2015). Bacterial recognition pathways that lead to inflammasome activation. Immunol. Rev. *265*, 112–129. https://doi.org/10.1111/imr.12289.

Sun, H., Kamanova, J., Lara-Tejero, M., and Galán, J.E. (2016). A family of *Salmonella* type III secretion effector proteins selectively targets the NF-κB signaling pathway to preserve host homeostasis. PLOS Pathog. *12*, e1005484. https://doi.org/10.1371/journal.ppat.1005484.

Sundquist, M., and Wick, M.J. (2009). *Salmonella* induces death of CD8α^{+} dendritic cells but not CD11c(int)CD11b(+) inflammatory cells in vivo via MyD88 and TNFR1. J. Leukoc. Biol. *85*, 225–234. https://doi.org/10.1189/jlb.0708413.

Tardif, V., Riquelme, S.A., Remy, S., Carreño, L.J., Cortés, C.M., Simon, T., Hill, M., Louvet, C., Riedel, C.A., Blancou, P., et al. (2013). Carbon monoxide decreases endosome-lysosome fusion and inhibits soluble antigen presentation by dendritic cells to T cells. Eur. J. Immunol. *43*, 2832–2844.

Tenor, J.L., McCormick, B.A., Ausubel, F.M., and Aballay, A. (2004). *Caenorhabditis elegans*-based screen identifies *Salmonella* virulence factors required for conserved host-pathogen interactions. Curr. Biol. *14*, 1018–1024.

Tezcan-Merdol, D., Nyman, T., Lindberg, U., Haag, F., Koch-Nolte, F., and Rhen, M. (2001). Actin is ADP-ribosylated by the *Salmonella enterica* virulence-associated protein SpvB. Mol. Microbiol. *39*, 606–619.

Thiennimitr, P., Winter, S.E., Winter, M.G., Xavier, M.N., Tolstikov, V., Huseby, D.L., Sterzenbach, T., Tsolis, R.M., Roth, J.R., and Bäumler, A.J. (2011). Intestinal inflammation allows *Salmonella* to use ethanolamine to compete with the microbiota. Proc. Natl. Acad. Sci. U.S.A. *108*, 17480–17485. https://doi.org/10.1073/pnas.1107857108.

Thompson, L.J., Dunstan, S.J., Dolecek, C., Perkins, T., House, D., Dougan, G., Nguyen, T.D.H., Tran, T.H.T.P. La, Doan, C. Du, Le, T.P., et al. (2009). Transcriptional response in the peripheral blood of patients infected with *Salmonella enterica* serovar Typhi. Proc. Natl. Acad. Sci. U.S.A. *106*, 22433–22438.

Thurston, T.L., Boyle, K.B., Allen, M., Ravenhill, B.J., Karpiyevich, M., Bloor, S., Kaul, A., Noad, J., Foeglein, A., Matthews, S.A., et al. (2016). Recruitment of TBK1 to cytosol-invading *Salmonella* induces WIPI2-dependent antibacterial autophagy. EMBO J. *182*, e201694491.

Thurston, T.L., Ryzhakov, G., Bloor, S., von Muhlinen, N., and Randow, F. (2009). The TBK1 adaptor and autophagy receptor NDP52 restricts the proliferation of ubiquitin-coated bacteria. Nat. Immunol. *10*, 1215–1221. https://doi.org/10.1038/ni.1800.

Thurston, T.L., Wandel, M.P., von Muhlinen, N., Foeglein, A., and Randow, F. (2012). Galectin 8 targets damaged vesicles for autophagy to defend cells against bacterial invasion. Nature *482*, 414–418. https://doi.org/10.1038/nature10744.

Tindall, B.J., Grimont, P.A.D., Garrity, G.M., and Euzéby, J.P. (2005). Nomenclature and taxonomy of the genus *Salmonella*. Int. J. Syst. Evol. Microbiol. *55*, 521–524.

Tobar, J.A., Carreño, L.J., Bueno, S.M., González, P.A., Mora, J.E., Quezada, S.A., and Kalergis, A.M. (2006). Virulent *Salmonella enterica* serovar Typhimurium evades adaptive immunity by preventing dendritic cells from activating T cells. Infect. Immun. *74*, 6438–6448.

Tobar, J.A., González, P.A., and Kalergis, A.M. (2004). *Salmonella* escape from antigen presentation can be overcome by targeting bacteria to Fc gamma receptors on dendritic cells. J. Immunol. *173*, 4058–4065.

Tran, Q.T., Gomez, G., Khare, S., Lawhon, S.D., Raffatellu, M., Bäumler, A.J., Ajithdoss, D., Dhavala, S., and Adams, L.G. (2010). The *Salmonella enterica* serotype Typhi Vi capsular antigen is expressed after the bacterium enters the ileal mucosa. Infect. Immun. *78*, 527–535.

Ugrinovic, S., Ménager, N., Goh, N., and Mastroeni, P. (2003). Characterization and development of T-Cell immune responses in B-cell-deficient (Igh-6(-/-) mice with *Salmonella enterica* serovar Typhimurium infection. Infect. Immun. *71*, 6808–6819.

van der Velden, A.W., Copass, M.K., and Starnbach, M.N. (2005). *Salmonella* inhibit T cell proliferation by a direct, contact-dependent immunosuppressive effect. Proc. Natl. Acad. Sci. U.S.A. *102*, 17769–17774.

van der Velden, A.W., Dougherty, J.T., and Starnbach, M.N. (2008). Down-modulation of TCR expression by *Salmonella enterica* serovar Typhimurium. J. Immunol. *180*, 5569–5574.

Van Kesteren, M., Bibby, B.G., and Berry, G.P. (1942). Studies on the antibacterial factors of human saliva. J. Bacteriol. *43*, 573–583.

Vogt, S.L., Peña-Díaz, J., and Finlay, B.B. (2015). Chemical communication in the gut: Effects of microbiota-generated metabolites on gastrointestinal bacterial pathogens. Anaerobe *34*, 106–115.

van der Waaij, D., Berghuis-de Vries, J.M., and Lekkerkerk Lekkerkerk-v (1971). Colonization resistance of the digestive tract in conventional and antibiotic-treated mice. J. Hyg. (Lond). *69*, 405–411.

Watson, K.G., and Holden, D.W. (2010). Dynamics of growth and dissemination of *Salmonella in vivo*. Cell. Microbiol. *12*, 1389–1397. https://doi.org/10.1111/j.1462-5822.2010.01511.x.

Wehkamp, J., Schauber, J., and Stange, E.F. (2007). Defensins and cathelicidins in gastrointestinal infections. Curr. Opin. Gastroenterol. 23, 32–38. https://doi.org/10.1097/MOG.0b013e32801182c2.

Wei, L., Jin, Q., Chu, Y., Wu, S., and Huang, R. (2015). Suppression of dendritic cell and T-cell activation by the $pRST^{98}$ *Salmonella* plasmid. Mol. Med. Rep. *11*, 2306–2314. https://doi.org/10.3892/mmr.2014.2919.

Wilson, R.P., Winter, S.E., Spees, A.M., Winter, M.G., Nishimori, J.H., Sanchez, J.F., Nuccio, S.-P., Crawford, R.W., Tükel, Ç., and Bäumler, A.J. (2011). The Vi capsular polysaccharide prevents complement receptor 3-mediated clearance of *Salmonella enterica* serotype Typhi. Infect. Immun. 79, 830–837.

Winter, S.E., Thiennimitr, P., Winter, M.G., Butler, B.P., Huseby, D.L., Crawford, R.W., Russell, J.M., Bevins, C.L., Adams, L.G., Tsolis, R.M., *et al.* (2010a). Gut inflammation provides a respiratory electron acceptor for *Salmonella*. Nature *467*, 426–429. https://doi.org/10.1038/nature09415.

Winter, S.E., Winter, M.G., Godinez, I., Yang, H.-J., Rüssmann, H., Andrews-Polymenis, H.L., and Bäumler, A.J. (2010b). A rapid change in virulence gene expression during the transition from the intestinal lumen into tissue promotes systemic dissemination of *Salmonella*. PLOS Pathog. *6*, e1001060.

Worley, M.J., Nieman, G.S., Geddes, K., and Heffron, F. (2006). *Salmonella* Typhimurium disseminates within its host by manipulating the motility of infected cells. Proc. Natl. Acad. Sci. U.S.A. *103*, 17915–17920.

Wu, H., Jones, R.M., and Neish, A.S. (2012). The *Salmonella* effector AvrA mediates bacterial intracellular survival during infection *in vivo*. Cell. Microbiol. *14*, 28–39. https://doi.org/10.1111/j.1462-5822.2011.01694.x.

Wynosky-Dolfi, M.A., Snyder, A.G., Philip, N.H., Doonan, P.J., Poffenberger, M.C., Avizonis, D., Zwack, E.E., Riblett, A.M., Hu, B., Strowig, T., *et al.* (2014). Oxidative metabolism enables *Salmonella* evasion of the NLRP3 inflammasome. J. Exp. Med. *211*, 653–668. https://doi.org/10.1084/jem.20130627.
Yachie, A., Niida, Y., Wada, T., Igarashi, N., Kaneda, H., Toma, T., Ohta, K., Kasahara, Y., and Koizumi, S. (1999). Oxidative stress causes enhanced endothelial cell injury in human heme oxygenase-1 deficiency. J. Clin. Invest. *103*, 129–135. https://doi.org/10.1172/JCI4165.
Yang, Z., Soderholm, A., Lung, T.W., Giogha, C., Hill, M.M., Brown, N.F., Hartland, E., and Teasdale, R.D. (2015). SseK3 is a *Salmonella* effector that binds TRIM32 and modulates the host's NF-κB signalling activity. PLOS ONE *10*, e0138529. https://doi.org/10.1371/journal.pone.0138529.
Ye, Z., Petrof, E.O., Boone, D., Claud, E.C., and Sun, J. (2007). *Salmonella* effector AvrA regulation of colonic epithelial cell inflammation by deubiquitination. Am. J. Pathol. *171*, 882–892.
Yin, J., Xia, J., Tao, M., Xu, L., Li, Q., Geng, S., and Jiao, X. (2016). Construction and characterization of a *cigR* deletion mutant of *Salmonella enterica* serovar Pullorum. Avian Pathol. *45*, 569–575. https://doi.org/10.1080/03079457.2016.1187708.
Zaki, M.H., Fujii, S., Okamoto, T., Islam, S., Khan, S., Ahmed, K.A., Sawa, T., and Akaike, T. (2009). Cytoprotective function of heme oxygenase 1 induced by a nitrated cyclic nucleotide formed during murine salmonellosis. J. Immunol. *182*, 3746–3756. https://doi.org/10.4049/jimmunol.0803363.
Zgurskaya, H.I., and Nikaido, H. (1999). Bypassing the periplasm: reconstitution of the AcrAB multidrug efflux pump of *Escherichia coli*. Proc. Natl. Acad. Sci. U.S.A. *96*, 7190–7195.
Zhang, S., Santos, R.L., Tsolis, R.M., Stender, S., Hardt, W.-D., Bäumler, A.J., and Adams, L.G. (2002). The *Salmonella enterica* serotype Typhimurium effector proteins SipA, SopA, SopB, SopD, and SopE2 act in concert to induce diarrhea in calves. Infect. Immun. *70*, 3843–3855.
Zhou, D., Chen, L.M., Hernandez, L., Shears, S.B., and Galán, J.E. (2001a). A *Salmonella* inositol polyphosphatase acts in conjunction with other bacterial effectors to promote host cell actin cytoskeleton rearrangements and bacterial internalization. Mol. Microbiol. *39*, 248–259.
Zhou, Z., Ribeiro, A.A., Lin, S., Cotter, R.J., Miller, S.I., and Raetz, C.R. (2001b). Lipid A modifications in polymyxin-resistant *Salmonella* Typhimurium: PMRA-dependent 4-amino-4-deoxy-L-arabinose, and phosphoethanolamine incorporation. J. Biol. Chem. *276*, 43111–43121.

Immune-evasion Strategies of Mycobacteria and their Implications for the Protective Immune Response

7

Alexandra G. Fraga[1,2], Ana Margarida Barbosa[1,2], Catarina M. Ferreira[1,2], João Fevereiro[1,2], Jorge Pedrosa[1,2] and Egídio Torrado[1,2*]

[1]Life and Health Sciences Research Institute (ICVS), School of Medicine, University of Minho, Braga, Portugal.
[2]PT Government Associate Laboratory ICVS/3B's, University of Minho, Braga, Portugal.

*Correspondence: egidiotorrado@med.uminho.pt

http://doi.org/10.21775/9781910190692.07

Abstract

Mycobacteria are intracellular pathogens that have macrophages as their main host cells. However, macrophages are also the primary line of defence against invading microorganisms. To survive in the intracellular compartment, virulent mycobacteria have developed several strategies to modulate the activation and the effector functions of macrophages. Despite this, antigen-specific T cells develop during infection. While T-cell responses are critical for protection they can also contribute to the success of mycobacteria as human pathogens, as immunopathology associated with these responses facilitates transmission. Here, we provide a brief overview of different immune-evasion strategies of mycobacteria and their impact on the protective immune response. This understanding will further our knowledge on host–pathogen interactions and may provide critical insights for the development of novel host-specific therapies.

Introduction

Mycobacteria are one of the most important and well-adapted group of human pathogens. Among diseases caused by mycobacteria, tuberculosis, caused by *Mycobacterium tuberculosis*, is undeniably one of the oldest recorded human diseases, and still one of the biggest killers. While mortality has decreased significantly since the 1990s (WHO, 2015), in 2014 alone, 9.6 million people were estimated to fall ill with tuberculosis and 1.5 million died from the disease (WHO, 2015). After tuberculosis, leprosy and Buruli ulcer, caused by

Mycobacterium leprae and *Mycobacterium ulcerans*, respectively, are the second and third most common mycobacterial diseases in humans (Walsh *et al.*, 2010). Although leprosy was declared eliminated by the end of 2000 in most countries by the World Health Organization (WHO), incidence remains fairly stable, suggesting active transmission (Schreuder *et al.*, 2016; Walsh *et al.*, 2010). On the other hand, Buruli ulcer is a re-emerging disease mainly due to environmental factors including deforestation, artificial topographic alterations (such dams and irrigation systems), and increasing populations engaged in basic manual agriculture in wetlands (Merritt *et al.*, 2010; van Ravensway *et al.*, 2012).

The outcome of mycobacterial infections depends on their interaction with the host immune system. These interactions can result in control of infection, establishment of latent infection or development of disease (Casanova and Abel, 2002). Specifically for *M. tuberculosis*, genetic analyses suggest that this pathogen and humans coevolved for thousands of years (Comas *et al.*, 2013). During this time, *M. tuberculosis* has developed multiple and redundant strategies to manipulate macrophage activation and establish an intracellular niche to prevent elimination. Despite this, antigen-specific T cells are activated upon exposure to mycobacteria, and these cells are an essential component of the protective response (Cooper, 2009). However, the inflammatory environment generated during infection can have important consequences for T-cell function and control of infection. In some cases this environment may even promote T-cell-mediated immunopathology, which facilitates transmission of infection. Therefore, immune evasion by mycobacteria is a rather sophisticated process, wherein the highly adapted pathogen uses multiple and redundant strategies to manipulate the host response and establish a suitable environment for proliferation, while promoting the immune-mediated damage required for transmission.

Here, we discuss some of the strategies used by mycobacteria to evade the host immune system and how these strategies impact the protective response.

Mycobacteria use redundant strategies to inhibit the microbicidal activity of myeloid cells

Upon infection, resident macrophages and dendritic cells (DCs) are likely to be the first cells to interact with mycobacteria. Owing the success of mycobacteria as pathogens, it is not surprising that some of the most important adaptations that pathogenic mycobacteria have acquired throughout evolution are directed at modulating the activation and the effector functions of these cells. For example, one of the first mycobacterial virulence traits identified and studied in detail, was the ability of *M. tuberculosis* to inhibit the fusion of mycobacteria-containing phagosomes with acidified lysosomes (Armstrong and Hart, 1971). A process similar to phagosome–lysosome fusion is autophagy. Autophagy is controlled by autophagy-related genes (Atg) and relies on the formation of an autophagosome which engulfs components of the cytoplasm and delivers them to degradation in the lysosome (Mizushima *et al.*, 2011). It is now well established that *M. tuberculosis* can also interfere with autophagy (Romagnoli *et al.*, 2012; Shin *et al.*, 2010). Although phagocytosis and autophagy are independent processes, both phagosomes and autophagosomes mature through acidification and acquisition of proteolytic enzymes that degrade intracellular material, including mycobacteria. These processes are not only important to control intracellular pathogen growth, but also affect acquired immune responses. Indeed, proteins and peptides degraded in the phagolysosome or autophagolysosome can then be loaded in major histocompatibility complex (MHC) molecules thereby inducing T-cell activation (Vyas *et al.*, 2008). Here we discuss some of the mechanisms wherein mycobacteria interfere with phagosome–lysosome fusion and autophagy to prevent rapid elimination and establish an intracellular niche to thrive.

The highly characteristic cell wall (Brennan and Nikaido, 1995) and capsular layer (Sani *et al.*, 2010) of mycobacteria is composed of unique glycolipids, glucans and proteins that engage several host pathogen recognition receptors (PRRs) thereby shaping the response of myeloid cells in the infection site (Awuh and Flo, 2016). Phagocytosis of *M. tuberculosis* has been shown to be mediated through multiple receptors expressed by macrophages, such as Fcγ receptors, complement receptor type 3 and lectin receptors (Schäfer *et al.*, 2009). Fcγ receptors and complement receptor type 3 promote phagocytosis of the bacteria coated with IgG antibodies and complement protein fragment C3bi, respectively (Caron and Hall, 1998).

During primary infection, phagocytosis of *M. tuberculosis* appears to be mostly through non-opsonic mechanisms (Schäfer *et al.*, 2009). This occurs likely because serum and complement components are limited in the alveolar space when the first interaction between *M. tuberculosis* and macrophages occurs (Schäfer *et al.*, 2009; Schluger, 2001). On the other hand, opsonized uptake of *M. tuberculosis* may be more important at later stages of the infection (Schäfer *et al.*, 2009). The relevance of these mechanisms is highlighted by the fact that phagocytosis of opsonized mycobacteria does not appear to inhibit phagosome–lysosome fusion (Kang *et al.*, 2005) whereas phagocytosis of non-opsonized mycobacteria does (Hmama *et al.*, 2015; Kang *et al.*, 2005). For example, phagocytosis of mycobacteria through mannose receptor (MR) binding of the mannose-capped lipoarabinomannan (ManLAM), one of the most prominent molecular patterns found in the cell wall of pathogenic mycobacteria (Brennan and Nikaido, 1995), results in phagosome–lysosome fusion arrest (Kang *et al.*, 2005). Interestingly, however, ManLAM has also been shown to interact with other PRRs, namely the dendritic cell-specific intracellular adhesion molecule 3 grabbing nonintegrin (DC-SIGN) and, more recently, Dectin-2 expressed by DCs (Yonekawa *et al.*, 2014). While the engagement of ManLAM with Dectin-2, in what regards phagosome–lysosome fusion, has not been yet determined, engagement of ManLAM with DC-SIGN does not induce phagosome–lysosome arrest (Kang *et al.*, 2005).

The effect of ManLAM on phagocytic cells is however not limited to phagosome-lysosome arrest but it also stimulates the production of cytokines, co-stimulatory molecules and MHC-II expression, important for macrophage function and stimulation of acquired responses (Kallenius *et al.*, 2016). Indeed, stimulation of bone marrow-derived DCs with ManLAM induces the production of IL-10, IL-12p40, IL-6 and TNF-α, up-regulation of the co-stimulatory molecules CD80 and CD86 and MHC-II, in a Dectin-2-dependent manner (Yonekawa *et al.*, 2014). Importantly, mice deficient in Dectin-2 are more susceptible to disease after *M. avium* pulmonary infection, showing that recognition of ManLAM by Dectin-2 is important to control infection (Yonekawa *et al.*, 2014). On the other hand, a recent study compared the virulence *M. tuberculosis* and *M. bovis* BCG with isogenic strains lacking the *CapA* locus, which is involved in the biosynthesis of the mannose cap of LAM (Afonso-Barroso *et al.*, 2013). This study showed that mycobacteria lacking the mannose cap of LAM were not less virulent than the isogenic strains that have ManLAM (Afonso-Barroso *et al.*, 2013). Thus, taken together, these data suggest that during infection the effects of ManLAM may be redundant and in the absence of this molecular pattern other ligands are engaged by the same receptors in myeloid cells. In this regard, several molecular patterns have been identified to interact with DC-SIGN (Appelmelk *et al.*, 2008; Driessen *et al.*, 2009; Geijtenbeek *et al.*, 2003; Pitarque *et al.*, 2005). An example is α-glucan that has been shown to induce the production of IL-10 upon lipopolysaccharide (LPS) stimulation of DCs, without altering IL-12p40 production (Geurtsen *et al.*, 2009). In the context of the whole bacteria however, it has been shown that stimulation of monocyte-derived DCs with *M. bovis* BCG, grown in conditions that are permissive to the maintenance of the capsular layer (rich in α-glucan), produced significantly lower levels of IL-12 and TNF-α, when compared to *M. bovis* BCG that did not have the α-glucan-rich capsular layer (Sani *et al.*, 2010). A recent study identified PPE10 protein, secreted by Esx-5, as the main responsible for maintaining the mycobacterial capsular integrity (Ates *et al.*, 2016). Interestingly, *M. marinum* deficient in Esx-5 or PPE10 were less virulent that the complemented strains in zebrafish embryos (Ates *et al.*, 2016), highlighting the importance of the capsular layer, and α-glucan, in the immune response and control of infection.

The above described data highlight the complex interaction between mycobacteria and phagocytic cells during infection. Specifically for ManLAM, its recognition through MR results in phagosome-lysosome arrest (Kang *et al.*, 2005), but its recognition through other PRRs may result in a different response leading to the production of cytokines, including TNF-α, IL-6, IL-12, IL-10 (Kallenius *et al.*, 2016) that have multiple roles during infection (Domingo-Gonzalez *et al.*, 2016; Torrado and Cooper, 2013). When these data are taken together with other data showing that (i) several mycobacterial molecular patterns are recognized by host cells through multiple PRRs, including Toll-like receptors (TLR), NOD-like receptors, Dectin-1 and Mincle; and (ii) engagement of these receptors leads to the production cytokines important in macrophage function and in the infection process (Awuh and Flo, 2016; Killick *et al.*, 2013; Mortaz *et al.*, 2015; Rajaram *et al.*, 2014); it is plausible to conclude that most of the molecular patterns of mycobacteria have redundant effects on phagocytic cells (Killick *et al.*, 2013). Because of these redundant functions, in some cases absence of one receptor in the host, or one molecular pattern in the pathogen, does not translate in altered bacterial burdens or pathology to the host. The data described above for ManLAM are an example. On the host side, absence of TLR-2 or TLR-9 does not impact the ability of mice to control *M. tuberculosis* infection; however, mice deficient in both TLR-2 and TLR-9 revealed to be more susceptible to *M. tuberculosis* infection (Bafica *et al.*, 2005).

Together, these data suggest not only that there is some degree of redundancy in the effects of mycobacterial molecular patterns, but also that the targeting of specific PRRs is critical for the pathogen to establish an intracellular niche and survive in the infected host.

In addition to molecular patterns recognized by cell-surface receptors during phagocytosis, *M. tuberculosis* has also been shown to secrete effector molecules that interfere with intracellular vesicle trafficking. Two of the most well characterized effectors that interfere with phagosome–lysosome fusion are PtpA (Bach *et al.*, 2008) and SapM (Vergne *et al.*, 2005). The mechanisms by which these proteins interfere with phagosome–lysosome fusion have been recently reviewed (Hmama *et al.*, 2015).

Other important mycobacterial proteins that trigger phagosome-lysosome arrest include TB10.4 (encoded by *esxH*) and TB9.8 (encoded by *esxG*), which are secreted by the mycobacterial Esx-3 type VII secretion system (Mehra *et al.*, 2013). The TB10.4/TB9.8 heterodimer is able to interact with the hepatocyte growth factor-regulated tyrosine kinase substrate (Hgs/Hrs), a component of the Endosomal Sorting Complex Required for Transport (ESCRT) (Mehra *et al.*, 2013). The ESCRT transport system is composed by four protein complexes that are sequentially recruited to the endosomal membrane, promoting the delivery of phagocytosed mycobacteria to the lysosome, thus restricting intracellular bacterial growth (Mehra *et al.*, 2013). The heterodimer was shown to directly target Hrs, inhibiting delivery of bacteria to lysosomes (Mehra *et al.*, 2013). It is worth noting that TB10.4/TB9.8 are highly antigenic, and a high frequency of T cells specific for these antigens has been shown to correlate with protection against *M. tuberculosis* infection (Hervas-Stubbs *et al.*, 2006). In fact, the introduction of the *M. tuberculosis* Esx-3 locus into an *M. smegmatis* strain lacking the endogenous Esx-3 region generates highly protective immunity against *M. tuberculosis* (Sweeney *et al.*, 2011). There is also a role for Esx-3 of *M. tuberculosis* and *M. smegmatis* in iron and zinc acquisition (Serafini *et al.*, 2009; Siegrist *et al.*, 2014).

The Esx-1 type VII secretion system also secretes other highly antigenic T-cell proteins that interfere with intracellular vesicle trafficking, such as the early secretory antigen target 6 kDa (ESAT-6, encoded by *esxA*) and the culture filtrate antigen 10 kDa (CFP-10, encode by *esxB*) (Tan *et al.*, 2006). However, a more important biological role of these proteins in mycobacterial pathogenesis is their phagosomal permeabilization properties implicated in rupture and translocation of *M. tuberculosis* and *M. leprae* from the phagosome into the cytosol (van der Wel *et al.*, 2007), activation of autophagy (Watson *et al.*, 2012) as well as the inflammasome (Mishra *et al.*, 2013). The importance of these proteins in infection is highlighted by several studies showing that, mutations affecting the synthesis or secretion of ESAT-6/CFP10 attenuate the virulence of mycobacteria in murine models of infection (Hsu *et al.*, 2003; Lewis *et al.*, 2003; Pym *et al.*, 2002, 2003).

Together, these data show that mycobacteria use redundant mechanisms to prevent macrophage activation and functions, highlighting the importance of these strategies to prevent rapid elimination.

Initial studies on macrophages infected with *M. tuberculosis* or *M. bovis* BCG suggested that autophagy induction could overcome phagososme–lysosome arrest imposed by mycobacteria (Gutierrez *et al.*, 2004). This and following studies, established autophagy as an important antimycobacterial defence mechanism, not only by overcoming phagosome-lysosome arrest, but also by promoting the generation of antimicrobial peptides with activity against mycobacteria (Ponpuak *et al.*, 2010). The critical role of autophagy during *M. tuberculosis* infection is underlined by the acute susceptibility of mice conditionally deficient for Atg5 in phagocytic cells (Castillo *et al.*, 2012; Kimmey *et al.*, 2015; Watson *et al.*, 2012), a phenotype similar to that of IFN-γ-deficient mice (Cooper *et al.*, 1993; Flynn *et al.*, 1993; Nandi and Behar, 2011). However, more recent data showed that deficiency in other Atg genes important for autophagy did not impact control of *M. tuberculosis* infection (Kimmey *et al.*, 2015), suggesting that Atg5 participates in unique functions not provided by other essential Atg proteins. Indeed, Atg5 deficiency in alveolar macrophages or DCs did not significantly impact control of infection; as opposed to Atg5 deficiency in neutrophils, which resulted in uncontrolled recruitment of these cells to the infection site (Kimmey *et al.*, 2015). The role of neutrophils in the susceptibility of Atg5 deficient mice was determined by antibody depletion studies that reverted the acute susceptible phenotype of Atg5 deficient mice (Kimmey *et al.*, 2015).

Although the data discussed above suggest autophagy-independent roles of Atg5 in controlling resistance to *M. tuberculosis* infection, several factors produced by *M. tuberculosis* have been identified to interfere with the autophagic process. The ε-aminoacetyltransferase Eis, for instance, has been shown to acetylate Lys-55 in a JNK-specific phosphatase and is critical to activate the autophagy regulator Beclin 1 (Kim *et al.*, 2012). Similar to phagosome–lysosome fusion, the Esx-1 secretion system has also been implicated in inhibiting autophagy in human DCs (Romagnoli *et al.*, 2012). Likewise, the PE_PGRS47 protein has been recently shown to inhibit autophagy with impact on antigen presentation, but without affecting MHC-II expression (Saini *et al.*, 2016). This protein was identified in a screening of random genomic DNA fragments of *M. tuberculosis* capable of inhibiting antigen processing and loading into MHC-II (Saini *et al.*, 2016). Compared to the wild-type or PE_PGRS47 complemented strains, macrophages infected with the mutant PE_PGRS47 strain showed an increase in microtubule-associated light chain 3 (LC3), a central component of autophagosome formation, and p62, which is recruited to newly formed autophagosomes (Saini *et al.*, 2016). These findings thus suggest that PE_PGRS47 prevents the formation of autophagosomes containing intracellular *M. tuberculosis*. Moreover, *in vivo* experiments show that mice infected with the PE_PGRS47 mutant strain of *M. tuberculosis* were more resistant to infection, maintaining lower bacterial burdens throughout chronic infection when compared to mice infected with the wild-type or complemented strains (Saini *et al.*, 2016). Interestingly, while there were no differences in the frequency of antigen-specific CD4 T cells at day 9 post infection (when T cells are primed in the lymph nodes, discussed below), there were more antigen-specific CD4 T cells at day 24 post infection in mice infected with the *M. tuberculosis* strain deficient in PE_PGRS47 than in mice infected with the wild-type or complemented strains (Saini *et al.*, 2016). These data show that by interfering with intracellular vesicle trafficking, *M. tuberculosis* impairs antigen-processing and subsequent loading into MHC-II molecules.

Since there were no differences in the frequency of CD4 T cell at day 9 post infection (Saini *et al.*, 2016), the data discussed above also suggest that inhibition of autophagy may not impact T-cell priming. In this regard, several studies have shown that mycobacteria can modulate antigen presentation through different mechanisms, including inhibition of MHC-II expression or antigen-processing and loading (Hmama *et al.*, 2015). However, these mechanisms may not fully explain the delayed T-cell priming during mycobacterial infection (discussed below) but may rather be more important for T-cell function in the infection site. A recent study took advantage of MHC-II mixed bone-marrow chimeras to compare the bacterial burdens in myeloid cells that were capable (MHC-II-expressing), or not (MHC-II deficient), of interacting with CD4 T cells (Srivastava and Ernst, 2013). Results showed that myeloid cells expressing MHC-II had reduced bacterial burdens when compared to myeloid cells that did not express MHC-II (Srivastava and Ernst, 2013). Therefore, direct recognition of infected cells by CD4 T cells appears to be required for controlling intracellular *M. tuberculosis* in the infection site. Accordingly, depletion of CD4 T cells resulted in increased numbers of bacteria in MHC-II-expressing myeloid cells but not MHC-II deficient cells (Srivastava and Ernst, 2013). These data suggest that, by interfering with autophagy and phagosome–lysosome fusion, *M. tuberculosis* may prevent the interaction between infected macrophages and antigen-specific CD4 T cells at the infection site, thereby limiting the ability of CD4 T cells to activate *M. tuberculosis*-infected macrophages.

Collectively, these studies show that during the first interaction with myeloid cells in the infection site different mycobacterial molecular patterns are recognized by distinct host receptors, resulting in signals leading to phagosome-lysosome arrest, but also the production of pro- and anti-inflammatory cytokines and chemokines with important roles during infection (Domingo-Gonzalez *et al.*, 2016; Torrado and Cooper, 2013). The different and redundant strategies used by mycobacteria to inhibit phagosome–lysosome fusion and autophagy highlight the key role of these strategies in the early establishment of an intracellular niche for the pathogen to proliferate (Fig. 7.1). During the course of infection, the balance between these signals may play key roles in determining the early control of infection and induction of T-cell responses (Fig. 7.1).

Mycobacteria modulate the microenvironment to subvert infected-cell death pathways and recruit permissive macrophages

Although mycobacteria are well equipped to prevent and escape phagosome–lysosome fusion and autophagy, macrophages can still activate apoptotic cell death programmes

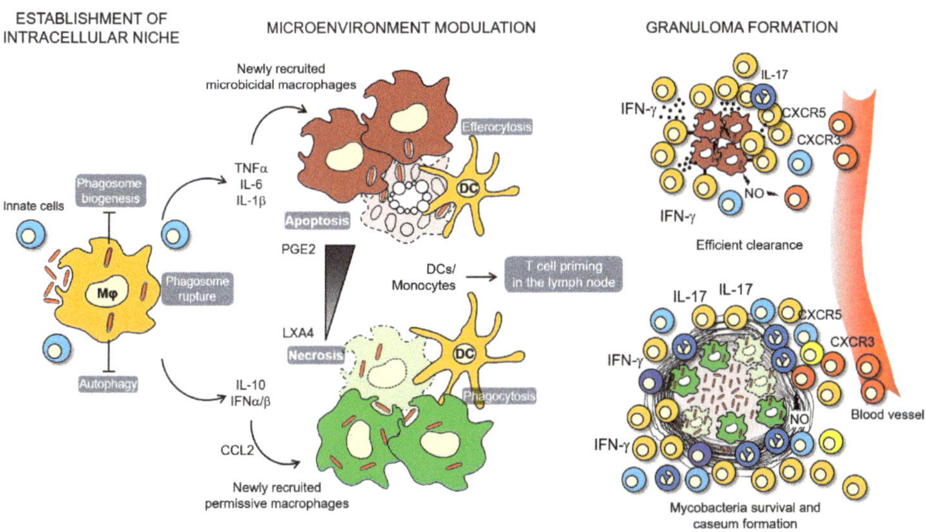

Figure 7.1 Progression of the immune response to *M. tuberculosis* infection. Early interaction with phagocytes through PRRs results in the phagocytosis of the bacteria. Virulent *M. tuberculosis* interferes with intracellular vesicle trafficking and escapes the phagosome to prevent early elimination and establish an intracellular niche. At the same time macrophages, and other inflammatory cells in the site of infection, produce inflammatory mediators that can impact the cell death modality of the infected-cell and induce the recruitment of other inflammatory cells. High levels of IFN-αβ induce the recruitment of CCR2-expressing monocytes with limited ability to control bacterial growth. As infected cells die newly recruited macrophages and monocytes efferocytose or phagocytose the bacteria and a granulomatous reaction begins to form. After the onset of acquired immunity, T cells begin to accumulate and a granuloma is formed. The efficient interaction between T cells and macrophages together with a balanced production of cytokines and chemokines results in a non-damaging inflammatory environment and control of infection. On the other hand, phagocytosis of the bacteria by permissive macrophages with a weak capacity to interact with T cells results in progression of infection. As the granuloma coalesces, the inflammatory environment generated may promote the constant accumulation of T cells contributing to immunopathological consequences to the host.

to prevent further microbial intracellular replication and increase exposure of microbial antigens. Apoptosis is a tightly regulated process of cellular deconstruction that minimizes inflammation and pathology by containing the dismembered cellular contents of dying cells within membrane-bound vesicles, called apoptotic bodies (Taylor *et al.*, 2008). These apoptotic bodies emit signals to be engulfed and digested by phagocytes in a process called efferocytosis (Martin *et al.*, 2014). On the other hand, necrosis is a more inflammatory process when compared to apoptosis, as it is defined by loss of membrane integrity with the release of intracellular contents to the extracellular space (Moraco and Kornfeld, 2014).

A growing body of evidence points to the ability of *M. tuberculosis* to inhibit apoptotic and promote necrotic cell death of macrophages as an important determinant of virulence and pathogenesis (Behar *et al.*, 2010; Chen *et al.*, 2008; Chen *et al.*, 2006; Divangahi *et al.*, 2009; Divangahi *et al.*, 2010). Early studies suggested apoptosis, but not necrosis, of the mycobacterial-infected cells to have direct bactericidal effects on intracellular bacterial viability. However, more recent data suggest that apoptosis itself is not intrinsically bactericidal but instead requires efferocytosis (Martin *et al.*, 2012).

One key mechanism wherein virulent *M. tuberculosis* prevents apoptosis and promotes necrotic death of infected macrophages is by regulating eicosanoid production (Chen *et al.*, 2008). Mechanistically, avirulent *M. tuberculosis* H37Ra strain has been shown to stimulate the production of prostaglandin E_2 (PGE_2), which is required for plasma membrane protection by regulating synaptotagmin 7, the calcium sensor involved in the lysosome-mediated repair mechanism (Divangahi *et al.*, 2009). On the other hand, virulent *M. tuberculosis* H37Rv and Erdman strains stimulate the production of lipoxins, such as lipoxin A4 (LXA4), which in turn down-regulate cyclooxygenase (COX)2 mRNA, thereby reducing PGE2 production and preventing its membrane protective effects, resulting in necrotic death of the infected cell (Chen *et al.*, 2008; Divangahi *et al.*, 2009). The relevance of these pathways is demonstrated by the observation that Alox5-deficient macrophages, which cannot synthesize LXA4, undergo more apoptosis and are better able to control *M. tuberculosis* H37Rv growth when compared to wild-type macrophages. On the other hand, PGE2-deficient macrophages undergo more necrosis and do not to control intracellular growth of either virulent or avirulent *M. tuberculosis* (Divangahi *et al.*, 2009). Therefore, apoptosis appears to be the fate of *M. tuberculosis*-infected cells if mitochondrial and plasma membrane integrity is protected by PGE2, whereas necrosis occurs if LXA4 predominates. How *M. tuberculosis* induces LXA4 production is not yet determined. Likely, the Esx-1 secretion system-induced rupture of the mycobacteria vacuole (van der Wel *et al.*, 2007), as well as the pore forming activity of ESAT-6 (Smith *et al.*, 2008), are important in the process. Another candidate is CpnT, a NAD^+-glycohydrolase which depletes cellular NAD^+ pools, resulting in host necrotic cell death (Danilchanka *et al.*, 2014; Sun *et al.*, 2015).

The ability of *M. tuberculosis* to prevent apoptosis of its host cells can also affect the T-cell response (Divangahi *et al.*, 2010). Indeed, when alveolar macrophages from *M. tuberculosis*-infected wild-type or Alox5-deficient mice were transferred into wild-type mice, the Alox5-deficient macrophages were able to stimulate pulmonary and systemic CD4 and CD8 T-cell responses, in contrast to wild-type macrophages (Divangahi *et al.*, 2010). These enhanced T-cell responses were not directly induced by the transferred macrophages. Instead, the antigen cargo contained within the apoptotic vesicles of Alox5-deficient macrophages was phagocytosed by DCs that, in turn, stimulated T-cell responses (Divangahi *et al.*, 2010).

The role of lipid mediators can also directly influence T-cell responses. For example, PGE2 has been shown to directly promote the differentiation and proinflammatory functions of human and murine Th17 cells (Boniface *et al.*, 2009; Gopal *et al.*, 2012). In this regard, IL-17 and Th17 cells are an important component of the protective immune response to highly virulent strains of *M. tuberculosis* (Gopal *et al.*, 2014), but uncontrolled Th17 responses can result in important immunopathological consequences to the host (Cruz *et al.*, 2010; Torrado and Cooper, 2010). More recently, IL-17 was found to promote the intracellular growth of *Mycobacterium bovis* BCG and *M. tuberculosis* by inhibiting the intrinsic apoptotic pathway (Cruz *et al.*, 2015a). This study showed that IL-17 induces a down-regulation of the apoptosis master regulator p53 resulting in a reduction in the percentage of caspase-3-positive macrophages, as well as an increase in the ratio of Bcl2/Bax proteins (Cruz *et al.*, 2015a). Together, these data suggest that during the early stages of infection the balance between eicosanoids plays a key role in determining the cell death modality of the infected cell; as the infection progresses the inflammatory environment generated in the infection site may alter this balance (Fig. 7.1). Future research is required to address these issues.

In addition to regulating the infected-cell death modality, PGE2 has been recently shown to regulate the production of type I interferons (IFN-α/β) (Mayer-Barber *et al.*, 2014). IFN-α/β has been shown to have a largely detrimental role in tuberculosis (Berry *et al.*, 2010; Manca *et al.*, 2001; Ordway *et al.*, 2007), with more virulent strains (such as Beijing strains) inducing the highest production of IFN-α/β (Carmona *et al.*, 2013; Manca *et al.*, 2001). The induction of IFN-α/β is likely associated with the phagosome membrane rupture mediated by the Esx-1 secretion system (van der Wel *et al.*, 2007) triggering a cytosolic DNA sensor and activating the Sting/Tbk1/Irf3 axis (Manzanillo *et al.*, 2012). This pathway is also activated by multiple PRRs, including TLRs (Barber, 2011). In this regard, *M. tuberculosis* Beijing strains have been shown to induce IFN-α/β through a TLR-4-dependent pathway (Carmona *et al.*, 2013).

The detrimental role of IFN-α/β during mycobacterial infections is complex and likely involves different mechanisms. One of these mechanisms is associated with the recruitment of permissive macrophages to the infection site. Indeed, induction of IFN-α/β production by intranasal challenge of *M. tuberculosis*-infected mice with polyinosinic-polycytidylic acid condensed with poly-l-lysine and carboxymethylcellulose (Poly-ICLC) results in elevated bacterial burdens associated with increased accumulation of a $CD11b^{+}Gr1^{int}$ myeloid population recruited via the CCL2-CCR2 pathway (Antonelli *et al.*, 2010). In Poly-ICLC treated mice, $CD11b^{+}Gr1^{int}$ cells displayed decreased expression of MHC-II and elevated bacterial levels when compared to the same subset of cells purified from infected, untreated controls (Antonelli *et al.*, 2010).

In another recent paper, it was shown that both *M. tuberculosis* and *M. marinum* actively promote the recruitment of permissive macrophages by using cell-surface-associated phthiocerol dimycoceroserate (PDIM) lipids through a host CCR2-mediated pathway (Cambier *et al.*, 2014). Interestingly, these PDIM appear to mask underlying mycobacterial molecular patterns. In the absence of PDIM, these molecular patterns signal a TLR-dependent recruitment of macrophages that produce nitric oxide and control intracellular *M. tuberculosis* and *M. marinum* growth (Cambier *et al.*, 2014). Whether or not this CCR2 pathway is also activated by IFN-α/β, was not determined. This intricate immune evasion strategy also provides an explanation for the observation that *M. tuberculosis* initiates infection in the lower airways, a relatively sterile environment. Indeed, in the upper airways the normal flora provides the TLR stimulation required to recruit microbicidal macrophages that are capable of controlling *M. tuberculosis* infection (Cambier *et al.*, 2014).

The other detrimental role of IFN-α/β is by suppressing IFN-γ-dependent macrophage activation (Teles *et al.*, 2013). This role of IFN-α/β is particularly evident in leprosy, which presents in two distinct clinical manifestations: a disseminated form (leprematous) or a self-healing form (tuberculoid). Self-healing tuberculoid lesions were shown to be associated with IFN-γ and its downstream vitamin D-dependent antimicrobial genes, whereas disseminated and progressive leprematous lesions were associated with an IFN-α/β gene programme, including IL-10 (Teles *et al.*, 2013). Importantly, IFN-γ-induced antimicrobial activity against *M. leprae* in monocytes was abrogated by IFN-α/β or IL-10 (Teles *et al.*, 2013). Recently, IFN-α/β was also shown to induce IL-10 production, which in turn inhibits the production of TNF-α, IL-12, and IL-1β and impairs the ability of IFN-γ-activated macrophages to control *M. tuberculosis* (McNab *et al.*, 2014). Taking into account that tuberculosis also consists of a spectrum of disease, even though not as distinct as leprosy, it is possible that IFN-α/β responses are involved in defining this spectrum.

Despite its largely detrimental role, IFN-α/β may also have protective functions during *M. tuberculosis* infection. Indeed, mice lacking both IFN-α/β and IFN-γ signalling are more susceptible to infection, when compared to single deficient mice (Desvignes *et al.*, 2012; Moreira-Teixeira *et al.*, 2016). In this model, the function of IFN-α/β receptor signalling was suggested to limit the early recruitment of permissive macrophages (Desvignes *et al.*, 2012). In accordance with this protective role is the observation that IFN-α/β may promote expression of the inducible Nitric Oxide Synthase (iNOS, encoded by *Nos2*) during the early stages of *M. tuberculosis* infection (Moreira-Teixeira *et al.*, 2016). Therefore, low amounts of IFN-α/β may be protective by activating macrophages to restrict intracellular mycobacterial growth; however, production of too much IFN-α/β results in the recruitment of large numbers of permissive macrophages, preventing control of infection.

Collectively, these data show that mycobacteria modulate the inflammatory environment of the infection site to interfere with the cell death modality of its host cells while promoting the recruitment of permissive macrophages to the infection site (Fig. 7.1).

Delayed T-cell priming during mycobacterial infections

Antigen-specific T cells are an essential component of the protective response to mycobacteria (Cooper, 2009). During tuberculosis, T-cell effector responses are dominated by Th1 (defined by IFN-γ production) and, to a lesser extent, Th17 (defined by IL-17 production) phenotypes, as well as CD8 T cells. These populations play key roles in the protective and pathological responses (Behar, 2013; Cooper, 2009; Lin and Flynn, 2015; Torrado and Cooper, 2010; Torrado *et al.*, 2011).

One of the most intriguing aspects of the host response against mycobacteria is the delayed priming of the T-cell response. As priming and initiation of acquired immune responses are difficult to address in humans, due to limitations in sampling and determining time of exposure, the aerosol mouse model of infection has been used to address this issue (Gallegos *et al.*, 2008; Reiley *et al.*, 2008; Wolf *et al.*, 2008). Using this model, the bacteria is deposited in the lung lower airways where *M. tuberculosis* grows in an apparent uncontrolled manner, until day 21 post infection (Cooper and Khader, 2008; Khader *et al.*, 2007). Notwithstanding, several studies have shown that antigen-specific T cells only become activated after day 7–10 of infection and that T-cell priming occurs in the lung draining mediastinal lymph nodes, but not in the lung (Gallegos *et al.*, 2008; Reiley *et al.*, 2008; Wolf *et al.*, 2008). This slow onset in the acquired response is in marked contrast to the kinetics typically observed for other aerogenic infections, such as influenza (Flynn *et al.*, 1998) and *Francisella tularensis* (Baron et *al.*, 2007), or with other intracellular bacteria, such as *Listeria monocytogenes* (Kursar *et al.*, 2002). In these infections, T-cell priming and expression of T-cell immunity in the infection site occurs as early as 3–7 days post infection and results in a rapid control of viral or bacterial proliferation (Baron *et al.*, 2007; Flynn *et al.*, 1998; Kursar *et al.*, 2002).

The slow pace at which the T-cell response is initiated could be, at least partially, explained by the reduced antigen production, associated with the slow proliferation rate of *M. tuberculosis.* However, during *in vitro* growth *M. tuberculosis* rapidly secretes T-cell antigenic proteins to the culture medium (Andersen *et al.*, 1992; Andersen *et al.*, 1991). Therefore, if the slow mycobacterial growth accounted for the insufficient antigen for CD4 T-cell activation, then the time required to initiate the immune response should be reduced by increasing the dose of infection. However, increasing the dose of infection by a factor of 10 does not significantly enhance T-cell priming (Reiley *et al.*, 2008; Wolf *et al.*, 2008). On the other hand, the observations that (i) T-cell priming in the draining mediastinal lymph node after aerosol challenge coincides with the arrival of viable bacteria to this location (Reiley *et al.*, 2008; Wolf *et al.*, 2008); and that (ii) the delay in the control of *M. tuberculosis* infection in the lungs after aerosol infection contrasts with the immediate control in the liver upon intravenous challenge, despite the fact that in both cases control of infection was associated with the accumulation of IFN-γ-producing CD4 T cells (Cooper, 2009; Cooper *et al.*, 1997a); suggest that dissemination of the bacteria to the draining lymph node is likely essential to T-cell priming. This hypothesis is also consistent with the observations that the lung environment is not optimal for T-cell priming, as myeloid cells isolated from lungs of *M. tuberculosis*-infected mice induce poor T-cell cytokine responses, when compared to the same populations isolated from the lymph nodes (Wolf *et al.*, 2007). Therefore to initiate T-cell responses, *M. tuberculosis*-infected myeloid cells need to migrate to the lymph node.

Still the question remains on why these cells do not migrate quicker to the draining lymph nodes. While there is still no definitive answer to this question, as referred above it is known that increasing the infection dose does not significantly enhance this response (Reiley *et al.*, 2008; Wolf *et al.*, 2008). As such, the bacteria itself does not appear to inhibit the migration of the infected cells; instead, primary infected cells are likely non-migratory cells. Therefore, despite the ability of *M. tuberculosis* to interfere with myeloid cell mechanisms required for T-cell activation, such as MHC-II expression (Hmama *et al.*, 2015), the delay in T-cell priming is largely attributed to the microenvironment of the lung and the transport of *M. tuberculosis*, or its antigens, to the draining lymph nodes.

In this regard, the transport of *M. tuberculosis* from the lung to the draining lymph nodes has been recently shown to be made by Ly6C monocytes, in a CCR2-dependent way (Samstein *et al.*, 2013). It is interesting to note, however, that these cells are not lung resident but migrate from the bone-marrow to the infection site in response to CCR2-binding chemokines, CCL2 and CCL7, where they presumably differentiate into macrophages or DCs (Serbina and Pamer, 2006). How these cells uptake *M. tuberculosis*, or its antigens, is still not known. The function of Ly6C monocytes in T-cell priming appears to be limited to shuttling the bacteria, or its antigens, to the draining lymph nodes as they do not prime

T cells directly (Samstein *et al.*, 2013). Perhaps this is the result of *M. tuberculosis'* ability to interfere with the differentiation of monocytes. Indeed, it has been suggested that *M. tuberculosis* infection impairs the differentiation of monocytes into DCs, inducing instead, the generation of macrophage-like cells that fail to up-regulate co-stimulatory and MHC-II molecules (Remoli *et al.*, 2011).

As T-cell priming is delayed, one could speculate that if primed T cells were already in the lung and ready to respond, control of infection would be more efficient. Indeed, mice infected with *M. tuberculosis* and subsequently treated with antibiotics to generate memory-immune mice (Cooper *et al.*, 1997a) or vaccinated with BCG (Cruz *et al.*, 2015b) control *M. tuberculosis* faster than naive mice; however, the kinetics of the protective response is still slow and control of infection is only observed at day 14 post infection (Cooper *et al.*, 1997a; Cruz *et al.*, 2015b). One possible caveat of these experiments is that the memory T-cell response generated by live bacteria does not populate the lung. To overcome this, transgenic CD4 T cells primed *in vitro* in Th1 differentiating conditions (to promote the development of IFN-γ-producing T cell) were transferred into naive mice before aerosol *M. tuberculosis* infection (Gallegos *et al.*, 2008). These transferred effector cells were shown to populate the lung and produce IFN-γ and, as expected, provided enhanced protection (Gallegos *et al.*, 2008); however, this enhanced protection was only detected after day 7 of infection (Gallegos *et al.*, 2008). Thus, delayed T-cell priming is not the only issue in preventing the rapid control of infection, but the ability of the effector T cells to sense antigen in the lung is also critical (Srivastava and Ernst, 2013). The establishment of an intracellular niche, described above, is therefore important for *M. tuberculosis* to shelter for the T-cell response.

The impact of the inflammatory environment in accumulating T cells

As T cells accumulate in the infected organ, bacterial proliferation is controlled (Cooper and Khader, 2008; Khader *et al.*, 2007; Winslow *et al.*, 2003). In the tuberculosis mouse model, transition from the exponential increase in bacterial numbers to a plateau phase occurs at day 21 of infection, and correlates with the accumulation of IFN-γ-producing CD4 T cells (Cooper and Khader, 2008; Khader *et al.*, 2007). The critical role of CD4 T cells and IFN-γ in the control of mycobacterial infections has been demonstrated in different animal models as well as in humans, particularly in those with acquired immunodeficiencies affecting CD4 T-cell immunity, such as HIV infection, which is one of the most important risk factors for development of active tuberculosis (Geldmacher *et al.*, 2012). Therefore, understanding the mechanisms that govern T-cell migration and accumulation within the infected tissue are important.

After priming, T cells exit the draining lymph nodes, enter into the peripheral blood circulation, and migrate to the lung tissue. Although the mechanisms that orchestrate the T-cell egress from the blood stream into the lung tissue are not completely understood, recent papers have identified key chemokine receptors that are important for T-cell migration into the lung tissue. One of these studies used intravascular flow cytometry, as previously described (Anderson *et al.*, 2014), to distinguish between *M. tuberculosis*-specific T cells that are in the lung vasculature or lung parenchyma (Sakai *et al.*, 2014). This approach identified two major subsets of CD4 T cells based on their location and phenotype. The CD4 T-cell population found mostly associated with the lung vasculature produced the highest amounts of IFN-γ and expressed killer cell lectin-like receptor subfamily G member 1 (KLRG1), CX3CR1 and high levels of the Th1 transcription factor T-bet. On the other hand, the lung parenchyma was enriched with CD4 T cells that produced lower amounts of IFN-γ, expressed CXCR3 and intermediate levels of T-bet, but did not express KLRG1 (Sakai *et al.*, 2014). To determine their protective role during infection, each population was purified and adoptively transferred into T-cell-deficient naive mice that were subsequently challenged with *M. tuberculosis* through the aerosol route. This approach showed that the population better able to induce control of infection produced lower levels of IFN-γ, but was more capable of migrating to the lung parenchyma (Sakai *et al.*, 2014).

In this context, the chemokine receptor CXCR5 was also shown to have an important role in CD4 T-cell migration (Slight *et al.*, 2013). CXCL13 is the ligand for CXCR5 and during *M. tuberculosis* infection both CXCL13 and CXCR5 are required for the formation of ectopic lymphoid follicles containing B and T cells (Khader *et al.*, 2011; Slight *et al.*, 2013). These follicles have also been identified in tuberculosis patients (Ulrichs *et al.*, 2004) and their absence has been associated with uncontrolled disease (Ulrichs *et al.*, 2005). In mice, CXCR5 expression is required to control *M. tuberculosis* infection (Slight *et al.*, 2013). However, the enhanced susceptibility of CXCR5 was not associated with delayed CD4 T-cell priming, accumulation of CD4 T cells in the lung tissue, or altered cytokine production (Slight *et al.*, 2013). Instead, T cells from CXCR5-deficient mice locate in distinct perivascular cuffs, a phenotype also observed in wild-type mice that received CXCL13 (the ligand for CXCR5) neutralizing antibody (Slight *et al.*, 2013).

Together, these data suggest that the coordinated expression of CXCR3 and CXCR5 is critical for *M. tuberculosis*-specific cells to migrate to the lung parenchyma and into the *M. tuberculosis* lesion where they interact with infected macrophages. Nevertheless, whether these receptors mediate the extravasation across the lung blood vessel into the lung parenchyma or the migration within the parenchyma into the lesion is yet to be determined.

The mechanism underlying the accumulation of T-cell phenotypes associated with the lung vasculature or parenchyma is still not well understood. While we cannot exclude the possibility that these phenotypes are imprinted during priming, recent data suggest that the lung inflammatory environment generated during infection plays a role. In this regard, it has been previously shown that mice deficient in IL-27 signalling are more resistant to *M. tuberculosis* (Holscher *et al.*, 2005; Pearl *et al.*, 2004). More recently, we have found that the enhanced resistance of IL-27R-deficient mice is replicated in mice deficient in IL-27R only in the T-cell compartment (Torrado *et al.*, 2015). Therefore, we asked whether T cells differentiated differently in the absence or presence of IL-27-mediated signalling. Our data show that in IL-27R deficient mice, CD4 T cells were enriched in the lung parenchyma migratory populations, displaying reduced T-bet expression, CXCR3 and produced more IL-2 when compared to wild-type mice (Torrado *et al.*, 2015). Importantly, in competitive bone-marrow chimeras IL-27R-deficient CD4 T cells were more fit to survive within the lung lesion site, when compared to wild-type CD4 T cells (Torrado *et al.*, 2015).

Another important modulator of the T-cell response is nitric oxide produced by IFN-γ-activated macrophages in the lesion site (Cooper *et al.*, 2002). To test the hypothesis that the more activated cells within the lesion site were more sensitive to the inflammatory environment we resorted to the *M. avium* model wherein iNOS is not required to control infection and, indeed, it appears to benefit the host (Gomes *et al.*, 1999). Using this model, we also found an increased accumulation of CD4 T cells expressing high levels of T-bet in iNOS deficient when compared to wild-type mice (Pearl *et al.*, 2012). Importantly, within the T-bet-expressing effector population, CD4 T cells expressing low levels of the early activation marker CD69 accumulate during chronic infection in higher numbers in iNOS deficient mice when compared to wild-type mice (Pearl *et al.*, 2012). These data suggest that the terminally differentiated population (based on the expression of high T-bet and reduced CD69) displays an enhanced sensitivity to the regulatory effects of nitric oxide (Pearl *et al.*, 2012). Therefore, while nitric oxide in the phagosome may restrain intracellular bacterial growth, it also regulates T-cell responses, preventing the accumulation of strongly activated and potentially pathological T cells at the site of infection (Fig. 7.1).

Since iNOS expression is required to control *M. tuberculosis*, the role of nitric oxide in this model is more difficult to address. To overcome this issue, a recent study used a genetically manipulated strain of *M. tuberculosis* that only proliferates in the presence of streptomycin (Mishra *et al.*, 2013). Without the confounding effect of bacterial proliferation, this study shows that iNOS deficient mice were still highly susceptibility to disease, displaying increased accumulation of neutrophils (Mishra *et al.*, 2013). This phenotype was associated with increased activation of the NLRP3 inflammasome, a process dependent on Esx-1-induced phagosome rupture, which resulted in elevated production of IL-1β (Mishra *et al.*, 2013). The protective role of nitric oxide was shown to be the inhibition of the inflammasome activation through protein nitrosilation (Mishra *et al.*, 2013). It is however important to note that, in the absence of IL-1R-mediated signalling mice are acutely susceptible to infection (Mayer-Barber *et al.*, 2010). As for the IFN-α/β discussed above, these data show that an adequate level of IL-1 is required to control infection but in excess, IL-1 causes important immunopathological consequences to the host.

Together, these data suggest that the protective capacity of CD4 T cells is defined by their ability to rapidly migrate to the lung parenchyma and interact with infected macrophages, rather than by producing high levels of IFN-γ. The protective function of IFN-γ is mediated by activating macrophages to control intracellular growth but also by restraining immunopathological responses (Nandi and Behar, 2011). The inflammatory environment generated in the lesion site also regulates T-cell responses. More research is required to discriminate between the pathological and protective inflammatory mediators and their impact in the control of infection.

The impact of chronic infection in T-cell function and granuloma formation

During chronic *M. tuberculosis* infection in mice, effector T-cell numbers remain relatively stable (Moguche *et al.*, 2015; Winslow *et al.*, 2003), suggesting that these cells are exposed to chronic antigen stimulation (Moguche *et al.*, 2015). Over the last decade, several papers have shown that T cells under repetitive exposure to their cognate peptide may display defects in their ability to proliferate and produce cytokines, a condition known as exhaustion (Barber *et al.*, 2006; Joshi *et al.*, 2007). In tuberculosis, CD4 T-cell activation (as measured by production of IFN-γ) is suboptimal in the lungs of infected animals, even at the peak of the response (Bold *et al.*, 2011; Egen *et al.*, 2011). As discussed above, mycobacteria can promote the recruitment of permissive macrophages that express low levels of MHC-II (Antonelli *et al.*, 2010; Cambier *et al.*, 2014), but can also interfere with the expression and loading of antigens in MHC-II molecules (Hmama *et al.*, 2015). Therefore, the suboptimal activation of T cells may be associated with the reduced interaction with MHC-II molecules within the infection site (Srivastava and Ernst, 2013). Alternatively, T cells may develop an exhaustion phenotype. Indeed, the proliferative capacity of antigen-specific T cells reduces as the infection progresses (Reiley *et al.*, 2010), and this is a hallmark of exhausted T cells (Barber *et al.*, 2006).

One of the first papers to address T-cell exhaustion in tuberculosis determined the expression of two well-known markers of T-cell exhaustion in viral infections (Joshi *et al.*, 2007): Programmed Death 1 (PD-1) and KLRG1 (Reiley *et al.*, 2010). This study showed that, although during chronic *M. tuberculosis* infection T cells acquired the expression of either PD-1 or KLRG1, these are not markers of exhaustion during tuberculosis (Reiley *et al.*, 2010). Indeed, both populations produced IFN-γ and TNF-α upon restimulation, with the KLRG1 population producing the highest cytokine levels (Reiley *et al.*, 2010). Upon adoptive transfer into *M. tuberculosis* infected mice, the purified PD-1-expressing population maintained proliferation whereas the KLRG1-expressing population did not proliferate and their numbers declined over time (Reiley *et al.*, 2010). These data suggest that PD-1-expressing CD4 T cells represent a proliferative population, likely responsible for the maintenance of the antigen-specific T-cell pool throughout infection, while KLRG1-expressing CD4 T cells represent a terminally differentiated effector population (Reiley *et al.*, 2010). Recent data support these observations and show that when transferred into uninfected animals, PD-1-expressing CD4 T cells are better able to persist and to provide enhanced protection to *M. tuberculosis* infection when compared to terminally differentiated, KLRG1-expressing, CD4 T cells (Moguche *et al.*, 2015). Importantly, the PD-1-expressing CD4 T-cell population was shown to reside preferentially in the lung parenchyma, exhibit higher capacity to produce IL-2 and a small fraction expressed the chemokine receptor CXCR5 (Moguche *et al.*, 2015).

While these data show that PD-1-expressing CD4 T cells are required for protection against *M. tuberculosis* infection, mice deficient in PD-1 are extremely susceptible to infection, developing wasting disease associated with a massive accumulation of mycobacteria-specific T cells (Barber *et al.*, 2011; Lázár-Molnár *et al.*, 2010). On the other hand, KLRG1 deficient mice are more resistant to *M. tuberculosis* infection, associated with increased number of CD4 T cells (Cyktor *et al.*, 2013). Not only does this evidence the potential pathological role of the T-cell response, but also reveals that the PD-1 pathway is required to restrain pathological T-cell responses and prevent lethal disease.

Although PD-1 by itself may not be a marker of exhaustion, T cells express multiple inhibitory receptors during chronic *M. tuberculosis* infection, including TIM3, LAG-3, and 2B4 (Jayaraman *et al.*, 2016). Interestingly, T cells that coexpress TIM3 and PD1 produced less IFN-γ, IL-2 and TNF-α, but more IL-10 when compared with T cells expressing only TIM3, suggesting that T cells that coexpress multiple inhibitory receptors are more likely to be exhausted (Jayaraman *et al.*, 2016). Accordingly, TIM3 blockade resulted in increased frequencies of T cells producing IFN-γ, TNF-α and IL-2 and more importantly reduced bacterial burdens (Jayaraman *et al.*, 2016). These data show that T-cell exhaustion does develop during tuberculosis, and this may be an important mechanism to maintain chronic infection.

As T cells accumulate in the infection site, the granuloma begins to acquire a more intact structure, with a macrophage-rich centre surrounded by T and B cells. With time, some granulomata can undergo remodelling and accumulate necrotic material in the centre, leading to the formation of caseum. The caseum can then undergo liquefaction resulting in the fusion with an adjacent airway, a process known as cavitation, which facilitates bacterial dissemination (Marakalala *et al.*, 2016; Ndlovu and Marakalala, 2016).

Classically, the granulomatous response has been associated with a protective response that contains infection and prevents spreading of the bacteria to other organs. The fact that granuloma formation associates with control of mycobacterial infections has been supported by the hypersusceptibility to *M. tuberculosis* infection of humans and mice under various immunocompromising conditions, such as TNF-α blockade therapies, deficiency in the myeloid differentiation primary response protein 88 (MyD88) or the IL-12/IFN-γ axis (Bean *et al.*, 1999; Chakravarty *et al.*, 2008; Cooper *et al.*, 1993, 1997b; Flynn *et al.*, 1993; Fremond *et al.*, 2004; Lawn *et al.*, 2002). As discussed above, infections by mycobacteria result in complex bacteria–host interactions wherein the bacteria can subvert phagocyte function. Therefore, it is possible that the hypersusceptibility to mycobacterial infections in immunodeficient individuals is a combination of a decreased microbicidal capacity of the infected macrophage and poorly formed granulomata. Furthermore, the role of granulomata in preventing spread of the disease to peripheral organs is also questionable. As discussed in the sections above, most of the *M. tuberculosis* modulatory functions are associated with the lung environment and we cannot exclude the possibility that the protective response in peripheral organs is more efficient in controlling bacterial growth, than it is in the lung. In support of this hypothesis is the fact that control of *M. tuberculosis* infection is more efficient after intravenous challenge than after aerosol challenge (Cooper, 2009; Cooper *et al.*, 1997a). Therefore, whether granulomata formation is host-protective or represents an evolutionary adaption wherein the bacteria promote the formation of these immunological structures to spread to new host is an area of intense research. In fact, these issues were the subject of a recent review (Ndlovu and Marakalala, 2016).

While the benefit of granulomata to the host is questionable, granuloma formation is likely critical for *M. tuberculosis* to spread to new hosts. Caseation followed by liquefaction of the granuloma can lead to rupture of the lesion and the release of infectious material into the airways. This material can then be aerosolized in cough droplets, thus facilitating bacterial transmission to new hosts (Cardona, 2011). However, not all granulomata evolve into caseating lesions. Indeed, these lesions can be highly heterogeneous in a single host, despite the fact that they are likely founded by a single bacterium (Lin *et al.*, 2014). Importantly, a substantial number of granulomata in animals with active disease and latent infection can become sterile after the onset of adaptive immunity (Lin *et al.*, 2014). Therefore, the local microenvironment of each granuloma determines control or progression of infection.

Despite heterogeneity in granulomata from a single host, a recent study used laser-capture microdissection, mass spectrometry, and confocal microscopy in human intact, caseous, and cavitary granulomata samples to design molecular maps of different regions of each granuloma type (Marakalala *et al.*, 2016). This study identified more alterations in protein and lipid signatures associated with different regions of the granuloma than between different granulomata (Marakalala *et al.*, 2016). Interestingly, the centres of caseous and cavitary granulomata (presumably more advanced lesions) were enriched in pro-inflammatory pathways with the presence of antimicrobial peptides, reactive oxygen species and pro-inflammatory eicosanoids (Marakalala *et al.*, 2016). In contrast, the most exterior tissue surrounding the granuloma displayed an anti-inflammatory signature (Marakalala *et al.*, 2016).

The impact of this structural organization of the granuloma in the T-cell response, and vice versa, is not yet determined. Future work will be critical to determine whether the anti-inflammatory rim of the granuloma is an evolutionary adaption of the host to prevent further immunopathological consequences or an intricate strategy of the pathogen to prevent the entrance of antigen-specific T cells or modulate their phenotype and function. Whatever the answer, understanding how T cells accumulate and organize within these structures will be critical to define T cell correlates of protection. As granulomata are essential for *M. tuberculosis* to spread to new hosts, T cells are also essential for the success of *M. tuberculosis* as a human pathogen. Indeed, *M. tuberculosis* epitopes recognized by T cells are evolutionarily conserved (Comas *et al.*, 2010; Coscolla *et al.*, 2015), suggesting that *M. tuberculosis* benefits from T-cell recognition.

Taken together with the data discussed above, pro-inflammatory eicosanoids appear to play a key role in the T-cell response but also in the progression of the granulomatous response. During the early stages of infection pro-inflammatory eicosanoids induce necrotic cell death of infected-macrophages, thus delaying T-cell responses and promoting phagocytosis of the bacteria by newly recruited macrophages. After the onset of acquired responses, these macrophages are likely defective in interacting with T cells and more permissive to intracellular growth. As these cells die, a necrotic centre begins to form promoting its fusion with the airways, facilitating transmission of infection (Fig. 7.1).

Mycobacterium ulcerans infection: an extreme among mycobacteria

Although *M. ulcerans* and *M. tuberculosis* display a high degree of genetic similarity (Demangel *et al.*, 2009), the phenotypic clinical manifestations associated to infection with these mycobacteria differ greatly. This is, at least in part, due to the fact that *M. ulcerans* produces mycolactone, a dermonecrotic exotoxin that is responsible for the extensive destruction of the subcutaneous tissue and adjacent epidermis observed in Buruli ulcer patients (George *et al.*, 1999). Early clinical manifestations of Buruli ulcer include the non-ulcerative nodules, plaques and oedemas. However, due to the indolent nature of the lesions, there is often a delay in health-care seeking, which allows non-ulcerative manifestations to progress to more severe ulcerative forms (Walsh *et al.*, 2008).

Mycolactone is a small hydrophobic macrolide produced by a group of polyketide synthases and polyketide-modifying enzymes, encoded by a giant virulence plasmid pMUM001 (Stinear *et al.*, 2004). This toxin has been shown to passively diffuse through the cell membrane (Snyder and Small, 2003) rather than via a transporter-mediated mechanism and concentrates in the cytoplasm (Snyder and Small, 2003).

Despite the unique ability of *M. ulcerans* to secrete a toxin, the mycobacterial nature of *M. ulcerans* plays an important role in the development of a cell mediated immune response. Like other pathogenic mycobacteria, *M. ulcerans* is an intracellular pathogen that elicits a persistent recruitment of inflammatory cells to the site of infection, namely neutrophils and macrophages (Oliveira *et al.*, 2005). Initial recognition of *M. ulcerans* appears to be mediated through TLR-2, TLR-4 and Dectin-1 that actively participate in the sensing and internalization of the pathogen (Lee *et al.*, 2009), contributing to an antimycobacterial response with the release of reactive oxygen species (Lee *et al.*, 2009), reactive nitrogen species (Torrado *et al.*, 2010) and inflammatory cytokines and chemokines (Torrado *et al.*, 2007). However, the bacterial proliferation associated to the progressively increasing concentrations of mycolactone at the site of infection, eventually starts to hamper the potentially protective immune response mounted by the host.

Among the deleterious effects of mycolactone, the toxin has been associated to structural alterations of actin- and tubulin-cytoskeleton, that disturbs the cell shape as well as multiple cytoskeleton-dependent cellular functions. In fact, mycolactone-induced cytoskeletal alterations can be mediated through the hyperactivation of the actin-cytoskeleton regulator Wiskott-Aldrich syndrome protein (WASP) (Guenin-Macé *et al.*, 2013), a nucleation promoting factor that activates the Arp2/3 complex in order to stimulate nucleation of branched actin filaments (Burianek and Soderling, 2013). By hijacking and disrupting WASP autoinhibition, mycolactone leads to uncontrolled activation of WASP and results in defective cell adhesion and migration due to the accumulation of small, highly branched actin filaments, rather than long microfilaments (Guenin-Macé *et al.*, 2013). Mycolactone also affects cytoskeleton-dependent functions with important consequences for cellular homeostasis. A recent proteomic study revealed several regulators and structural constituents of the cytoskeleton that are affected upon exposure to mycolactone (Gama *et al.*, 2014). For example, the cytoplasmic dynein 1 intermediate chain 2 (Dync1i2), a non-catalytic subunit of the microtubule-associated molecular motor dynein, was shown to be down-regulated in mycolactone-treated cells (Gama *et al.*, 2014). Several cellular processes dependent on cytoskeleton architecture and dynein-driven transport have been reported to be affected by mycolactone. Specifically, the toxin can affect the transport of vesicles and organelles throughout the cytoskeleton tracks, allowing the pathogen to subvert innate microbicidal defences. Indeed, increasing concentrations of mycolactone impair phagosome–lysosome fusion in IFN-γ activated macrophages, which results in a deficient control in bacterial proliferation (Torrado *et al.*, 2010), as well an accumulation of autophagosomes, given their inability to fuse with lysosomes (Gama *et al.*, 2014).

Mycolactone does not act only upon the cytoskeleton. Another well-known effect of mycolactone is the down-regulation of specific proteins implicated in host immunity, such as cytokines and chemokines (Baron *et al.*, 2016; Hall *et al.*, 2014; Simmonds *et al.*, 2009; Torrado *et al.*, 2007). Interestingly, however, the reduced cytokine production during *M. ulcerans* infection does not appear to result from decreased gene transcription or even protein translation, but rather a blockage in the process of protein export (Hall *et al.*, 2014; Simmonds *et al.*, 2009). Indeed, mycolactone has been shown to directly target the α subunit of the Sec61 translocon (Baron *et al.*, 2016) and inhibit the cotranslational translocation of nascent proteins into the endoplasmatic reticulum due to a conformational change in the Sec61 translocation complex, to a stabilized closed conformation (Hall *et al.*, 2014; McKenna *et al.*, 2016). The mycolactone-dependent blockade of Sec61 translocation ultimately results in the accumulation and degradation of newly synthesized ER-transiting proteins in the cytoplasm (Hall *et al.*, 2014; Simmonds *et al.*, 2009). Accordingly, it was recently shown that there is an excessive accumulation of ubiquitnated proteins in fibroblasts treated with mycolactone (Gama *et al.*, 2014).

Despite the impact of mycolactone on the innate response, an acquired immune response is mounted, with the accumulation and proliferation of antigen-specific T cells in the draining lymph node (Fraga *et al.*, 2011). However, similar to what is observed in *M. tuberculosis* infection, there is a delay in the generation of acquired immunity, taking up to 7 days post infection to develop a specific T-cell response. This delay has been attributed, at least in part, to the ability of mycolactone to inhibit the maturation and migration of DCs to the draining lymph node (Coutanceau *et al.*, 2007). This delay allows exponential bacterial proliferation with the further build-up of mycolactone in the infected tissue, amplifying its deleterious effects. For example, the toxin has been shown to have a direct effect on T-cell responsiveness to stimulation by affecting the expression of the T-cell homing receptor CD62L, thus impairing their capacity to reach peripheral lymph nodes and respond to chemotactic signals (Guenin-Macé *et al.*, 2011). Moreover, mycolactone also induces the hyperactivation of the Src-family kinase, Lck, which results in a deregulation of intracellular signalling pathways involve in T-cell receptor activation (Boulkroun *et al.*, 2010). Ultimately, the uncontrolled proliferation of *M. ulcerans* leads to the complete destruction of the draining lymph node (Fraga *et al.*, 2011), either due to bacterial colonization as a result of dissemination of *M. ulcerans* via afferent lymphatic drainage or due to diffusion of mycolactone from the site of infection.

The end result of the effect of mycolactone is cell death. While *M. tuberculosis* can clearly manipulate the fate of the infected cell and determine the modality of cell death to its own advantage, *M. ulcerans* takes this trait to a whole new level. Indeed, the first identified function of mycolactone was the induction of cell death. It has been shown that mycolactone-treated cells present an arrest in the G0/G1-phase as early as 48 hours post incubation (Gama *et al.*, 2014; George *et al.*, 1999). Longer periods of incubation lead to an accumulation of the sub-G0/G1 population, compatible with apoptotic cells (Gama *et al.*, 2014; George *et al.*, 1999). Interestingly, the arrest of the G1 cell cycle progression has been shown to occur after drug-induced inhibition of actin assembly (Bohmer *et al.*, 1996; Huang *et al.*, 1998; Iwig *et al.*, 1995; Maness and Walsh, 1982; Reshetnikova *et al.*, 2000), further implicating the cytoskeleton in the mycolactone-mediated cytopathic effect. Additional evidence for the role of mycolactone as an apoptotic inducer are *in vivo* reports describing the presence of apoptotic cells in human Buruli ulcer lesions (Walsh *et al.*, 2005) and in infected lesions of experimental animal models (George *et al.*, 1999, 2000; Oliveira *et al.*, 2005; Silva-Gomes *et al.*, 2015). However, the continuous build-up of mycolactone during progressive *M. ulcerans* infection results in an overwhelming accumulation of apoptotic cells in the infected tissue. The inability of scavenger phagocytic cells to efficiently clear the tissue of these apoptotic cells eventually leads to secondary necrosis followed by autolysis. This post-apoptotic secondary necrosis is behind the progressively expanding necrotic areas observed in Buruli ulcer lesions (Silva *et al.*, 2009).

Despite the size of these necrotic lesions, with some exceeding 15 cm in diameter, Buruli ulcer is an indolent disease. Mycolactone is described to have an analgesic effect to annul the pain caused during *M. ulcerans* infection (Marion *et al.*, 2014). Hypoesthesia is achieved through the binding of mycolactone to type 2 angiotensin II receptors (AT2R), which triggers phospholipase A2 (PLA2) activation and arachidonic acid release (Marion *et al.*, 2014). Arachidonic acid is then metabolized by COX-1 into PGE2, leading to the activation of TRAAK channels and a consequent potassium-dependent hyperpolarization of neurons (Marion *et al.*, 2014). This strategy seems to provide an evolutionary advantage for *M. ulcerans*, in which the lack of pain could possibly allow an uncontrolled expansion of necrotic areas with numerous extracellular bacilli and the potential shedding of bacteria to the environment, perpetuating its infectious cycle.

Collectively, these data show that, like other mycobacteria, *M. ulcerans* is an intracellular pathogen and induces T-cell responses which are required to control infection. However, the pathophysiology of Buruli ulcer is dominated by the effect of mycolactone, which makes *M. ulcerans* unique within the genus.

Conclusion

The outcome of mycobacterial infections is dependent on the complex interaction between the intracellular pathogen and the immune system. Through this interaction, mycobacteria have evolved to adapt to its host by developing intricate mechanisms to subvert the immune response. However, the immune response may also contribute decisively for the success of mycobacteria as human pathogens. In this regard, *M. tuberculosis* has developed several mechanisms to modulate the activation and the effector function of the phagocytic cell, likely to prevent its rapid elimination rather than to completely shut down the immune system. The initial manipulation of the phagocyte response then allows the bacteria to survive within the host cell, while an antigen-specific immune response is still being mounted. This response eventually culminates in the formation of an inflammatory granulomatous structure that prevents the spreading of the bacteria to peripheral organs. However, its formation and breakdown has also proven to be essential for the bacteria to spread to new hosts. Understanding this duality will be important for the development of host-specific therapies. As an extreme example of mycobacterial survival, the immune evasion mechanisms used by *M. ulcerans* mainly depend on the actions of the necrotizing toxin mycolactone.

Acknowledgements

Our work is funded by the project NORTE-01-0145-FEDER-000013, supported by the Northern Portugal Regional Operational Programme (NORTE 2020), under the Portugal 2020 Partnership Agreement, through the European Regional Development Fund (FEDER); ERA-NET Infect-ERA (2015) grant BU_SPONT_HEAL; the Fundação para a Ciência e Tecnologia (FCT) through the FCT investigator grant IF/01390/2014 to E.T. and the postdoctoral grant SFRH/BPD/112903/2015 to A.G.F; and the doctoral fellowship PD/BD/113699/2015 to J.F.

References

Afonso-Barroso, A., Clark, S.O., Williams, A., Rosa, G.T., Nobrega, C., Silva-Gomes, S., Vale-Costa, S., Ummels, R., Stoker, N., Movahedzadeh, F., *et al.* (2013). Lipoarabinomannan mannose caps do not affect mycobacterial virulence or the induction of protective immunity in experimental animal models of infection and have minimal impact on in vitro inflammatory responses. Cell. Microbiol. *15*, 660–674. https://doi.org/10.1111/cmi.12065.

Andersen, P., Askgaard, D., Gottschau, A., Bennedsen, J., Nagai, S., and Heron, I. (1992). Identification of immunodominant antigens during infection with *Mycobacterium tuberculosis*. Scand. J. Immunol. *36*, 823–831. https://doi.org/10.1111/j.1365-3083.1992.tb03144.x.

Andersen, P., Askgaard, D., Ljungqvist, L., Bennedsen, J., and Heron, I. (1991). Proteins released from *Mycobacterium tuberculosis* during growth. Infect. Immun. *59*, 1905–1910.

Anderson, K.G., Mayer-Barber, K., Sung, H., Beura, L., James, B.R., Taylor, J.J., Qunaj, L., Griffith, T.S., Vezys, V., Barber, D.L., *et al.* (2014). Intravascular staining for discrimination of vascular and tissue leukocytes. Nat. Protoc. *9*, 209–222. https://doi.org/10.1038/nprot.2014.005.

Antonelli, L.R., Gigliotti Rothfuchs, A., Goncalves, R., Roffe, E., Cheever, A.W., Bafica, A., Salazar, A.M., Feng, C.G., and Sher, A. (2010). Intranasal Poly-IC treatment exacerbates tuberculosis in mice through the pulmonary recruitment of a pathogen-permissive monocyte/macrophage population. J. Clin. Invest. *120*, 1674–1682. https://doi.org/10.1172/JCI40817.

Appelmelk, B.J., den Dunnen, J., Driessen, N.N., Ummels, R., Pak, M., Nigou, J., Larrouy-Maumus, G., Gurcha, S.S., Movahedzadeh, F., Geurtsen, J., *et al.* (2008). The mannose cap of mycobacterial lipoarabinomannan does not dominate the *Mycobacterium*-host interaction. Cell. Microbiol. *10*, 930–944.

Armstrong, J.A., and Hart, P.D. (1971). Response of cultured macrophages to *Mycobacterium tuberculosis*, with observations on fusion of lysosomes with phagosomes. J. Exp. Med. *134*, 713–740. https://doi.org/10.1084/jem.134.3.713.

Ates, L.S., van der Woude, A.D., Bestebroer, J., van Stempvoort, G., Musters, R.J., Garcia-Vallejo, J.J., Picavet, D.I., Weerd, R.v., Maletta, M., Kuijl, C.P., *et al.* (2016). The ESX-5 System of Pathogenic Mycobacteria Is Involved In Capsule Integrity and Virulence through Its Substrate PPE10. PLOS Pathog. *12*, e1005696. https://doi.org/10.1371/journal.ppat.1005696.

Awuh, J.A., and Flo, T.H. (2016). Molecular basis of mycobacterial survival in macrophages. Cell. Mol. Life Sci. [Epub ahead of print]. https://doi.org/10.1007/s00018-016-2422-8.

Bach, H., Papavinasasundaram, K.G., Wong, D., Hmama, Z., and Av-Gay, Y. (2008). *Mycobacterium tuberculosis* virulence is mediated by PtpA dephosphorylation of human vacuolar protein sorting 33B. Cell Host Microbe *3*, 316–322. https://doi.org/10.1016/j.chom.2008.03.008.

Bafica, A., Scanga, C.A., Feng, C.G., Leifer, C., Cheever, A., and Sher, A. (2005). TLR9 regulates Th1 responses and cooperates with TLR2 in mediating optimal resistance to *Mycobacterium tuberculosis*. J. Exp. Med. *202*, 1715–1724.

Barber, D.L., Mayer-Barber, K.D., Feng, C.G., Sharpe, A.H., and Sher, A. (2011). CD4 T cells promote rather than control tuberculosis in the absence of PD-1-mediated inhibition. J. Immunol. *186*, 1598–1607. https://doi.org/10.4049/jimmunol.1003304.

Barber, D.L., Wherry, E.J., Masopust, D., Zhu, B., Allison, J.P., Sharpe, A.H., Freeman, G.J., and Ahmed, R. (2006). Restoring function in exhausted CD8 T cells during chronic viral infection. Nature *439*, 682–687.

Barber, G.N. (2011). Innate immune DNA sensing pathways: STING, AIMII and the regulation of interferon production and inflammatory responses. Curr. Opin. Immunol. *23*, 10–20. https://doi.org/10.1016/j.coi.2010.12.015.

Baron, L., Paatero, A.O., Morel, J.D., Impens, F., Guenin-Macé, L., Saint-Auret, S., Blanchard, N., Dillmann, R., Niang, F., Pellegrini, S., *et al.* (2016). Mycolactone subverts immunity by selectively blocking the Sec61 translocon. J. Exp. Med. *213*, 2885–2896.

Baron, S.D., Singh, R., and Metzger, D.W. (2007). Inactivated *Francisella tularensis* live vaccine strain protects against respiratory tularemia by intranasal vaccination in an immunoglobulin A-dependent fashion. Infect. Immun. *75*, 2152–2162. https://doi.org/10.1128/IAI.01606-06.

Bean, A.G., Roach, D.R., Briscoe, H., France, M.P., Korner, H., Sedgwick, J.D., and Britton, W.J. (1999). Structural deficiencies in granuloma formation in TNF gene-targeted mice underlie the heightened susceptibility to aerosol *Mycobacterium tuberculosis* infection, which is not compensated for by lymphotoxin. J. Immunol. *162*, 3504–3511.

Behar, S.M. (2013). Antigen-specific CD8(+) T cells and protective immunity to tuberculosis. Adv. Exp. Med. Biol. *783*, 141–163. https://doi.org/10.1007/978-1-4614-6111-1_8.

Behar, S.M., Divangahi, M., and Remold, H.G. (2010). Evasion of innate immunity by *Mycobacterium tuberculosis*: is death an exit strategy? Nat. Rev. Microbiol. *8*, 668–674. https://doi.org/10.1038/nrmicro2387.

Berry, M.P., Graham, C.M., McNab, F.W., Xu, Z., Bloch, S.A., Oni, T., Wilkinson, K.A., Banchereau, R., Skinner, J., Wilkinson, R.J., *et al.* (2010). An interferon-inducible neutrophil-driven blood transcriptional signature in human tuberculosis. Nature *466*, 973–977. https://doi.org/10.1038/nature09247.

Bohmer, R.M., Scharf, E., and Assoian, R.K. (1996). Cytoskeletal integrity is required throughout the mitogen stimulation phase of the cell cycle and mediates the anchorage-dependent expression of cyclin D1. Mol. Biol. Cell *7*, 101–111. https://doi.org/10.1091/mbc.7.1.101.

Bold, T.D., Banaei, N., Wolf, A.J., and Ernst, J.D. (2011). Suboptimal activation of antigen-specific CD4+ effector cells enables persistence of *M. tuberculosis* in vivo. PLOS Pathog. *7*, e1002063. https://doi.org/10.1371/journal.ppat.1002063.

Boniface, K., Bak-Jensen, K.S., Li, Y., Blumenschein, W.M., McGeachy, M.J., McClanahan, T.K., McKenzie, B.S., Kastelein, R.A., Cua, D.J., and de Waal Malefyt, R. (2009). Prostaglandin E2 regulates Th17 cell differentiation and function through cyclic AMP and EP2/EP4 receptor signaling. J. Exp. Med. *206*, 535–548. https://doi.org/10.1084/jem.20082293.

Boulkroun, S., Guenin-Macé, L., Thoulouze, M.I., Monot, M., Merckx, A., Langsley, G., Bismuth, G., Di Bartolo, V., and Demangel, C. (2010). Mycolactone suppresses T cell responsiveness by altering both early signaling and posttranslational events. J. Immunol. *184*, 1436–1444. https://doi.org/10.4049/jimmunol.0902854.

Brennan, P.J., and Nikaido, H. (1995). The envelope of mycobacteria. Annu. Rev. Biochem. *64*, 29–63. https://doi.org/10.1146/annurev.bi.64.070195.000333.

Burianek, L.E., and Soderling, S.H. (2013). Under lock and key: spatiotemporal regulation of WASP family proteins coordinates separate dynamic cellular processes. Semin. Cell. Dev. Biol. *24*, 258–266. https://doi.org/10.1016/j.semcdb.2012.12.005.

Cambier, C.J., Takaki, K.K., Larson, R.P., Hernandez, R.E., Tobin, D.M., Urdahl, K.B., Cosma, C.L., and Ramakrishnan, L. (2014). Mycobacteria manipulate macrophage recruitment through coordinated use of membrane lipids. Nature *505*, 218-222. https://doi.org/10.1038/nature12799.

Cardona, P.J. (2011). A spotlight on liquefaction: evidence from clinical settings and experimental models in tuberculosis. Clin. Dev. Immunol. *2011*, 868246. https://doi.org/10.1155/2011/868246.

Carmona, J., Cruz, A., Moreira-Teixeira, L., Sousa, C., Sousa, J., Osorio, N.S., Saraiva, A.L., Svenson, S., Kallenius, G., Pedrosa, J., *et al.* (2013). *Mycobacterium tuberculosis* strains are differentially recognized by TLRs with an impact on the immune response. PLOS ONE *8*, e67277. https://doi.org/10.1371/journal.pone.0067277.

Caron, E., and Hall, A. (1998). Identification of two distinct mechanisms of phagocytosis controlled by different Rho GTPases. Science *282*, 1717–1721.

Casanova, J.L., and Abel, L. (2002). Genetic dissection of immunity to mycobacteria: the human model. Annu. Rev. Immunol. *20*, 581–620. https://doi.org/10.1146/annurev.immunol.20.081501.125851.

Castillo, E.F., Dekonenko, A., Arko-Mensah, J., Mandell, M.A., Dupont, N., Jiang, S., Delgado-Vargas, M., Timmins, G.S., Bhattacharya, D., Yang, H., *et al.* (2012). Autophagy protects against active tuberculosis by suppressing bacterial burden and inflammation. Proc. Natl. Acad. Sci. U.S.A. *109*, E3168–76. https://doi.org/10.1073/pnas.1210500109.

Chakravarty, S.D., Zhu, G., Tsai, M.C., Mohan, V.P., Marino, S., Kirschner, D.E., Huang, L., Flynn, J., and Chan, J. (2008). Tumor necrosis factor blockade in chronic murine tuberculosis enhances granulomatous inflammation and disorganizes granulomas in the lungs. Infect. Immun. *76*, 916–926. https://doi.org/10.1128/IAI.01011-07.

Chen, M., Divangahi, M., Gan, H., Shin, D.S., Hong, S., Lee, D.M., Serhan, C.N., Behar, S.M., and Remold, H.G. (2008). Lipid mediators in innate immunity against tuberculosis: opposing roles of PGE2 and LXA4 in the induction of macrophage death. J. Exp. Med. *205*, 2791–2801. https://doi.org/10.1084/jem.20080767.

Chen, M., Gan, H., and Remold, H.G. (2006). A mechanism of virulence: virulent *Mycobacterium tuberculosis* strain H37Rv, but not attenuated H37Ra, causes significant mitochondrial inner membrane disruption in macrophages leading to necrosis. J Immunol *176*, 3707-3716. https://doi.org/10.4049/jimmunol.176.6.3707.

Comas, I., Chakravartti, J., Small, P.M., Galagan, J., Niemann, S., Kremer, K., Ernst, J.D., and Gagneux, S. (2010). Human T cell epitopes of *Mycobacterium tuberculosis* are evolutionarily hyperconserved. Nat. Genet. *42*, 498–503. https://doi.org/10.1038/ng.590.

Comas, I., Coscolla, M., Luo, T., Borrell, S., Holt, K.E., Kato-Maeda, M., Parkhill, J., Malla, B., Berg, S., Thwaites, G., *et al.* (2013). Out-of-Africa migration and Neolithic coexpansion of *Mycobacterium tuberculosis* with modern humans. Nat. Genet. *45*, 1176–1182. https://doi.org/10.1038/ng.2744.

Cooper, A.M. (2009). Cell-mediated immune responses in tuberculosis. Annu. Rev. Immunol. *27*, 393–422. https://doi.org/10.1146/annurev.immunol.021908.132703.

Cooper, A.M., Adams, L.B., Dalton, D.K., Appelberg, R., and Ehlers, S. (2002). IFN-gamma and NO in mycobacterial disease: new jobs for old hands. Trends Microbiol. *10*, 221–226.

Cooper, A.M., Callahan, J.E., Keen, M., Belisle, J.T., and Orme, I.M. (1997a). Expression of memory immunity in the lung following re-exposure to *Mycobacterium tuberculosis*. Tuber. Lung Dis. *78*, 67–73.

Cooper, A.M., Dalton, D.K., Stewart, T.A., Griffin, J.P., Russell, D.G., and Orme, I.M. (1993). Disseminated tuberculosis in interferon gamma gene-disrupted mice. J. Exp. Med. *178*, 2243–2247.

Cooper, A.M., and Khader, S.A. (2008). The role of cytokines in the initiation, expansion, and control of cellular immunity to tuberculosis. Immunol. Rev. *226*, 191–204. https://doi.org/10.1111/j.1600-065X.2008.00702.x.

Cooper, A.M., Magram, J., Ferrante, J., and Orme, I.M. (1997b). Interleukin 12 (IL-12) is crucial to the development of protective immunity in mice intravenously infected with *Mycobacterium tuberculosis*. J. Exp. Med. *186*, 39–45.

Coscolla, M., Copin, R., Sutherland, J., Gehre, F., de Jong, B., Owolabi, O., Mbayo, G., Giardina, F., Ernst, J.D., and Gagneux, S. (2015). *M. tuberculosis* T Cell Epitope Analysis Reveals Paucity of Antigenic Variation and Identifies Rare Variable TB Antigens. Cell Host Microbe *18*, 538–548.

Coutanceau, E., Decalf, J., Martino, A., Babon, A., Winter, N., Cole, S.T., Albert, M.L., and Demangel, C. (2007). Selective suppression of dendritic cell functions by *Mycobacterium ulcerans* toxin mycolactone. J. Exp. Med. *204*, 1395–1403.

Cruz, A., Fraga, A.G., Fountain, J.J., Rangel-Moreno, J., Torrado, E., Saraiva, M., Pereira, D.R., Randall, T.D., Pedrosa, J., Cooper, A.M., *et al.* (2010). Pathological role of interleukin 17 in mice subjected to repeated BCG vaccination after infection with *Mycobacterium tuberculosis*. J. Exp. Med. *207*, 1609–1616. https://doi.org/10.1084/jem.20100265.

Cruz, A., Ludovico, P., Torrado, E., Gama, J.B., Sousa, J., Gaifem, J., Appelberg, R., Rodrigues, F., Cooper, A.M., Pedrosa, J., *et al.* (2015a). IL-17A promotes intracellular growth of *Mycobacterium* by inhibiting apoptosis of infected macrophages. Front. Immunol. *6*, 498. https://doi.org/10.3389/fimmu.2015.00498.

Cruz, A., Torrado, E., Carmona, J., Fraga, A.G., Costa, P., Rodrigues, F., Appelberg, R., Correia-Neves, M., Cooper, A.M., Saraiva, M., *et al.* (2015b). BCG vaccination-induced long-lasting control of *Mycobacterium tuberculosis* correlates with the accumulation of a novel population of $CD4^{+}IL\text{-}17^{+}TNF^{+}IL\text{-}2^{+}$ T cells. Vaccine 33, 85–91. https://doi.org/10.1016/j.vaccine.2014.11.013.

Cyktor, J.C., Carruthers, B., Stromberg, P., Flano, E., Pircher, H., and Turner, J. (2013). Killer cell lectin-like receptor G1 deficiency significantly enhances survival after *Mycobacterium tuberculosis* infection. Infect. Immun. *81*, 1090–1099. https://doi.org/10.1128/IAI.01199-12.

Danilchanka, O., Sun, J., Pavlenok, M., Maueröder, C., Speer, A., Siroy, A., Marrero, J., Trujillo, C., Mayhew, D.L., Doornbos, K.S., *et al.* (2014). An outer membrane channel protein of *Mycobacterium tuberculosis* with exotoxin activity. Proc. Natl. Acad. Sci. U.S.A. *111*, 6750–6755. https://doi.org/10.1073/pnas.1400136111.

Demangel, C., Stinear, T.P., and Cole, S.T. (2009). Buruli ulcer: reductive evolution enhances pathogenicity of *Mycobacterium ulcerans*. Nat. Rev. Microbiol. 7, 50–60. https://doi.org/10.1038/nrmicro2077.

Desvignes, L., Wolf, A.J., and Ernst, J.D. (2012). Dynamic roles of type I and type II IFNs in early infection with *Mycobacterium tuberculosis*. J. Immunol. *188*, 6205–6215. https://doi.org/10.4049/jimmunol.1200255.

Divangahi, M., Chen, M., Gan, H., Desjardins, D., Hickman, T.T., Lee, D.M., Fortune, S., Behar, S.M., and Remold, H.G. (2009). *Mycobacterium tuberculosis* evades macrophage defenses by inhibiting plasma membrane repair. Nat. Immunol. *10*, 899–906. https://doi.org/10.1038/ni.1758.

Divangahi, M., Desjardins, D., Nunes-Alves, C., Remold, H.G., and Behar, S.M. (2010). Eicosanoid pathways regulate adaptive immunity to *Mycobacterium tuberculosis*. Nat. Immunol. *11*, 751–758. https://doi.org/10.1038/ni.1904.

Domingo-Gonzalez, R., Prince, O., Cooper, A., and Khader, S.A. (2016). Cytokines and Chemokines in *Mycobacterium tuberculosis* Infection. Microbiol. Spectr. *4*. https://doi.org/10.1128/microbiolspec.TBTB2-0018-2016.

Driessen, N.N., Ummels, R., Maaskant, J.J., Gurcha, S.S., Besra, G.S., Ainge, G.D., Larsen, D.S., Painter, G.F., Vandenbroucke-Grauls, C.M., Geurtsen, J., *et al.* (2009). Role of phosphatidylinositol mannosides in the interaction between mycobacteria and DC-SIGN. Infect. Immun. *77*, 4538–4547. https://doi.org/10.1128/IAI.01256-08.

Egen, J.G., Rothfuchs, A.G., Feng, C.G., Horwitz, M.A., Sher, A., and Germain, R.N. (2011). Intravital imaging reveals limited antigen presentation and T cell effector function in mycobacterial granulomas. Immunity *34*, 807–819. https://doi.org/10.1016/j.immuni.2011.03.022.

Flynn, J.L., Chan, J., Triebold, K.J., Dalton, D.K., Stewart, T.A., and Bloom, B.R. (1993). An essential role for interferon gamma in resistance to *Mycobacterium tuberculosis* infection. J. Exp. Med. *178*, 2249–2254.

Flynn, K.J., Belz, G.T., Altman, J.D., Ahmed, R., Woodland, D.L., and Doherty, P.C. (1998). Virus-specific CD8+ T cells in primary and secondary influenza pneumonia. Immunity *8*, 683–691.

Fraga, A.G., Cruz, A., Martins, T.G., Torrado, E., Saraiva, M., Pereira, D.R., Meyers, W.M., Portaels, F., Silva, M.T., Castro, A.G., *et al.* (2011). *Mycobacterium ulcerans* triggers T-cell immunity followed by local and regional but not systemic immunosuppression. Infect. Immun. *79*, 421–430. https://doi.org/10.1128/IAI.00820-10.

Fremond, C.M., Yeremeev, V., Nicolle, D.M., Jacobs, M., Quesniaux, V.F., and Ryffel, B. (2004). Fatal *Mycobacterium tuberculosis* infection despite adaptive immune response in the absence of MyD88. J. Clin. Invest. *114*, 1790–1799. https://doi.org/10.1172/JCI21027.

Gallegos, A.M., Pamer, E.G., and Glickman, M.S. (2008). Delayed protection by ESAT-6-specific effector CD4+ T cells after airborne *M. tuberculosis* infection. J. Exp. Med. *205*, 2359–2368. https://doi.org/10.1084/jem.20080353.

Gama, J.B., Ohlmeier, S., Martins, T.G., Fraga, A.G., Sampaio-Marques, B., Carvalho, M.A., Proenca, F., Silva, M.T., Pedrosa, J., and Ludovico, P. (2014). Proteomic analysis of the action of the *Mycobacterium ulcerans* toxin mycolactone: targeting host cells cytoskeleton and collagen. PLOS Negl. Trop. Dis. *8*, e3066. https://doi.org/10.1371/journal.pntd.0003066.

Geijtenbeek, T.B., Van Vliet, S.J., Koppel, E.A., Sanchez-Hernandez, M., Vandenbroucke-Grauls, C.M., Appelmelk, B., and Van Kooyk, Y. (2003). Mycobacteria target DC-SIGN to suppress dendritic cell function. J. Exp. Med. *197*, 7–17.

Geldmacher, C., Zumla, A., and Hoelscher, M. (2012). Interaction between HIV and *Mycobacterium tuberculosis*: HIV-1-induced CD4 T-cell depletion and the development of active tuberculosis. Curr. Opin. HIV AIDS 7, 268–275. https://doi.org/10.1097/COH.0b013e3283524e32.

George, K.M., Chatterjee, D., Gunawardana, G., Welty, D., Hayman, J., Lee, R., and Small, P.L. (1999). Mycolactone: a polyketide toxin from *Mycobacterium ulcerans* required for virulence. Science *283*, 854–857.

George, K.M., Pascopella, L., Welty, D.M., and Small, P.L. (2000). A *Mycobacterium ulcerans* toxin, mycolactone, causes apoptosis in guinea pig ulcers and tissue culture cells. Infect. Immun. *68*, 877–883. https://doi.org/10.1128/IAI.68.2.877-883.2000.

Geurtsen, J., Chedammi, S., Mesters, J., Cot, M., Driessen, N.N., Sambou, T., Kakutani, R., Ummels, R., Maaskant, J., Takata, H., *et al.* (2009). Identification of mycobacterial alpha-glucan as a novel ligand for DC-SIGN: involvement of mycobacterial capsular polysaccharides in host immune modulation. J. Immunol. *183*, 5221–5231. https://doi.org/10.4049/jimmunol.0900768.

Gomes, M.S., Flórido, M., Pais, T.F., and Appelberg, R. (1999). Improved clearance of *Mycobacterium avium* upon disruption of the inducible nitric oxide synthase gene. J. Immunol. *162*, 6734–6739.

Gopal, R., Lin, Y., Obermajer, N., Slight, S., Nuthalapati, N., Ahmed, M., Kalinski, P., and Khader, S.A. (2012). IL-23-dependent IL-17 drives Th1-cell responses following *Mycobacterium bovis* BCG vaccination. Eur. J. Immunol. *42*, 364–373. https://doi.org/10.1002/eji.201141569.

Gopal, R., Monin, L., Slight, S., Uche, U., Blanchard, E., Fallert Junecko, B.A., Ramos-Payan, R., Stallings, C.L., Reinhart, T.A., Kolls, J.K., *et al.* (2014). Unexpected role for IL-17 in protective immunity against hypervirulent *Mycobacterium tuberculosis* HN878 infection. PLOS Pathog. *10*, e1004099. https://doi.org/10.1371/journal.ppat.1004099.

Guenin-Macé, L., Carrette, F., Asperti-Boursin, F., Le Bon, A., Caleechurn, L., Di Bartolo, V., Fontanet, A., Bismuth, G., and Demangel, C. (2011). Mycolactone impairs T cell homing by suppressing microRNA control of L-selectin expression. Proc. Natl. Acad. Sci. U.S.A. *108*, 12833–12838. https://doi.org/10.1073/pnas.1016496108.

Guenin-Macé, L., Veyron-Churlet, R., Thoulouze, M.I., Romet-Lemonne, G., Hong, H., Leadlay, P.F., Danckaert, A., Ruf, M.T., Mostowy, S., Zurzolo, C., *et al.* (2013). Mycolactone activation of Wiskott-Aldrich syndrome proteins underpins Buruli ulcer formation. J. Clin. Invest. *123*, 1501–1512. https://doi.org/10.1172/JCI66576.

Gutierrez, M.G., Master, S.S., Singh, S.B., Taylor, G.A., Colombo, M.I., and Deretic, V. (2004). Autophagy is a defense mechanism inhibiting BCG and *Mycobacterium tuberculosis* survival in infected macrophages. Cell *119*, 753–766. https://doi.org/10.1016/j.cell.2004.11.038.

Hall, B.S., Hill, K., McKenna, M., Ogbechi, J., High, S., Willis, A.E., and Simmonds, R.E. (2014). The pathogenic mechanism of the *Mycobacterium ulcerans* virulence factor, mycolactone, depends on blockade of protein translocation into the ER. PLOS Pathog. *10*, e1004061. https://doi.org/10.1371/journal.ppat.1004061.

Hervas-Stubbs, S., Majlessi, L., Simsova, M., Morova, J., Rojas, M.J., Nouze, C., Brodin, P., Sebo, P., and Leclerc, C. (2006). High frequency of CD4+ T cells specific for the TB10.4 protein correlates with protection against *Mycobacterium tuberculosis* infection. Infect. Immun. *74*, 3396–3407. https://doi.org/10.1128/IAI.02086-05.

Hmama, Z., Peña-Díaz, S., Joseph, S., and Av-Gay, Y. (2015). Immunoevasion and immunosuppression of the macrophage by *Mycobacterium tuberculosis*. Immunol. Rev. *264*, 220–232. https://doi.org/10.1111/imr.12268.

Holscher, C., Holscher, A., Ruckerl, D., Yoshimoto, T., Yoshida, H., Mak, T., Saris, C., and Ehlers, S. (2005). The IL-27 receptor chain WSX-1 differentially regulates antibacterial immunity and survival during experimental tuberculosis. J. Immunol. *174*, 3534–3544. https://doi.org/10.4049/jimmunol.174.6.3534.

Hsu, T., Hingley-Wilson, S.M., Chen, B., Chen, M., Dai, A.Z., Morin, P.M., Marks, C.B., Padiyar, J., Goulding, C., Gingery, M., *et al.* (2003). The primary mechanism of attenuation of bacillus Calmette-Guerin is a loss of secreted lytic function required for invasion of lung interstitial tissue. Proc. Natl. Acad. Sci. U.S.A. *100*, 12420–12425. https://doi.org/10.1073/pnas.1635213100.

Huang, S., Chen, C.S., and Ingber, D.E. (1998). Control of cyclin D1, p27(Kip1), and cell cycle progression in human capillary endothelial cells by cell shape and cytoskeletal tension. Mol. Biol. Cell *9*, 3179–3193.

Iwig, M., Czeslick, E., Müller, A., Gruner, M., Spindler, M., and Glaesser, D. (1995). Growth regulation by cell shape alteration and organization of the cytoskeleton. Eur. J. Cell Biol. *67*, 145–157.

Jayaraman, P., Jacques, M.K., Zhu, C., Steblenko, K.M., Stowell, B.L., Madi, A., Anderson, A.C., Kuchroo, V.K., and Behar, S.M. (2016). TIM3 Mediates T Cell Exhaustion during *Mycobacterium tuberculosis* Infection. PLOS Pathog. *12*, e1005490. https://doi.org/10.1371/journal.ppat.1005490.

Joshi, N.S., Cui, W., Chandele, A., Lee, H.K., Urso, D.R., Hagman, J., Gapin, L., and Kaech, S.M. (2007). Inflammation directs memory precursor and short-lived effector CD8(+) T cell fates via the graded expression of T-bet transcription factor. Immunity *27*, 281–295. https://doi.org/10.1016/j.immuni.2007.07.010.

Kallenius, G., Correia-Neves, M., Buteme, H., Hamasur, B., and Svenson, S.B. (2016). Lipoarabinomannan, and its related glycolipids, induce divergent and opposing immune responses to *Mycobacterium tuberculosis* depending on structural diversity and experimental variations. Tuberculosis (Edinb.) *96*, 120–130. https://doi.org/10.1016/j.tube.2015.09.005.

Kang, P.B., Azad, A.K., Torrelles, J.B., Kaufman, T.M., Beharka, A., Tibesar, E., DesJardin, L.E., and Schlesinger, L.S. (2005). The human macrophage mannose receptor directs *Mycobacterium tuberculosis* lipoarabinomannan-mediated phagosome biogenesis. J. Exp. Med. *202*, 987–999. https://doi.org/10.1084/jem.20051239.

Khader, S.A., Bell, G.K., Pearl, J.E., Fountain, J.J., Rangel-Moreno, J., Cilley, G.E., Shen, F., Eaton, S.M., Gaffen, S.L., Swain, S.L., *et al.* (2007). IL-23 and IL-17 in the establishment of protective pulmonary CD4+ T cell responses after vaccination and during *Mycobacterium tuberculosis* challenge. Nat. Immunol. *8*, 369–377. https://doi.org/10.1038/ni1449.

Khader, S.A., Guglani, L., Rangel-Moreno, J., Gopal, R., Junecko, B.A., Fountain, J.J., Martino, C., Pearl, J.E., Tighe, M., Lin, Y.Y., *et al.* (2011). IL-23 is required for long-term control of *Mycobacterium tuberculosis* and B cell follicle formation in the infected lung. J. Immunol. *187*, 5402–5407. https://doi.org/10.4049/jimmunol.1101377.

Killick, K.E., Ní Cheallaigh, C., O'Farrelly, C., Hokamp, K., MacHugh, D.E., and Harris, J. (2013). Receptor-mediated recognition of mycobacterial pathogens. Cell. Microbiol. *15*, 1484–1495. https://doi.org/10.1111/cmi.12161.

Kim, K.H., An, D.R., Song, J., Yoon, J.Y., Kim, H.S., Yoon, H.J., Im, H.N., Kim, J., Kim, D.J., Lee, S.J., *et al.* (2012). *Mycobacterium tuberculosis* Eis protein initiates suppression of host immune responses by acetylation of DUSP16/MKP-7. Proc. Natl. Acad. Sci. U.S.A. *109*, 7729–7734. https://doi.org/10.1073/pnas.1120251109.

Kimmey, J.M., Huynh, J.P., Weiss, L.A., Park, S., Kambal, A., Debnath, J., Virgin, H.W., and Stallings, C.L. (2015). Unique role for ATG5 in neutrophil-mediated immunopathology during *M. tuberculosis* infection. Nature *528*, 565–569. https://doi.org/10.1038/nature16451.

Kursar, M., Bonhagen, K., Köhler, A., Kamradt, T., Kaufmann, S.H., and Mittrücker, H.W. (2002). Organ-specific CD4+ T cell response during *Listeria monocytogenes* infection. J. Immunol. *168*, 6382–6387.

Lawn, S.D., Butera, S.T., and Shinnick, T.M. (2002). Tuberculosis unleashed: the impact of human immunodeficiency virus infection on the host granulomatous response to *Mycobacterium tuberculosis*. Microbes Infect. *4*, 635-646. https://doi.org/10.1016/S1286-4579(02)01582-4.

Lázár-Molnár, E., Chen, B., Sweeney, K.A., Wang, E.J., Liu, W., Lin, J., Porcelli, S.A., Almo, S.C., Nathenson, S.G., and Jacobs, W.R. (2010). Programmed death-1 (PD-1)-deficient mice are extraordinarily sensitive to tuberculosis. Proc. Natl. Acad. Sci. U.S.A. *107*, 13402–13407. https://doi.org/10.1073/pnas.1007394107.

Lee, H.M., Shin, D.M., Choi, D.K., Lee, Z.W., Kim, K.H., Yuk, J.M., Kim, C.D., Lee, J.H., and Jo, E.K. (2009). Innate immune responses to *Mycobacterium ulcerans* via toll-like receptors and dectin-1 in human keratinocytes. Cell. Microbiol. *11*, 678–692. https://doi.org/10.1111/j.1462-5822.2009.01285.x.

Lewis, K.N., Liao, R., Guinn, K.M., Hickey, M.J., Smith, S., Behr, M.A., and Sherman, D.R. (2003). Deletion of RD1 from *Mycobacterium tuberculosis* mimics bacille Calmette-Guerin attenuation. J. Infect. Dis. *187*, 117–123. https://doi.org/10.1086/345862.

Lin, P.L., and Flynn, J.L. (2015). CD8 T cells and *Mycobacterium tuberculosis* infection. Semin. Immunopathol. *37*, 239–249. https://doi.org/10.1007/s00281-015-0490-8.

Lin, P.L., Ford, C.B., Coleman, M.T., Myers, A.J., Gawande, R., Ioerger, T., Sacchettini, J., Fortune, S.M., and Flynn, J.L. (2014). Sterilization of granulomas is common in active and latent tuberculosis despite within-host variability in bacterial killing. Nat. Med. *20*, 75–79. https://doi.org/10.1038/nm.3412.

Manca, C., Tsenova, L., Bergtold, A., Freeman, S., Tovey, M., Musser, J.M., Barry, C.E., 3rd, Freedman, V.H., and Kaplan, G. (2001). Virulence of a *Mycobacterium tuberculosis* clinical isolate in mice is determined by failure to induce Th1 type immunity and is associated with induction of IFN-alpha /beta. Proc. Natl. Acad. Sci. U.S.A. *98*, 5752–5757. https://doi.org/10.1073/pnas.091096998.

Maness, P.F., and Walsh, R.C. (1982). Dihydrocytochalasin B disorganizes actin cytoarchitecture and inhibits initiation of DNA synthesis in 3T3 cells. Cell *30*, 253–262.

Manzanillo, P.S., Shiloh, M.U., Portnoy, D.A., and Cox, J.S. (2012). *Mycobacterium tuberculosis* activates the DNA-dependent cytosolic surveillance pathway within macrophages. Cell Host Microbe *11*, 469-480. https://doi.org/10.1016/j.chom.2012.03.007.

Marakalala, M.J., Raju, R.M., Sharma, K., Zhang, Y.J., Eugenin, E.A., Prideaux, B., Daudelin, I.B., Chen, P.Y., Booty, M.G., Kim, J.H., *et al.* (2016). Inflammatory signaling in human tuberculosis granulomas is spatially organized. Nat. Med. *22*, 531–538. https://doi.org/10.1038/nm.4073.

Marion, E., Song, O.R., Christophe, T., Babonneau, J., Fenistein, D., Eyer, J., Letournel, F., Henrion, D., Clere, N., Paille, V., *et al.* (2014). Mycobacterial toxin induces analgesia in buruli ulcer by targeting the angiotensin pathways. Cell *157*, 1565–1576. https://doi.org/10.1016/j.cell.2014.04.040.

Martin, C.J., Booty, M.G., Rosebrock, T.R., Nunes-Alves, C., Desjardins, D.M., Keren, I., Fortune, S.M., Remold, H.G., and Behar, S.M. (2012). Efferocytosis is an innate antibacterial mechanism. Cell Host Microbe *12*, 289–300. https://doi.org/10.1016/j.chom.2012.06.010.

Martin, C.J., Peters, K.N., and Behar, S.M. (2014). Macrophages clean up: efferocytosis and microbial control. Curr. Opin. Microbiol. *17*, 17–23. https://doi.org/10.1016/j.mib.2013.10.007.

Mayer-Barber, K.D., Andrade, B.B., Oland, S.D., Amaral, E.P., Barber, D.L., Gonzales, J., Derrick, S.C., Shi, R., Kumar, N.P., Wei, W., *et al.* (2014). Host-directed therapy of tuberculosis based on interleukin-1 and type I interferon cross-talk. Nature *511*, 99–103. https://doi.org/10.1038/nature13489.

Mayer-Barber, K.D., Barber, D.L., Shenderov, K., White, S.D., Wilson, M.S., Cheever, A., Kugler, D., Hieny, S., Caspar, P., Núñez, G., *et al.* (2010). Caspase-1 independent IL-1beta production is critical for host resistance to *Mycobacterium tuberculosis* and does not require TLR signaling in vivo. J. Immunol. *184*, 3326–3330. https://doi.org/10.4049/jimmunol.0904189.

McKenna, M., Simmonds, R.E., and High, S. (2016). Mechanistic insights into the inhibition of Sec61-dependent co- and post-translational translocation by mycolactone. J. Cell Sci. *129*, 1404–1415. https://doi.org/10.1242/jcs.182352.

McNab, F.W., Ewbank, J., Howes, A., Moreira-Teixeira, L., Martirosyan, A., Ghilardi, N., Saraiva, M., and O'Garra, A. (2014). Type I IFN induces IL-10 production in an IL-27-independent manner and blocks responsiveness to IFN-gamma for production of IL-12 and bacterial killing in *Mycobacterium tuberculosis*-infected macrophages. J. Immunol. *193*, 3600–3612. https://doi.org/10.4049/jimmunol.1401088.

Mehra, A., Zahra, A., Thompson, V., Sirisaengtaksin, N., Wells, A., Porto, M., Köster, S., Penberthy, K., Kubota, Y., Dricot, A., *et al.* (2013). *Mycobacterium tuberculosis* type VII secreted effector EsxH targets host ESCRT to impair trafficking. PLOS Pathog. *9*, e1003734. https://doi.org/10.1371/journal.ppat.1003734.

Merritt, R.W., Walker, E.D., Small, P.L., Wallace, J.R., Johnson, P.D., Benbow, M.E., and Boakye, D.A. (2010). Ecology and transmission of Buruli ulcer disease: a systematic review. PLOS Negl. Trop. Dis. *4*, e911. https://doi.org/10.1371/journal.pntd.0000911.

Mishra, B.B., Rathinam, V.A., Martens, G.W., Martinot, A.J., Kornfeld, H., Fitzgerald, K.A., and Sassetti, C.M. (2013). Nitric oxide controls the immunopathology of tuberculosis by inhibiting NLRP3 inflammasome-dependent processing of IL-1β. Nat. Immunol. *14*, 52–60. https://doi.org/10.1038/ni.2474.

Mizushima, N., Yoshimori, T., and Ohsumi, Y. (2011). The role of Atg proteins in autophagosome formation. Annu. Rev. Cell Dev. Biol. *27*, 107–132. https://doi.org/10.1146/annurev-cellbio-092910-154005.

Moguche, A.O., Shafiani, S., Clemons, C., Larson, R.P., Dinh, C., Higdon, L.E., Cambier, C.J., Sissons, J.R., Gallegos, A.M., Fink, P.J., *et al.* (2015). ICOS and Bcl6-dependent pathways maintain a CD4 T cell population with memory-like properties during tuberculosis. J. Exp. Med. *212*, 715–728. https://doi.org/10.1084/jem.20141518.

Moraco, A.H., and Kornfeld, H. (2014). Cell death and autophagy in tuberculosis. Semin. Immunol. *26*, 497–511. https://doi.org/10.1016/j.smim.2014.10.001.

Moreira-Teixeira, L., Sousa, J., McNab, F.W., Torrado, E., Cardoso, F., Machado, H., Castro, F., Cardoso, V., Gaifem, J., Wu, X., *et al.* (2016). Type I IFN Inhibits alternative macrophage activation during *Mycobacterium tuberculosis* infection and leads to enhanced protection in the absence of IFN-gamma signaling. J. Immunol. *197*, 4714–4726. https://doi.org/10.4049/jimmunol.1600584.

Mortaz, E., Adcock, I.M., Tabarsi, P., Masjedi, M.R., Mansouri, D., Velayati, A.A., Casanova, J.L., and Barnes, P.J. (2015). Interaction of Pattern Recognition Receptors with *Mycobacterium* Tuberculosis. J. Clin. Immunol. *35*, 1–10. https://doi.org/10.1007/s10875-014-0103-7.

Nandi, B., and Behar, S.M. (2011). Regulation of neutrophils by interferon-γ limits lung inflammation during tuberculosis infection. J. Exp. Med. *208*, 2251–2262. https://doi.org/10.1084/jem.20110919.

Ndlovu, H., and Marakalala, M.J. (2016). Granulomas and inflammation: host-directed therapies for tuberculosis. Front. Immunol. 7, 434. https://doi.org/10.3389/fimmu.2016.00434.

Oliveira, M.S., Fraga, A.G., Torrado, E., Castro, A.G., Pereira, J.P., Filho, A.L., Milanezi, F., Schmitt, F.C., Meyers, W.M., Portaels, F., *et al.* (2005). Infection with *Mycobacterium ulcerans* induces persistent inflammatory responses in mice. Infect. Immun. *73*, 6299–6310. https://doi.org/10.1128/IAI.73.10.6299-6310.2005.

Ordway, D., Henao-Tamayo, M., Harton, M., Palanisamy, G., Troudt, J., Shanley, C., Basaraba, R.J., and Orme, I.M. (2007). The hypervirulent *Mycobacterium tuberculosis* strain HN878 induces a potent TH1 response followed by rapid down-regulation. J. Immunol. *179*, 522–531. https://doi.org/10.4049/jimmunol.179.1.522.

Pearl, J.E., Khader, S.A., Solache, A., Gilmartin, L., Ghilardi, N., deSauvage, F., and Cooper, A.M. (2004). IL-27 signaling compromises control of bacterial growth in mycobacteria-infected mice. J. Immunol. *173*, 7490–7496.

Pearl, J.E., Torrado, E., Tighe, M., Fountain, J.J., Solache, A., Strutt, T., Swain, S., Appelberg, R., and Cooper, A.M. (2012). Nitric oxide inhibits the accumulation of CD4+CD44hiTbet+CD69lo T cells in mycobacterial infection. Eur. J. Immunol. *42*, 3267–3279. https://doi.org/10.1002/eji.201142158.

Pitarque, S., Herrmann, J.L., Duteyrat, J.L., Jackson, M., Stewart, G.R., Lecointe, F., Payre, B., Schwartz, O., Young, D.B., Marchal, G., *et al.* (2005). Deciphering the molecular bases of *Mycobacterium tuberculosis* binding to the lectin DC-SIGN reveals an underestimated complexity. Biochem. J. *392*, 615–624. https://doi.org/10.1042/BJ20050709.

Ponpuak, M., Davis, A.S., Roberts, E.A., Delgado, M.A., Dinkins, C., Zhao, Z., Virgin, H.W.t., Kyei, G.B., Johansen, T., Vergne, I., *et al.* (2010). Delivery of cytosolic components by autophagic adaptor protein p62 endows autophagosomes with unique antimicrobial properties. Immunity *32*, 329–341. https://doi.org/10.1016/j.immuni.2010.02.009.

Pym, A.S., Brodin, P., Brosch, R., Huerre, M., and Cole, S.T. (2002). Loss of RD1 contributed to the attenuation of the live tuberculosis vaccines *Mycobacterium bovis* BCG and *Mycobacterium microti*. Mol. Microbiol. *46*, 709–717.

Pym, A.S., Brodin, P., Majlessi, L., Brosch, R., Demangel, C., Williams, A., Griffiths, K.E., Marchal, G., Leclerc, C., and Cole, S.T. (2003). Recombinant BCG exporting ESAT-6 confers enhanced protection against tuberculosis. Nat. Med. *9*, 533–539. https://doi.org/10.1038/nm859.

Rajaram, M.V., Ni, B., Dodd, C.E., and Schlesinger, L.S. (2014). Macrophage immunoregulatory pathways in tuberculosis. Semin. Immunol. *26*, 471–485. https://doi.org/10.1016/j.smim.2014.09.010.

Reiley, W.W., Calayag, M.D., Wittmer, S.T., Huntington, J.L., Pearl, J.E., Fountain, J.J., Martino, C.A., Roberts, A.D., Cooper, A.M., Winslow, G.M., *et al.* (2008). ESAT-6-specific CD4 T cell responses to aerosol *Mycobacterium tuberculosis* infection are initiated in the mediastinal lymph nodes. Proc. Natl. Acad. Sci. U.S.A. *105*, 10961–10966. https://doi.org/10.1073/pnas.0801496105.

Reiley, W.W., Shafiani, S., Wittmer, S.T., Tucker-Heard, G., Moon, J.J., Jenkins, M.K., Urdahl, K.B., Winslow, G.M., and Woodland, D.L. (2010). Distinct functions of antigen-specific CD4 T cells during murine *Mycobacterium tuberculosis* infection. Proc. Natl. Acad. Sci. U.S.A. *107*, 19408–19413. https://doi.org/10.1073/pnas.1006298107.

Remoli, M.E., Giacomini, E., Petruccioli, E., Gafa, V., Severa, M., Gagliardi, M.C., Iona, E., Pine, R., Nisini, R., and Coccia, E.M. (2011). Bystander inhibition of dendritic cell differentiation by *Mycobacterium tuberculosis*-induced IL-10. Immunol. Cell Biol. *89*, 437–446. https://doi.org/10.1038/icb.2010.106.

Reshetnikova, G., Barkan, R., Popov, B., Nikolsky, N., and Chang, L.S. (2000). Disruption of the actin cytoskeleton leads to inhibition of mitogen-induced cyclin E expression, Cdk2 phosphorylation, and nuclear accumulation of the retinoblastoma protein-related p107 protein. Exp. Cell Res. *259*, 35–53. https://doi.org/10.1006/excr.2000.4966.

Romagnoli, A., Etna, M.P., Giacomini, E., Pardini, M., Remoli, M.E., Corazzari, M., Falasca, L., Goletti, D., Gafa, V., Simeone, R., *et al.* (2012). ESX-1 dependent impairment of autophagic flux by *Mycobacterium tuberculosis* in human dendritic cells. Autophagy *8*, 1357–1370. https://doi.org/10.4161/auto.20881.

Saini, N.K., Baena, A., Ng, T.W., Venkataswamy, M.M., Kennedy, S.C., Kunnath-Velayudhan, S., Carreño, L.J., Xu, J., Chan, J., Larsen, M.H., *et al.* (2016). Suppression of autophagy and antigen presentation by *Mycobacterium tuberculosis* PE_PGRS47. Nat. Microbiol. *1*, 16133. https://doi.org/10.1038/nmicrobiol.2016.133.

Sakai, S., Kauffman, K.D., Schenkel, J.M., McBerry, C.C., Mayer-Barber, K.D., Masopust, D., and Barber, D.L. (2014). Cutting edge: control of *Mycobacterium tuberculosis* infection by a subset of lung parenchyma-homing CD4 T cells. J. Immunol. *192*, 2965–2969. https://doi.org/10.4049/jimmunol.1400019.

Samstein, M., Schreiber, H.A., Leiner, I.M., Susac, B., Glickman, M.S., and Pamer, E.G. (2013). Essential yet limited role for CCR2(+) inflammatory monocytes during *Mycobacterium tuberculosis*-specific T cell priming. Elife 2, e01086. https://doi.org/10.7554/eLife.01086.

Sani, M., Houben, E.N., Geurtsen, J., Pierson, J., de Punder, K., van Zon, M., Wever, B., Piersma, S.R., Jimenez, C.R., Daffe, M., *et al.* (2010). Direct visualization by cryo-EM of the mycobacterial capsular layer: a labile structure containing ESX-1-secreted proteins. PLOS Pathog. *6*, e1000794. https://doi.org/10.1371/journal.ppat.1000794.

Schäfer, G., Jacobs, M., Wilkinson, R.J., and Brown, G.D. (2009). Non-opsonic recognition of *Mycobacterium tuberculosis* by phagocytes. J. Innate Immun. *1*, 231–243. https://doi.org/10.1159/000173703.

Schluger, N.W. (2001). Recent advances in our understanding of human host responses to tuberculosis. Respir. Res. *2*, 157–163.

Schreuder, P.A., Noto, S., and Richardus, J.H. (2016). Epidemiologic trends of leprosy for the 21st century. Clin. Dermatol. *34*, 24–31. https://doi.org/10.1016/j.clindermatol.2015.11.001.

Serafini, A., Boldrin, F., Palù, G., and Manganelli, R. (2009). Characterization of a *Mycobacterium tuberculosis* ESX-3 conditional mutant: essentiality and rescue by iron and zinc. J. Bacteriol. *191*, 6340–6344. https://doi.org/10.1128/JB.00756-09.

Serbina, N.V., and Pamer, E.G. (2006). Monocyte emigration from bone marrow during bacterial infection requires signals mediated by chemokine receptor CCR2. Nat. Immunol. *7*, 311–317. https://doi.org/10.1038/ni1309.

Shin, D.M., Jeon, B.Y., Lee, H.M., Jin, H.S., Yuk, J.M., Song, C.H., Lee, S.H., Lee, Z.W., Cho, S.N., Kim, J.M., *et al.* (2010). *Mycobacterium tuberculosis* eis regulates autophagy, inflammation, and cell death through redox-dependent signaling. PLOS Pathog. *6*, e1001230. https://doi.org/10.1371/journal.ppat.1001230.

Siegrist, M.S., Steigedal, M., Ahmad, R., Mehra, A., Dragset, M.S., Schuster, B.M., Philips, J.A., Carr, S.A., and Rubin, E.J. (2014). mycobacterial Esx-3 requires multiple components for iron acquisition. MBio *5*, e01073–14. https://doi.org/10.1128/mBio.01073-14.

Silva-Gomes, R., Marcq, E., Trigo, G., Gonçalves, C.M., Longatto-Filho, A., Castro, A.G., Pedrosa, J., and Fraga, A.G. (2015). Spontaneous healing of *Mycobacterium ulcerans* lesions in the guinea pig model. PLOS Negl. Trop. Dis. *9*, e0004265. https://doi.org/10.1371/journal.pntd.0004265.

Silva, M.T., Portaels, F., and Pedrosa, J. (2009). Pathogenetic mechanisms of the intracellular parasite *Mycobacterium ulcerans* leading to Buruli ulcer. Lancet Infect. Dis. *9*, 699–710. https://doi.org/10.1016/S1473-3099(09)70234-8.

Simmonds, R.E., Lali, F.V., Smallie, T., Small, P.L., and Foxwell, B.M. (2009). Mycolactone inhibits monocyte cytokine production by a posttranscriptional mechanism. J. Immunol. *182*, 2194–2202. https://doi.org/10.4049/jimmunol.0802294.

Slight, S.R., Rangel-Moreno, J., Gopal, R., Lin, Y., Fallert Junecko, B.A., Mehra, S., Selman, M., Becerril-Villanueva, E., Baquera-Heredia, J., Pavon, L., *et al.* (2013). CXCR5(+) T helper cells mediate protective immunity against tuberculosis. J. Clin. Invest. *123*, 712–726. https://doi.org/10.1172/JCI65728.

Smith, J., Manoranjan, J., Pan, M., Bohsali, A., Xu, J., Liu, J., McDonald, K.L., Szyk, A., LaRonde-LeBlanc, N., and Gao, L.Y. (2008). Evidence for pore formation in host cell membranes by ESX-1-secreted ESAT-6 and its role in *Mycobacterium marinum* escape from the vacuole. Infect. Immun. *76*, 5478–5487. https://doi.org/10.1128/IAI.00614-08.

Snyder, D.S., and Small, P.L. (2003). Uptake and cellular actions of mycolactone, a virulence determinant for *Mycobacterium ulcerans*. Microb. Pathog. *34*, 91–101.

Srivastava, S., and Ernst, J.D. (2013). Cutting edge: Direct recognition of infected cells by CD4 T cells is required for control of intracellular *Mycobacterium tuberculosis* in vivo. J. Immunol. *191*, 1016–1020. https://doi.org/10.4049/jimmunol.1301236.
Stinear, T.P., Mve-Obiang, A., Small, P.L., Frigui, W., Pryor, M.J., Brosch, R., Jenkin, G.A., Johnson, P.D., Davies, J.K., Lee, R.E., *et al.* (2004). Giant plasmid-encoded polyketide synthases produce the macrolide toxin of *Mycobacterium ulcerans*. Proc. Natl. Acad. Sci. U.S.A. *101*, 1345–1349. https://doi.org/10.1073/pnas.0305877101.
Sun, J., Siroy, A., Lokareddy, R.K., Speer, A., Doornbos, K.S., Cingolani, G., and Niederweis, M. (2015). The tuberculosis necrotizing toxin kills macrophages by hydrolyzing NAD. Nat. Struct. Mol. Biol. *22*, 672–678. https://doi.org/10.1038/nsmb.3064.
Sweeney, K.A., Dao, D.N., Goldberg, M.F., Hsu, T., Venkataswamy, M.M., Henao-Tamayo, M., Ordway, D., Sellers, R.S., Jain, P., Chen, B., *et al.* (2011). A recombinant *Mycobacterium smegmatis* induces potent bactericidal immunity against *Mycobacterium tuberculosis*. Nat. Med. *17*, 1261–1268. https://doi.org/10.1038/nm.2420.
Taylor, R.C., Cullen, S.P., and Martin, S.J. (2008). Apoptosis: controlled demolition at the cellular level. Nat. Rev. Mol. Cell Biol. *9*, 231–241.
Teles, R.M., Graeber, T.G., Krutzik, S.R., Montoya, D., Schenk, M., Lee, D.J., Komisopoulou, E., Kelly-Scumpia, K., Chun, R., Iyer, S.S., *et al.* (2013). Type I interferon suppresses type II interferon-triggered human anti-mycobacterial responses. Science *339*, 1448–1453. https://doi.org/10.1126/science.1233665.
Torrado, E., Adusumilli, S., Fraga, A.G., Small, P.L., Castro, A.G., and Pedrosa, J. (2007). Mycolactone-mediated inhibition of tumor necrosis factor production by macrophages infected with *Mycobacterium ulcerans* has implications for the control of infection. Infect. Immun. *75*, 3979–3988. https://doi.org/10.1128/IAI.00290-07.
Torrado, E., and Cooper, A.M. (2010). IL-17 and Th17 cells in tuberculosis. Cytokine Growth Factor Rev. *21*, 455–462. https://doi.org/10.1016/j.cytogfr.2010.10.004.
Torrado, E., and Cooper, A.M. (2013). Cytokines in the balance of protection and pathology during mycobacterial infections. Adv. Exp. Med. Biol. *783*, 121–140. https://doi.org/10.1007/978-1-4614-6111-1_7.
Torrado, E., Fountain, J.J., Liao, M., Tighe, M., Reiley, W.W., Lai, R.P., Meintjes, G., Pearl, J.E., Chen, X., Zak, D.E., *et al.* (2015). Interleukin 27R regulates CD4+ T cell phenotype and impacts protective immunity during Mycobacterium *tuberculosis* infection. J. Exp. Med. *212*, 1449–1463. https://doi.org/10.1084/jem.20141520.
Torrado, E., Fraga, A.G., Logarinho, E., Martins, T.G., Carmona, J.A., Gama, J.B., Carvalho, M.A., Proenca, F., Castro, A.G., and Pedrosa, J. (2010). IFN-gamma-dependent activation of macrophages during experimental infections by *Mycobacterium ulcerans* is impaired by the toxin mycolactone. J. Immunol. *184*, 947–955. https://doi.org/10.4049/jimmunol.0902717.
Torrado, E., Robinson, R.T., and Cooper, A.M. (2011). Cellular response to mycobacteria: balancing protection and pathology. Trends Immunol. *32*, 66–72. https://doi.org/10.1016/j.it.2010.12.001.
Ulrichs, T., Kosmiadi, G.A., Jörg, S., Pradl, L., Titukhina, M., Mishenko, V., Gushina, N., and Kaufmann, S.H. (2005). Differential organization of the local immune response in patients with active cavitary tuberculosis or with nonprogressive tuberculoma. J. Infect. Dis. *192*, 89–97.
Ulrichs, T., Kosmiadi, G.A., Trusov, V., Jorg, S., Pradl, L., Titukhina, M., Mishenko, V., Gushina, N., and Kaufmann, S.H. (2004). Human tuberculous granulomas induce peripheral lymphoid follicle-like structures to orchestrate local host defence in the lung. J. Pathol. *204*, 217–228. https://doi.org/10.1002/path.1628.
van der Wel, N., Hava, D., Houben, D., Fluitsma, D., van Zon, M., Pierson, J., Brenner, M., and Peters, P.J. (2007). *M. tuberculosis* and *M. leprae* translocate from the phagolysosome to the cytosol in myeloid cells. Cell *129*, 1287–1298.
van Ravensway, J., Benbow, M.E., Tsonis, A.A., Pierce, S.J., Campbell, L.P., Fyfe, J.A., Hayman, J.A., Johnson, P.D., Wallace, J.R., and Qi, J. (2012). Climate and landscape factors associated with Buruli ulcer incidence in Victoria, Australia. PLOS ONE *7*, e51074. https://doi.org/10.1371/journal.pone.0051074.
Vergne, I., Chua, J., Lee, H.H., Lucas, M., Belisle, J., and Deretic, V. (2005). Mechanism of phagolysosome biogenesis block by viable *Mycobacterium tuberculosis*. Proc. Natl. Acad. Sci. U.S.A. *102*, 4033–4038.
Vyas, J.M., Van der Veen, A.G., and Ploegh, H.L. (2008). The known unknowns of antigen processing and presentation. Nat. Rev. Immunol. *8*, 607–618. https://doi.org/10.1038/nri2368.

Walsh, D.S., Meyers, W.M., Portaels, F., Lane, J.E., Mongkolsirichaikul, D., Hussem, K., Gosi, P., and Myint, K.S. (2005). High rates of apoptosis in human *Mycobacterium ulcerans* culture-positive buruli ulcer skin lesions. Am. J. Trop. Med. Hyg. *73*, 410–415.

Walsh, D.S., Portaels, F., and Meyers, W.M. (2008). Buruli ulcer (*Mycobacterium ulcerans* infection). Trans. R. Soc. Trop. Med. Hyg. *102*, 969–978. https://doi.org/10.1016/j.trstmh.2008.06.006.

Walsh, D.S., Portaels, F., and Meyers, W.M. (2010). Recent advances in leprosy and Buruli ulcer (*Mycobacterium ulcerans* infection). Curr. Opin. Infect. Dis. *23*, 445–455. https://doi.org/10.1097/QCO.0b013e32833c2209.

Watson, R.O., Manzanillo, P.S., and Cox, J.S. (2012). Extracellular *M. tuberculosis* DNA targets bacteria for autophagy by activating the host DNA-sensing pathway. Cell *150*, 803–815. https://doi.org/10.1016/j.cell.2012.06.040.

W.H.O. (2015). Global tuberculosis report. The World Health Organization.

Winslow, G.M., Roberts, A.D., Blackman, M.A., and Woodland, D.L. (2003). Persistence and turnover of antigen-specific CD4 T cells during chronic tuberculosis infection in the mouse. J. Immunol. *170*, 2046–2052.

Wolf, A.J., Desvignes, L., Linas, B., Banaiee, N., Tamura, T., Takatsu, K., and Ernst, J.D. (2008). Initiation of the adaptive immune response to *Mycobacterium tuberculosis* depends on antigen production in the local lymph node, not the lungs. J. Exp. Med. *205*, 105–115. https://doi.org/10.1084/jem.20071367.

Wolf, A.J., Linas, B., Trevejo-Nunez, G.J., Kincaid, E., Tamura, T., Takatsu, K., and Ernst, J.D. (2007). *Mycobacterium tuberculosis* infects dendritic cells with high frequency and impairs their function in vivo. J. Immunol. *179*, 2509–2519. https://doi.org/10.4049/jimmunol.179.4.2509.

Yonekawa, A., Saijo, S., Hoshino, Y., Miyake, Y., Ishikawa, E., Suzukawa, M., Inoue, H., Tanaka, M., Yoneyama, M., Oh-Hora, M., *et al.* (2014). Dectin-2 is a direct receptor for mannose-capped lipoarabinomannan of mycobacteria. Immunity *41*, 402–413. https://doi.org/10.1016/j.immuni.2014.08.005

Role of Cyclic di-GMP in the Bacterial Virulence and Evasion of the Plant Immunity

8

Marta Martínez-Gil* and Cayo Ramos

Área de Genética, Facultad de Ciencias, Instituto de Hortofruticultura Subtropical y Mediterránea 'La Mayora', Universidad de Málaga-Consejo Superior de Investigaciones Científicas (IHSM-UMA-CSIC), Málaga, Spain.

*Correspondence: martamgv@uma.es

http://doi.org/10.21775/9781910190692.08

Abstract

Plant pathogenic bacteria are responsible for the loss of hundreds of millions of dollars each year, impacting a wide range of economically relevant agricultural crops. The plant immune system detects conserved bacterial molecules and deploys an arsenal of effective defence measures at different levels; however, during compatible interactions, some pathogenic bacteria suppress and manipulate the host immunity and colonize and infect the plant host. Different bacteria employ similar strategies to circumvent plant innate immunity, while other tactics are specific to certain bacterial species. Recent studies have highlighted the secondary messenger c-di-GMP as a key molecule in the transmission of environmental cues in an intracellular regulatory network that controls virulence traits in many plant pathogenic bacteria. In this review, we focus on the recent knowledge of the molecular basis of c-di-GMP signalling mechanisms that promote or prevent the evasion of bacterial phytopathogens from the plant immune system. This review will highlight the considerable diversity of mechanisms evolved in plant-associated bacteria to elude plant immunity.

An overview of bacterial plant diseases

Plants coexist with a myriad of microorganisms, bacteria, viruses or fungi, some of which deplete plant energy and resources. In particular, more than 100 bacterial species are responsible for many serious plant diseases worldwide, and many of these bacteria are responsible for important economic losses in agriculture, reducing the quantity and quality of produce. The bacteria *Xanthomonas oryzae* pv. *oryzae* is responsible for rice leaf blight, a disease that is estimated to cause a 50% economic loss in the annual rice yields worldwide (He *et al.*, 2010). The fire blight infection of apples, pears and other species from the

Rosaceae family results from the pathogen *Erwinia amylovora* and is estimated to cost €11.8 million in annual inspection in nurseries and garden centres in the EU (Caspari, 2011). The bacteria *Xylella fastidiosa* causes different diseases in many economically important crops, including Pierce's disease of grapevines, citrus variegated chlorosis and, more recently, olive quick decline syndrome (Purcell, 2013; Guan *et al.*, 2015). The Centre for Wine Economics predicted that the annual cost of Pierce's disease is US$61 million, and Pierce's disease control programmes have resulted in savings of US$261 million (Fuller *et al.*, 2012). Not surprisingly, *X. fastidiosa* was the first phytopathogen sequenced (Simpson *et al.*, 2016). *Xanthomonas arboricola* pv. pruni is the causal agent for the canker of many stone fruits, and the disease outbreak, affecting 30% of the produce, is estimated to cause economic losses of approximately €11.200 per ha (Stefani, 2010). The olive knot disease is one of the most serious diseases of olive trees, causing major damage on young trees and affecting the vigour, growth and total yield. It is caused by the pathogen *Pseudomonas savastanoi*, a bacterium belonging to the *Pseudomonas syringae* complex (Mansfield *et al.*, 2012). Although it is considered one of the most relevant diseases affecting the olive crop, economic losses as a result of the disease have not ben accurately estimated (Ramos *et al.*, 2012). This critical situation, considering the quality and quantity of the crop yield elicited by pathogens, is amplified with the increasing human population and the imperative need for an adequate food supply. A report by the Food and Agriculture Organization of the United Nations estimated a 50% increase in the global demand for food supply for 2030–2050 (World agriculture: towards 2030/2050, FAO, 2006).

The acknowledgement that microorganisms were ultimately a cause and not a consequence of plant diseases increased the efforts towards understanding the mechanisms behind plant diseases. In the early twentieth century, the first specific enzymes and toxins involved in the pathogenicity of plant pathogens were identified. In 1950, some plant pathogens that produce the plant hormone indolacetic acid (IAA), which is associated with bacterial pathogenicity, were identified. Since the 80s, the plant pathology field has benefited from the progress and development of molecular biology techniques. In particular, molecular tools have contributed to the detection, identification, isolation and modification of genes involved in disease resistance and pathogen virulence. Currently, the plant bacterial pathology field has enormously gained from the discovery of novel mechanisms that bacterial phytopathogens have evolved to interact with plants to successfully manipulate the host response and cause disease. Among these mechanisms, microbial pathogens produce diverse molecules, including nucleic acids, chitooligosaccharides and a great variety of effector proteins, that ultimately impact the host immune system (Pel and Pieterse, 2013; Le Fevre *et al.*, 2015).

The understanding of the plant defence machinery and the different strategies used by plant pathogens to overcome these barriers will contribute to the production of foods in the following two complementary ways: the protection of the food supply by combating pathogen infections and the development of highly disease-resistant plants in the near future.

Innate versus adaptive immunity in plants

Plants have evolved a great variety of mechanisms to protect themselves against pathogens, including the production of antimicrobial compounds, such as reactive oxygen species

(ROS), phytoalexins, chitinases, denfensins, oxylipins, and proteinases, and the induction of local and/or systemic acquired defence (Freeman, 2008). In contrast to mammals, plants lack an adaptive immune system and mobile defender cells, which enable the organism to have an extraordinary memory for future infections together with an infinite diversity of protection. However, plants and animals share some commonalities in the components of the innate immunity, such as the production of antimicrobial compounds and programmed cell death (Iriti and Faoro, 2007). Despite the absence of acquired or adaptive immunity, plants possess sophisticated and complex surveillance mechanisms to efficiently recognize and respond against putative pathogens. Plant immunity is based on the following two aspects: (i) a cell-autonomous basal resistance, also called innate immunity, determined by each individual cell (gene-for-gene response) and (ii) an unspecific acquired systemic resistance (SAR), where signals originating in the infection site spread to the uninfected parts of the plant. Given the localized response of plant innate immunity, compared with the animal immune system, the majority of plant cell types produce a broad arsenal of antimicrobial defences (Spoel and Dong, 2012).

The plant innate immunity has two different and consecutive levels (Jones and Dangl, 2006). The first more ancient and generalist level consistently remains ready to respond and involves cell surface proteins, called pattern recognition receptors (PRRs), which recognize specific microbial- or pathogen-associated molecular patterns (MAMPs or PAMPs), such as flagellin, peptidoglycan, elongation factor Tu and lipoteichoic acid among many others (Gust *et al.*, 2007; Pel and Pieterse, 2013; Newman *et al.*, 2013). At least two types of PRRs in plants are involved in bacterial MAMP recognition as follows: (i) the receptor-like kinases (RLKs) with an N-terminal extracellular domain, a single transmembrane region and a cytoplasmic kinase domain and (ii) the receptor-like proteins (RLPs), which differ from RLKs as they lack the intracellular kinase domain (Newman *et al.*, 2013). This recognition system is also present in animal innate immunity; however, plants have no clear Toll and Toll-like receptors, which is an important group of PRRs in animals (Hayashi *et al.*, 2001). Bacterial molecular patterns are generally immutable and conserved; however, the protein receptors involved in the recognition of similar molecular patterns are different in animals and plants. For example, the Toll-like receptor in mammals detects the bacterial flagellin, whereas the transmembrane receptor kinase FLS2 in plants directly interacts with flagellin (Gómez-Gómez and Boller, 2000; Smith *et al.*, 2003). The fact that PAMPs are highly conserved structures (essential in microbial recognition) and that their recognition varies depending on the organism suggests that microbial detection by different innate immunity systems has potentially converged (Nürnberger *et al.*, 2004; Zipfel and Felix, 2005). An immense list of immune receptors compensates for the lack of a pathogen-specific immune response in plants (Ausubel, 2005), particularly in *Arabidopsis thaliana*, which possesses more than 600 RLKs, whereas there are approximately 20 TLRs in mammals (Hopkins and Sriskandan, 2005; Ausubel, 2005; Willmann *et al.*, 2011).

Early pathogen perception elicits PAMP-triggered immunity (PTI), which is the first plant defence response and involves callose deposition and ROS production to restrict bacterial colonization and dispersal among the plant tissues (Trda *et al.*, 2015). However, some pathogens are able to mask this PTI response with the translocation of a large number of different effectors via the type III secretion system (T3SS), resulting in effector-triggered susceptibility (ETS) (Fig. 8.1) (Jones and Dangl, 2006; Win *et al.*, 2013; Castañeda-Ojeda

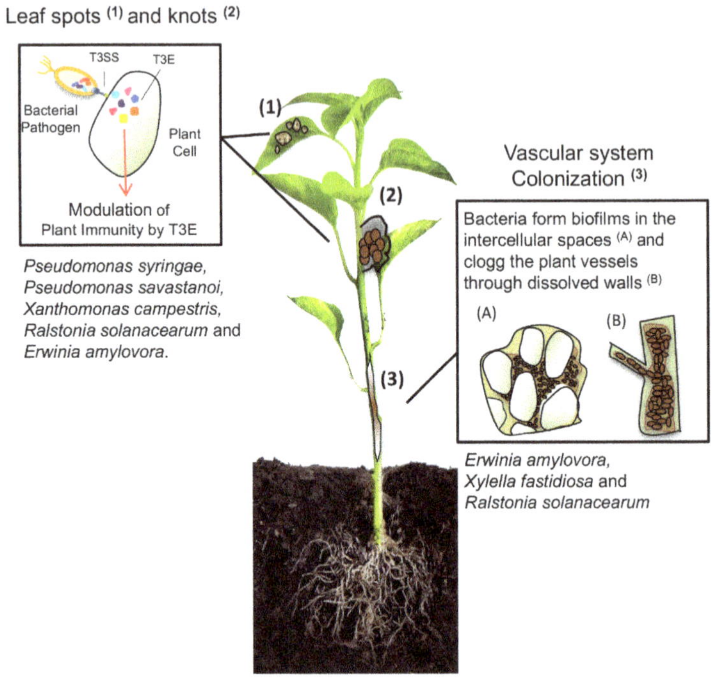

Figure 8.1 Virulence mechanisms in relevant plant pathogenic bacteria and the kinds of symptoms they cause. Bacteria as *Xanthomonas campestris*, *Ralstonia solanacearum*, *Erwinia amylovora* and bacteria belonging to the *Pseudomonas syringae* complex among others, cause leave spots and knots, primarily through the secretion of type III secretion system (T3SS) effectors (T3E) which interfere with the plant immune system, allowing further colonization of the plant tissues and survival within the plant. Some bacteria as *E. amylovora*, *R. solanacearum* and *Xylella fastidiosa* besides other virulence factors produce a variety of degradative enzymes and extracellular polysaccharides that will promote biofilm formation within the plant apoplast and further colonization of the plant vascular system protecting bacteria from the plant defences and causing disease.

et al., 2016; Washington *et al.*, 2016). Then, plants activate a second level of resistance that is mediated by the plant cytoplasmic NB-LRR receptors (encoded by *R genes*), which directly or indirectly recognize, bind and, therefore, block some of these bacterial effectors, causing effector-triggered immunity (ETI), a disease resistance condition that leads to a hypersensitive cell-death response (HR) during incompatible interactions (Dodds and Rathjen, 2010; Ishikawa *et al.*, 2014). However, this description has been simplified, as pathogens are constantly acquiring new strategies to overcome plant defences, and simultaneously plants co-evolve to recognize and respond to new bacterial equipment in a fascinating endless 'tug of war' game between plants, bacteria and natural selection.

Both pathogenic and commensal/beneficial bacteria on the plant surface are initially recognized as harmful organisms similar to animal innate immunity (Pel and Pieterse, 2013). Interestingly, the microbiome of plants and animals plays a key role in the adaptation of the host to its environment and contributes to the development of the immune system (Guttman *et al.*, 2014; Sun *et al.*, 2015). Consequently, some bacteria utilize different strategies to counterattack plant immunity at different levels. This strategy is well exemplified in

P. syringae or *Xanthomonas campestris,* two bacterial species that reduce the expression of flagellar genes or present polymorphisms of the flg22 epitope (Wenxian Sun *et al.*, 2006; Lan *et al.*, 2006; Wang *et al.*, 2015). In animals, important pathogens, such as *Helicobacter pylori* or *Campylobacter jejuni,* modify the entire flagellin recognition site to evade perception through the Toll-like receptor 5 (TLR5) (Purcell, 2013; Guan *et al.*, 2015; Broz and Monack, 2013). In general, plant pathogenic bacteria use different strategies to suppress plant defences during infection, including the synthesis of lipopolysaccharides and degradative enzymes, phytotoxins, antioxidants against ROS reactions, the secretion of effectors through the T3SS or manipulating hormone synthesis (Abramovitch *et al.*, 2006; Macho, 2016; Pfeilmeier *et al.*, 2016a; Albus *et al.*, 2001).

In addition to the knowledge acquired in recent years concerning how plant pathogenic bacteria evade plant immunity, there is still a gap in the current understanding of how these evasion mechanisms are regulated and the hierarchical and temporal control of these mechanisms (Pel and Pieterse, 2013; Le Fevre *et al.*, 2015; Pfeilmeier *et al.*, 2016a). Recent studies have demonstrated a role for the secondary messenger cyclic di-GMP (c-di-GMP) in the regulation of the elicitation of the plant defences (Freeman, 2008; Pfeilmeier *et al.*, 2016a; Pfeilmeier *et al.*, 2016b;). Secondary messengers function as signalling molecules and play important roles in the control of bacterial virulence (Fig. 8.2) (Iriti and Faoro, 2007; Liang, 2015; Pfeilmeier *et al.*, 2016a). In the next sections, we will discuss the role of the secondary messenger c-di-GMP in the virulence of plant pathogenic bacteria with a special emphasis on the impact of c-di-GMP on the bacterial evasion of the plant immunity. We will also discuss the relevance of c-di-GMP in the integration of environmental cues into regulatory networks governing numerous behaviours, such as motility, biofilm formation, degradative enzyme secretion and T3SS expression, that ultimately help to overcome plant immunity.

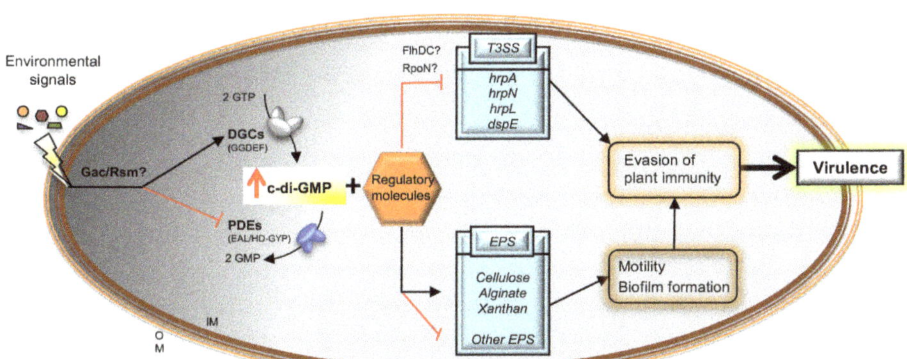

Figure 8.2 Regulation cascade of virulence factors by the secondary messenger cyclic di-GMP. Diguanylate cyclases (DGC) and phosphodiesterases (PDE) respond to different environmental signals through global regulatory two-component signal transduction systems, balancing intracellular levels of c-di-GMP. Specific regulatory molecules (proteins or RNAs riboswitches) bind, interact or sense c-di-GMP levels affecting the regulatory output (gene expression or enzymatic activity) of direct targets as, for example, the T3SS and exopolysaccharides, resulting in changes in the bacteria lifestyle and virulence.

Cyclic di-GMP: a relevant secondary messenger in pathogen–host interactions

C-di-GMP is a small diffusible secondary messenger discovered in 1987 as an allosteric activator of cellulose synthase in *Gluconacetobacter xylinum* (Ross *et al.*, 1987; Spoel and Dong, 2012). The history of c-di-GMP discovery has been well documented (Ross *et al.*, 1991; Jones and Dangl, 2006; Römling *et al.*, 2013). In the early 1980s, Benziman's group disclosed that cellulose synthase presented low activity without the secondary messenger (Aloni *et al.*, 1982). Further independent studies identified the cellulose synthase cofactor as bis (3′→5′) cyclic dimeric guanylic acid (c-di-GMP) (Ross *et al.*, 1987; Amikam and Benziman, 1989; Newman *et al.*, 2013). Several genetic studies identified, sequenced and characterized the genes encoding the enzymes responsible for c-di-GMP synthesis and degradation (Tal *et al.*, 1998; Hayashi *et al.*, 2001; Ausmees *et al.*, 2001; Simm *et al.*, 2004; Römling *et al.*, 2005; Tischler and Camilli, 2005). Specifically, two enzymes, the diguanylate cyclases (DGCs) and the phosphodiesterases (PDEs), which possess opposite activities, control c-di-GMP metabolism: (i) DGCs have a GGDEF domain and are responsible for the synthesis of c-di-GMP from two guanosine triphosphate molecules (2 GTP) and (ii) PDEs have an EAL or HD-GYP domain and generate two molecules of guanosine monophosphate (2 GMP) or a single phosphoguanylyl (3′–5′) guanosine molecule (pGpG) (Ross *et al.*, 1987; Ryan *et al.*, 2006; Sondermann *et al.*, 2012). The combined activity of these enzymes is responsible for the intracellular homeostasis of this secondary messenger, and numerous studies have inquired into the domains, function, activity and biochemistry of these two proteins (Fig. 8.2) (Chan *et al.*, 2004; Christen *et al.*, 2005; Malone *et al.*, 2007; Paul *et al.*, 2007; Henry and Crosson, 2011; Römling *et al.*, 2013; Wassmann *et al.*, 2007).

The c-di-GMP-signalling network connects specific environmental conditions (nutrient availability, oxygen, blue light, antibiotics, bile salts, among others) to the specific adaptive responses of bacteria, including the modification of gene expression, metabolism and bacterial lifestyle switches. The broad number of environmental signals that bacterial could sense is reflected on the diversity of signal-input domains observed in DGCs and PDEs proteins (Galperin *et al.*, 2001; Nürnberger *et al.*, 2004; Jenal and Malone, 2006; Hengge, 2009). These domains could include those reacting to light, such as the Blrp1 from *Klebsiella pneumoniae*, or gases, such as DosC and DosP from *Escherichia coli* (Barends *et al.*, 2009; Tuckerman *et al.*, 2009). The regulation of DGC, PDE and multicomponent response regulators by different environmental signals is well established and has been extensively reviewed (Hengge *et al.*, 2009; Krasteva *et al.*, 2012; Sondermann *et al.*, 2012).

The downstream signal transduction reflects the existence of a large variety of cytoplasmic proteins also known as c-di-GMP receptors, which specifically bind to c-di-GMP and translate the intracellular changes into specific outputs (Fig. 8.2) (Ausubel, 2005; Hengge, 2009). Several types of c-di-GMP receptors have been discovered, and the molecular details of c-di-GMP activity are beginning to be clarified: the PilZ domain proteins, transcriptional factors, enzymatically inactive DGC/PDE proteins and RNA riboswitches, are activated or repressed upon interaction with this secondary messenger (Weinhouse *et al.*, 1997; Hopkins and Sriskandan, 2005; Christen *et al.*, 2007; Pratt *et al.*, 2007; Vincent T Lee *et al.*, 2007; Hickman and Harwood, 2008; Hengge, 2009; Newell *et al.*, 2011; Srivastava *et al.*, 2011; Martínez-Granero *et al.*, 2014). However, recent investigations have demonstrated that c-di-GMP is able to bind to a large number of proteins without predictable binding domains as the membrane-bound PgaCD EPS synthase, the flagellar/type III export-ATPases FliI and

HrcN, the glutamyl ligase RimK and the histidine phosphatase/kinase CckA (Steiner *et al.*, 2013; Trampari *et al.*, 2015; Dubey *et al.*, 2016; Little *et al.*, 2016). This great variability of c-di-GMP binding proteins suggests that bioinformatics predictions based on predictable domains will fail in the identification of numerous important targets, therefore making further biochemical analysis essential. Given this vast diversity of effectors, regulation by the secondary messenger could occur at different levels (Sonderman *et al.*, 2012). The expression of numerous genes is controlled by transcriptional factors that will sense and bind to c-di-GMP as FleQ, which controls expression of flagella and adhesins in *Pseudomonas* sp. and Clp, which controls bacteria attachment in *X. campestris* (Hickman *et al.*, 2008; Tao *et al.*, 2010; Martínez-Gil *et al.*, 2014). Other receptors function at a post-translational level, involving protein–protein interactions, as YcgR from *E. coli* which has been demonstrated to interact and regulate the flagella motor (Boehm *et al.*, 2010).

The availability of the whole genome sequences of human-, animal- and plant- pathogenic bacteria has revealed the presence of numerous GGDEF, EAL and HD-GYP domain-containing proteins involved in c-di-GMP turnover, reinforcing the ubiquity and important role of this messenger in bacterial adaptation and lifestyle (Table 8.1) (Ausubel, 2005; Kulasakara *et al.*, 2006; Ryan *et al.*, 2007; Tamayo *et al.*, 2007; Bobrov *et al.*, 2011; Willmann *et al.*, 2011; Matas *et al.*, 2012). This secondary messenger has been implicated in the regulation of numerous bacterial behaviours, including biofilm formation (synthesis of adhesins and exopolysaccharides), flagellum-mediated motility, and virulence (Table 8.2). Bacterial processes associated with virulence and modulated by c-di-GMP signalling are typically associated with motility, adherence, invasion, cytotoxicity, intracellular infection, virulence factor secretion and immune response stimulation (Fig. 8.2) (Tischler and Camilli, 2005; Kulasakara *et al.*, 2006; Dow *et al.*, 2006; Hwee Siang Lee *et al.*, 2010; Moscoso *et al.*, 2011; Huang *et al.*, 2013; Moscoso *et al.*, 2014; Trda *et al.*, 2015; Trampari *et al.*, 2015). In particular, the association between bacterial motility, virulence and c-di-GMP has been widely studied. Flagella motility plays an essential role in effective plant colonization, migration into the apoplast and the plant vascular system. However, given that bacterial flagellin contributes to the induction of the early plant immune response, plant-associated bacteria needs to balance flagellar motility and their evasion of plant immunity. In several Pseudomonas, it has been shown that elevated levels of c-di-GMP reduce flagellin levels, helping the bacteria to evade the FLS2-mediated immune response (Pfeilmeier *et al.*, 2016b). In *P. syringae* high levels of c-di-GMP decrease the expression of flagellin (Engl *et al.*, 2014). Interestingly, this evasion of the FLS2-mediated immunity does not entail increase virulence, indicating that c-di-GMP controls several virulence-related phenotypes and the importance of motility for bacterial virulence. Interestingly, a recent study in numerous *Pseudomonas* species showed evidences that c-di-GMP is able to bind to FliI, the flagellum export ATPase and to HrcN, the homologue of the T3SS, suggesting a direct allosteric regulation by c-di-GMP (Trampari *et al.*, 2015).

The first observations of the role for c-di-GMP in bacterial virulence were reported in the animal pathogenic bacterium *Vibrio cholerae*, where low intracellular levels of c-di-GMP activated cholera toxin, and high levels of the secondary messenger attenuated virulence in an infant mouse model (Tischler and Camilli, 2005). Since these pioneering observations, numerous studies using genetic screens, overexpression approaches, or transcriptional profiling have implicated c-di-GMP levels in virulence regulation in a wide number of human pathogens, such as *Pseudomonas aeruginosa*, *Legionella pneumophila*, *Burkholderia*

Table 8.1 Enzymes involved in cyclic di-GMP metabolism (containing GGDEF and EAL/HD-GYP domains) in representative plant pathogenic bacteria

Organism	Disease	Genome size (bp)	Total proteins	C-di-GMP metabolism proteins	Enzymes described (references)	GenBank reference
Erwinia amylovora ATCC49946	Fire blight in fruit trees	3,795,810	3315	13 (Edmunds *et al.*, 2013)	EdcA, EdcC and EdcE (Edmunds *et al.* 2013)	FN666575
Xylella fastidiosa 9a5c	Pierce's disease in grapevines	2,555,410	2141	5 (de Souza *et al.*, 2013)	CgsA, PD1671, EaL (Chatterjee *et al.*, 2010; Cursino *et al.*, 2015; de Souza *et al.*, 2013)	AE003849
Dickeya dadantii Ech586	Soft rot disease in crops	4,870,170	4105	22 (He and Chou, 2015)	EcpB, EcpC and CsrD (Yi *et al.*, 2010; Wu *et al.*, 2014; Yuan *et al.*, 2015)	CP002038.1
Ralstonia solanacearum GMI1000	Wilt disease in crops	5,567,800	4479	37 (He and Chou, 2015)	–	AL646052
Xanthomonas campestris pv. campestris ATCC33913	Black rot in crucifers	5,025,750	409	44 (Agrios, 2005)	RpfG (Ryan *et al.*, 2006; Ryan *et al.*, 2007)	AE008922
Xanthomonas oryzae pv. oryzae PXO99A	Leaf blight in rice	479,432	3736	30 (Agrios, 2005)	DgcA, GdpX1 (Su *et al.*, 2016; Yang *et al.*, 2016)	CP000967
Pseudomonas syringae pv. tomato DC3000	Bacterial speck disease	6,150,270	5255	56 (He and Chou, 2015)	BifA, Chp8 (Aragón *et al.*, 2015a; Engl *et al.*, 2014)	AE016853
Pseudomonas savastanoi pv. savastanoi NCPPB3335	Olive knot disease	6,052,810	5098	34 (Rodriguez-Palenzuela *et al.*, 2010)	PleD, BifA, DgcP (Pérez-Mendoza *et al.*, 2014; Aragón *et al.*, 2015a, 2015b	CM001834.1

Table 8.2 C-di-GMP specific sensors and effectors in representative plant pathogenic bacteria

Organism	Host	Sensor/effector	Motif/domain	Phenotypic output	Regulation level	Reference
Dickeya dadantii	*Solanaceae*	YcgR	PilZ	Regulation of T3SS (*hrpA*)	Transcriptional	Yuan *et al.*, 2015
		EcpC and EGcpB	PDE	Positive regulation of T3SS (*hrpA*, *hrpN* and *dspE*)	Transcriptional	Yi *et al.*, 2010
Pseudomonas syringae pv. tomato	Tomato plants	Chp8	DGC	Decreases flagellin and induces EPS	Unknown	Engl *et al.*, 2014
		PleD	DGC	Negative regulation of *hrpL* and *hrpA*	Transcriptional	Pérez-Mendoza *et al.*, 2014
Erwinia amylovora	Fruit trees	EdcC and EdcE	DGC	Induces amylovoran and biofilm formation; negative regulation of flagellar motility	Unknown	Edmunds *et al.*, 2013
				Negative regulation of *hrpA*	Transcriptional	Edmunds *et al.*, 2013
Pseudomonas savastanoi pv. savastanoi	Olive plants	DgcP	DGC	Negative regulation of motility and positive regulation of biofilm formation	Unknown	Aragón *et al.*, 2015b
		BifA	PDE	Positive regulation of swimming motility; negative regulation of biofilm formation	Unknown	Aragón *et al.*, 2015a
Xanthomonas oryzae	Rice	GdpX1	DGC	Negative regulation of T3SS, EPS and motility	Transcriptional	Yang *et al.*, 2016
		DgcA	DGC	Negative regulation of EPS and motility; positive regulation of biofilm formation	Unknown	Su *et al.*, 2016
Xanthomonas campestris	Crucifers	RsmA	Transcriptional regulator	Negative regulation of biofilm formation	Post transcriptional	Lu *et al.*, 2012
		YaqJ	C-di-GMP binding protein	Biofilm formation and virulence	Transcriptional/ protein–protein interaction	An *et al.*, 2014
Xylella fastidiosa	Fruit trees	CsgA	DGC	Negative impact in EPS and biofilm formation	Unknown	Chatterjee *et al.*, 2010

DGC, diguanylate cyclase; EPS, exopolysaccharides; PDE, phosphodiesterase; T3SS, type III secretion system.

pseudomallei, *Brucella melitensis* and pathogenic *E. coli* (Tamayo *et al.*, 2007; Hwee Siang Lee *et al.*, 2010; Dodds and Rathjen, 2010; Levi *et al.*, 2011; Römling *et al.*, 2013; Peterson and Artis, 2014; Ishikawa *et al.*, 2014).

The role of c-di-GMP signalling in pathogenicity and virulence of plant pathogenic bacteria has been poorly treated compared with animal pathogens. Recent investigations of the plant pathogenic bacteria *X. campestris* pv. *campestris*, *P. syringae* pv. *tomato* and *P. savastanoi* pv. *savastanoi* revealed that DGCs and PDEs proteins are linked to a wide range of virulence phenotypes. The deletion of PDEs is frequently associated with a decrease in virulence, whereas the deletion of DGCs is associated with an increase in virulence (Ryan *et al.*, 2007; Yi *et al.*, 2010; Chatterjee *et al.*, 2010; Pérez-Mendoza *et al.*, 2011; Pel and Pieterse, 2013; Edmunds *et al.*, 2013; Aragón *et al.*, 2015a,b; Chin *et al.*, 2010). Nevertheless, the relationship between c-di-GMP sensors and effectors and bacterial virulence is more complex than previously anticipated, and several authors have reported a reduction in the virulence of various pathogenic bacteria as a consequence of the overexpression of some DGCs or the loss of function of either PDEs or DGCs. Currently, there is abundant evidence supporting an important role for c-di-GMP in bacterial virulence through effects on diverse bacterial regulation pathways in a complex interaction between DGC, PDE and the regulatory targets of c-di-GMP.

Cyclic di-GMP controls the expression of the type III secretion system

Consistent with many Gram-negative animal pathogens, the virulence of numerous phytopathogens, such as *P. syringae*, *X. campestris*, *Erwinia* species, *P. savastanoi*, *Ralstonia solanacearum* and *Pantoea stewartii*, relies on the assembly of the T3SS, a needle-like appendage that delivers a variety of virulence factors, called effectors, into the host cells (Fig. 8.1) (Tang *et al.*, 2006; Perez-Martinez *et al.*, 2010; Wenting Li *et al.*, 2014; Matas *et al.*, 2014; Wu *et al.*, 2015; Castañeda-Ojeda *et al.*, 2016; Popov *et al.*, 2016). Genes encoding the T3SS have been identified in *hrp* (hypersensitive response and pathogenicity)/*hrc* (hypersensitive response and conserved) gene clusters, which are essential for the translocation of effectors and the development of symptoms and diseases in host plants or the induction of the HR in non-hosts (Arnold *et al.*, 2003). The majority of T3SS effectors suppress plant immunity, particularly the signalling pathways triggered by flagellin and the elongation factor EF-Tu (Block and Alfano, 2011; Guttman *et al.*, 2014; Sun *et al.*, 2015). For more information on the functionality of the T3SS and its effectors in the modulation of plants, animals and human immunity readers are encouraged to consult these other reviews (McKinney, 2004; Alfano and Collmer, 2004; Tang *et al.*, 2006; Win *et al.*, 2013; Galán *et al.*, 2014; Pearson *et al.*, 2015). In addition, recent studies have demonstrated that T3SS-related effectors might play additional roles in the direct suppression of the plant immune system, such as contributing to bacterial survival, replication and dissemination by the modulation of plant hormone signalling, or metabolism and organelle function, all of which ultimately manipulate the host response (Guangyong Li *et al.*, 2014; Kang *et al.*, 2014; Kazan and Lyons, 2014; Macho, 2016). In addition to the *hrp*/*hrc* and T3SS effector genes, bacteria co-express a variety of other virulence-related genes that help bacteria to survive under the conditions encountered within the host (Yang *et al.*, 2004; Lam *et al.*, 2014). The expression of these genes is tightly regulated by a variety of environmental conditions (temperature, medium composition, pH), host factors, surface sensing and plant-specific signals (Rahme *et al.*, 1992). In the phytopathogen *R. solanacearum* the expression of the T3SS is induced upon contact with host cells, similarly to the animal pathogen *Yersinia enterocolitica*. However, and differently from animal pathogens, this induction of the T3SS is not dependent on the T3SS pilus (Stainier *et al.*, 1997; Aldon *et al.*, 2000).

The effect of the secondary messenger c-di-GMP on T3SS expression was initially described in the human pathogen *P. aeruginosa* (Kulasakara *et al.*, 2006). Other studies in *P. aeruginosa* have demonstrated that c-di-GMP signalling, particularly the DGC SadC, coordinates the switch between the T3SS and T6SS via the RetS/GacS/LadS pathway (Moscoso *et al.*, 2011; Moscoso *et al.*, 2014). The mechanism of this regulation in *P. aeruginosa* is still under intense scrutiny, and novel results have demonstrated that c-di-GMP represses T3SS expression through intracellular levels of cAMP and the activity of Orn, a 3′ to 5′ exonuclease required for the full activation of the T3SS (Almblad *et al.*, 2015; Chen *et al.*, 2016). Further studies have revealed similar results in the human pathogens *V. cholerae* and *Salmonella enterica* serovar Typhimurium (*S.* Typhimurium) (Ahmad *et al.*, 2011; Roelofs *et al.*, 2015). In *S.* Typhimurium, high levels of c-di-GMP abolished the pro-inflammatory immune response and reduced the invasion phenotype in epithelial cells by the repression of the T3SS effector SipA (Lamprokostopoulou *et al.*, 2010; Ahmad *et al.*, 2011). Yi and colleagues (2010) demonstrated the role of c-di-GMP in T3SS expression in *Dickeya dadantii*, a plant phytopathogen responsible for the necrosis, blight and ‘soft rot’ disease in potato, celery, bulbs of many other vegetables and a variety of ornamental crops. Two PDE proteins, EGcpB and EcpC, from the 18 GGDEF/EAL domain proteins present in the genome, positively impact T3SS expression via the sigma factor HrpL, affecting *D. dadantii* virulence (Yi *et al.*, 2010). Particularly, c-di-GMP levels affect the transcription of the *hrpA*, *hrpN*, *dspE* and *hrpL* genes, which encode the major structural component of the T3SS pilus, a hairpin, an effector and an alternative sigma factor, respectively. Additional studies have demonstrated that $YcgR_{3937}$, a PilZ-domain protein from *D. dadantii* is required for T3SS regulation by EGcpB both *in vivo* and *in vitro* (Yuan *et al.*, 2015). Specifically, $YcgR_{3937}$ was demonstrated to bind c-di-GMP and impair *hrpA* expression. Simultaneously, *hrpL* expression is controlled by the flagella master regulator FlhDC, upon the induction of EcpC via the alternative sigma factor RpoN. Interestingly, EGcpB and EcpC contribute differently to T3SS expression, suggesting that each protein might play specific roles under certain environmental conditions (Yi *et al.*, 2010; Yuan *et al.*, 2015). The intracellular c-di-GMP effector YcgR is conserved in *Enterobacteriaceae*, although in *E. coli* and *S.* Typhimurium, this effector protein has been exclusively associated with bacterial motility (Ryjenkov *et al.*, 2006; Zorraquino *et al.*, 2013). YhjH, the EcpC homologue in *E. coli*, is activated through FliA differently from *D. dadantii* (Pesavento *et al.*, 2008).

Similar to *D. dadantii*, high levels of c-di-GMP inhibit the expression of *hrpA* in *E. amylovora*, the causative agent of fire blight disease in plants from the *Rosaceae* family. However, the global effect of the secondary messenger in *E. amylovora* virulence is complex and varies depending on the infection stage as follows: at early stages, low levels of c-di-GMP induce the expression of T3SS genes, which are required for initial host colonization; however, the subsequent bacterial establishment in the host vascular tissues requires high levels of c-di-GMP that increase the production of exopolysaccharide and the synthesis of degradative enzymes (Edmunds *et al.*, 2013; Piqué *et al.*, 2015).

The bacterial plant pathogen *P. syringae*, responsible for major disease outbreaks in fruit trees, rice, tomatoes, corn, cucumber and beans, also uses a T3SS to deliver virulence factors and promote colonization and infection (Anderson *et al.*, 2014). In *P. syringae* pv. tomato and pv. phaseolicola, the overexpression of the DGC PleD leads to a 2-fold decrease in the expression of *hrpL* and *hrpA* genes, encoding the alternative sigma factor and protein subunits of the T3SS, respectively. However, this reduction in T3SS expression apparently does not affect bacterial virulence *in vivo* (Pérez-Mendoza *et al.*, 2014). On the contrary, the overexpression of PleD in *P. savastanoi* does not inhibit *hrpL* and *hrpA* expression, although it clearly affects the symptomatology of the disease (Pérez-Mendoza *et al.*, 2014). The DGC DgcP, a conserved protein in Pseudomonads, has been associated with an increase in the intracellular levels of c-di-GMP and the reduction of the virulence of *P. aeruginosa* PAK and *P. savastanoi* pv. savastanoi NCPPB3335 in an acute lung injury model and 1-year-old olive plants, respectively (Aragón *et al.*, 2015b). Moreover, the overexpression of DgcP repressed the T3SS and induced the T6SS in *P. aeruginosa*, consistent with the previous observations of Moscoso and colleagues (2011). Recent findings in *X. oryzae* demonstrated the negative impact of the DGC GdpX1 in *hrpG*, *hrpX* and *hpa1* mRNA levels and other virulence factors, such as exopolysaccharide production and motility (Yang *et al.*, 2016).

These apparently contradictory findings suggest that a highly complex regulation pathway is mediated by the secondary messenger c-di-GMP and specific modulatory enzymes, which introduce diversification in the final regulation of T3SS expression in related and non-related bacteria.

Cyclic di-GMP coordinates exopolysaccharide production

Plant-associated bacteria form surface communities known as biofilms. Within a biofilm, bacterial cells are tightly associated via an extracellular matrix that provides protection against changing environmental conditions, such as desiccation and radiation (Flemming and Wingender, 2010). Any of the plants' surfaces, aerial tissues, vascular system and surrounding plant roots can potentially be covered with a biofilm of either epiphyte or endophyte bacteria, including pathogens or symbionts microbes (Ramey *et al.*, 2004; Danhorn and Fuqua, 2007; Castiblanco and Sundin, 2016). Indeed, the ability to form biofilms is a common strategy for the survival or transmission of pathogens (Koczan *et al.*, 2009; Chatterjee *et al.*, 2010), particularly in the xylem-dwelling pathogens *Erwinia amylovora*, *R. solanacearum*, *X. campestris* and *X. fastidiosa* (Fig. 8.1) (Koczan *et al.*, 2009; Chatterjee *et al.*, 2010; Castiblanco and Sundin, 2016).

Surface proteins, surface polysaccharides (SPSS), including exopolysaccharides (EPSs) and lipopolysaccharides (LPSs), and extracellular DNA are major structural components that constitute the extracellular matrix of bacterial biofilms (Flemming and Wingender, 2010). Interestingly, SPSS, EPSs and LPSs are among the major virulence factors in bacterial pathogens that help bacterial cells evade plant immunity (D'Haeze and Holsters, 2004). Plant pathogens produce a large number of different polysaccharides, as alginate, cellulose, levan, amylovoran, succinoglycan and xanthan (Sunish Kumar and Sakthivel, 2001; Kumar *et al.*, 2003; D'Haeze *et al.*, 2004; Laue, 2006; Koczan *et al.*, 2009; Vrancken *et al.*, 2013; Pérez-Mendoza *et al.*, 2014; Aragón *et al.*, 2015a; Castiblanco and Sundin, 2016; Ordax *et al.*, 2010). Reflecting their highly hydrated and anionic characteristics, EPSs contribute to bacterial fitness and interactions with plants in a multifaceted manner, involving resistance to stress tolerance, colonization, growth, and survival in plants and an increase in bacterial pathogenicity and virulence (Fig 8.1) (Denny, 1995; D'Haeze *et al.*, 2004; Ordax *et al.*, 2010). Moreover, recent findings have suggested that SPSS could contribute to the differential host specificity of closely related plant pathogenic bacteria (Kawaharada *et al.*, 2015). During infection, EPSs could help bacterial cells escape from plant defences or suppress basal plant responses (Pellock *et al.*, 2000; D'Haeze and Holsters, 2004; Büttner and Bonas, 2010; Aslam *et al.*, 2008). As previously described, plant responses involve the massive production of ROS, antibiotics and antibacterial cationic peptides, thus it has been proposed that both SPSS and EPSs might protect bacterial cells from the plant immunity in the following three different ways: (i) sequestering antibiotics and antimicrobial compounds in different compartments; (ii) forming electrostatic barriers to repel cationic antimicrobial compounds; and (iii) acting as a physical diffusion barrier against ROS (D'Haeze *et al.*, 2004).

Studies with *X. campestris* have demonstrated that xanthan and cyclic glucan EPSs suppress callose deposition in tobacco and *A. thaliana* plants, thereby increasing the susceptibility of plants and the progression of the diseases (Yun *et al.*, 2006; Rigano *et al.*, 2007). *Xanthomonas* polysaccharides also chelate calcium ions present in plant apoplasts, which is essential for the activation of the PTI response (Aslam *et al.*, 2008). The vascular pathogen *E. amylovora* produces two types of EPSs, levan and amylovoran, which are both involved in biofilm formation, virulence and pathogenicity (Koczan *et al.*, 2009). Both molecules also mediate survival under certain environmental conditions and likely contribute resistance to anti-microbial compounds produced from plants during infection (Ordax *et al.*, 2010). Interestingly, recent findings revealed that amylovoran production and T3SS expression are negatively regulated by two-component signal transduction systems EnvZ/OmpR and GrrS/GrrA (Wenting Li *et al.*, 2014). Interestingly, some EPSs induce the host immune system. Two examples of this phenomena are as follows: (i) the EPS I from the plant pathogen *R. solanacearum*, which specifically induces the defence response in wilt-resistant tomato plants; and (ii) the EPS from *X. campestris* pv. *vesicatoria*, which induces the production of the broad-spectrum antimicrobial molecule phytoalexin (Romeiro and Kimura, 1997; D'Haeze *et al.*, 2004; Milling *et al.*, 2011).

Phytopathogenic bacteria have evolved sophisticated regulatory systems to adjust the synthesis of EPSs to extracellular conditions, such as nutrient availability and the presence of plant-derived molecules (Vojnov *et al.*, 2001; He *et al.*, 2009). Two-component signal transduction systems, stress response mechanisms, sigma factors, and post-transcriptional regulators constitute the regulatory network of polysaccharide synthesis during plant infection (He and Zhang, 2008; Büttner and Bonas, 2010). In many pathogenic or commensal bacterial species, c-di-GMP levels control EPS production in different ways (Steiner *et al.*, 2013; Edmunds *et al.*, 2013; Cursino *et al.*, 2015; Pérez-Mendoza *et al.*, 2015; Aragón *et al.*, 2015b; Köseoğlu *et al.*, 2015; Pérez-Mendoza and Sanjuán, 2016; Yang *et al.*, 2016). Recent studies on bacterial plant pathogens from the *P. syringae* complex have implicated c-di-GMP signalling in the virulence of these phytopathogens (Pérez-Mendoza *et al.*, 2014; Aragón *et al.*, 2015a; Aragón *et al.*, 2015b; Pfeilmeier *et al.*, 2016a). Furthermore, the active DGC DgcP and the PDE BifA are highly conserved among bacteria from the genus *Pseudomonas*, including the opportunistic pathogen *P. aeruginosa*. Particularly, the overexpression of DgcP positively impacts EPS production, although *dgcP* mutants of either *P. savastanoi* and *P. aeruginosa*, are arrested in virulence (Aragón *et al.*, 2015b). The overexpression of PleD, a DGC enzyme, from *Caulobacter crescentus*, did not influence the virulence of *P. syringae* pv. tomato or *P. syringae* pv. phaseolicola; however, this enzyme increases the knot size and reduces the plant necrosis induced by *P. savastanoi* pv. savastanoi (Pérez-Mendoza *et al.*, 2014). The DGC Chp8 from *Pseudomonas syringae* pv. tomato DC3000 is responsible for the increase in c-di-GMP levels during early stages of infection, which decreases flagellin synthesis and increases EPS production, cooperatively contributing to effective bacterial evasion from the plant immunity (Engl *et al.*, 2014). Interestingly, Engl and collaborators (2014) showed that Chp8 is encoded within the *hrp* cluster regulon, although its expression is independent of HrpL. This study have also shown that phloretin, a plant flavonoid produced upon bacterial infection, induces *chp8* expression with the consequent increase in c-di-GMP levels and the production of the EPS (Engl *et al.*, 2014).

Although it is commonly assumed that the increase in the secondary messenger c-di-GMP induces EPS synthesis, biofilm formation and bacterial cell autoaggregation, the connection between these processes is complex and depends on the bacterial species and its adaptation to host signals and the environment. In *E. amylovora* pathogenesis, biofilm formation and the amylovoran EPS are essential for vascular tissue colonization and virulence. The synthesis of this EPS is regulated by two DGCs: EdcC and EdcE (Ayers *et al.*, 1979; Edmunds *et al.*, 2013). The double mutant *edcCE* showed low levels of c-di-GMP and was impaired in biofilm formation and amylovoran production, but surprisingly, showed increased virulence, most likely affecting the expression of additional virulence factors as discussed above (Edmunds *et al.*, 2013). This outcome could reflect the fact that *E. amylovora* pathogenesis is complex and still poorly understood. However, this finding also indicates that different virulence factors might be essential for early stages of infection, while other factors might be required only during subsequent periods. In *X. oryzae* pv. *oryzae*, the DGC DgcA negatively affects EPS production, motility, autoaggregation and virulence against two different rice cultivars, although positively contributing to biofilm formation (Su *et al.*, 2016). Nevertheless, high levels of c-di-GMP in *X. fastidiosa* repress both EPS synthesis and biofilm formation (Chatterjee *et al.*, 2010; de Souza *et al.*, 2013). Chatterjee and colleagues (2010) demonstrated that the DGC CsgA negatively impacts EPS production and biofilm formation and promotes bacterial migration through the vascular system of the plant, thereby causing disease.

Currently, c-di-GMP has been implicated in the regulation of more than ten different EPSs from plant and animal pathogens at different levels (Pérez-Mendoza and Sanjuán, 2016). Studies have highlighted that bacterial EPSs have diverse functions in bacterial cells during interactions with the host and are key elements in the cross talk between bacteria and plants or animal hosts, which determines the outcome of a given interaction.

Conclusions and future perspectives

Approximately 100 bacterial species cause plant diseases and a variety of symptoms from leaf spots and blights to soft rots, wilts, knots and cankers (Agrios, 2005). The diversity of the symptomatology relies on the arsenal of different virulence mechanisms and regulatory pathways engaged in pathogenesis. In recent years, the secondary messenger c-di-GMP has emerged as a universal key molecule in the control and regulation of virulence in plant pathogenic bacteria. However, the multiplicity of single species of DGCs, PDEs enzymes and c-di-GMP-binding effector proteins has cluttered the comprehension of c-di-GMP signalling. The multiple divergences observed in different bacterial species regarding c-di-GMP regulation could be explained by the research approach based on the overexpression or mutation of specific DGCs or PDEs and further evaluation of its functionality on global virulence, instead of particular virulence-related phenotypes in a temporal frame of reference.

To acknowledge the importance of c-di-GMP in the virulence of plant pathogenic bacteria, future investigations should commit to determining the specific molecular and physiological functions of all of the enzymes involved in c-di-GMP metabolism in a temporal-spatial context within the infection process. The following are some relevant questions that should be addressed to fully comprehend the intricate pathway of c-di-GMP signalling: (i) what is the nature of the signals responsible for the activation or repression of DGCs and PDEs within the infection process?; (ii) what is the specific function of these enzymes in the cell within a time and space during different infection stages; or (iii) how is the cross-talk between the diverse enzymes and the enzymatic network interconnected with other global regulatory networks operating in bacteria? The understanding of this close molecular relationship between bacterial pathogens and their hosts could contribute to the development of more pathogen resilient crops and improve strategies for crop development.

Conflicts of interest disclosure

The authors declare no competing financial interests.

Acknowledgements

This work was financially supported by grant AGL2014-53242-C2-1-R from the Ministerio de Economía y Competitividad (MINECO), Spain and co-financed by FEDER. The authors would like to thank Dr Diego Romero (Universidad de Malaga, Spain) for providing critical comments on the manuscript.

References

Abramovitch, R.B., Anderson, J.C., and Martin, G.B. (2006). Bacterial elicitation and evasion of plant innate immunity. Nat. Rev. Mol. Cell Biol. 7, 601–611.

Agrios, G.N. (2005). Plant Pathology. Waltham, MA (Eselvier Acad).

Ahmad, I., Lamprokostopoulou, A., Le Guyon, S., Streck, E., Barthel, M., Peters, V., *et al.* (2011). Complex c-di-GMP signaling networks mediate transition between virulence properties and biofilm formation in *Salmonella enterica* serovar Typhimurium. PLOS ONE 6, e28351.

Albus, U., Baier, R., Holst, O., Pühler, A., and Niehaus, K. (2001). Suppression of an elicitor-induced oxidative burst reaction in *Medicago sativa* cell cultures by *Sinorhizobium meliloti* lipopolysaccharides. New Phytol. 151, 597–606.

Aldon, D., Brito, B., Boucher, C., and Genin, S. (2000). A bacterial sensor of plant cell contact controls the transcriptional induction of *Ralstonia solanacearum* pathogenicity genes. EMBO J. 19, 2304–2314.

Alfano, J.R., and Collmer, A. (2004). Type III secretion system effector proteins: double agents in bacterial disease and plant defense. Annu. Rev. Phytopathol. *42*, 385–414. https://doi.org/10.1146/annurev.phyto.42.040103.110731.

Almblad, H., Harrison, J.J., Rybtke, M., Groizeleau, J., Givskov, M., Parsek, M.R., and Tolker-Nielsen, T. (2015). The cyclic AMP-Vfr signaling pathway in *Pseudomonas aeruginosa* is inhibited by cyclic di-GMP. J. Bacteriol. 197, 2190–2200.

Aloni, Y., Delmer, D.P., and Benziman, M. (1982). Achievement of high rates of *in vitro* synthesis of 1,4-beta-D-glucan: activation by cooperative interaction of the *Acetobacter xylinum* enzyme system with GTP, polyethylene glycol, and a protein factor. PNAS 79 (21), 6448–6452.

Amikam, D., and Benziman, M. (1989). Cyclic diguanylic acid and cellulose synthesis in *Agrobacterium tumefaciens*. J. Bacteriol. *171*, 6649–6655.

Anderson, J.C., Wan, Y., Kim, Y.-M., Pasa-Tolic, L., Metz, T.O., and Peck, S.C. (2014). Decreased abundance of type III secretion system-inducing signals in *Arabidopsis* mkp1 enhances resistance against *Pseudomonas syringae*. PNAS 111, 6846–6851.

An, S.Q., Caly, D.L., McCarthy, Y., Murdoch, S.L., Ward, J., Febrer, M., Dow, J.M., and Ryan, R.P. (2014). Novel cyclic di-GMP effectors of the YajQ protein family control bacterial virulence. PLOS Pathog. *10*, e1004429. https://doi.org/10.1371/journal.ppat.1004429.

Aragón, I.M., Pérez-Mendoza, D., Gallegos, M.T., and Ramos, C. (2015a). The c-di-GMP phosphodiesterase BifA is involved in the virulence of bacteria from the *Pseudomonas syringae* complex. Mol. Plant Pathol. *16*, 604–615. https://doi.org/10.1111/mpp.12218.

Aragón, I.M., Pérez-Mendoza, D., Moscoso, J.A., Faure, E., Guery, B., Gallegos, M.T., Filloux, A., and Ramos, C. (2015b). Diguanylate cyclase DgcP is involved in plant and human *Pseudomonas* spp. infections. Environ. Microbiol. *17*, 4332–4351. https://doi.org/10.1111/1462-2920.12856.

Arnold, D.L., Pitman, A., and Jackson, R.W. (2003). Pathogenicity and other genomic islands in plant pathogenic bacteria. Mol. Plant Pathol. *4*, 407–420. https://doi.org/10.1046/j.1364-3703.2003.00187.x.

Aslam, S.N., Newman, M.-A., Erbs, G., Morrissey, K.L., Chinchilla, D., Boller, T., *et al.* (2008). Bacterial polysaccharides suppress induced innate immunity by calcium chelation. Curr. Biol. 18, 1078–1083.

Ausmees, N., Mayer, R., Weinhouse, H., Volman, G., Amikam, D., Benziman, M., and Lindberg, M. (2001). Genetic data indicate that proteins containing the GGDEF domain possess diguanylate cyclase activity. FEMS Microbiol. Lett. 204, 163–167.

Ausubel, F.M. (2005). Are innate immune signaling pathways in plants and animals conserved? Nat. Immunol. *6*, 973–979.

Ayers, A.R., Ayers, S.B., and Goodman, R.N. (1979). Extracellular Polysaccharide of *Erwinia amylovora*: a Correlation with Virulence. Appl. Environ. Microbiol. *38*, 659–666.

Barends, T.R.M., Hartmann, E., Griese, J.J., Beitlich, T., Kirienko, N.V., Ryjenkov, D.A., *et al.* (2009). Structure and mechanism of a bacterial light-regulated cyclic nucleotide phosphodiesterase. Nature 459, 1015–1018.

Block, A., and Alfano, J.R. (2011). Plant targets for *Pseudomonas syringae* type III effectors: virulence targets or guarded decoys? Curr. Opin. Microbiol. *14*, 39–46. https://doi.org/10.1016/j.mib.2010.12.011.

Bobrov, A.G., Kirillina, O., and Ryjenkov, D.A. (2011). Systematic analysis of cyclic di-GMP signalling enzymes and their role in biofilm formation and virulence in *Yersinia pestis*. Mol. Microbiol. *79*, 533–551.

Boehm, A., Kaiser, M., Li, H., Spangler, C., Kasper, C.A., Ackermann, M., Kaever, V., Sourjik, V., Roth, V., and Jenal, U. (2010). Second messenger-mediated adjustment of bacterial swimming velocity. Cell *141*, 107–116. https://doi.org/10.1016/j.cell.2010.01.018.

Broz, P., and Monack, D.M. (2013). Newly described pattern recognition receptors team up against intracellular pathogens. Nat. Rev. Immunol. *13*, 551–565. https://doi.org/10.1038/nri3479.

Büttner, D., and Bonas, U. (2010). Regulation and secretion of *Xanthomonas* virulence factors. FEMS Microbiol. Rev. *34*, 107–133. https://doi.org/10.1111/j.1574-6976.2009.00192.x.

Caspari, C. *et al.*, (2011). Quantification of costs and benefits of amendments to the EU plant health regime: draft final report. DG Sanco Framework Contract for evaluation Lot3: Food Chain. European Commision Directorate General for Health and Consumers.

Castañeda-Ojeda, M.P., López-Solanilla, E., and Ramos, C. (2016). Differential modulation of plant immune responses by diverse members of the *Pseudomonas savastanoi* pv. savastanoi HopAF type III effector family. Mol. Plant Pathol. https://doi.org//mpp.12420.

Castiblanco, L.F., and Sundin, G.W. (2016). New insights on molecular regulation of biofilm formation in plant-associated bacteria. J. Integr. Plant Biol. *58*, 362–372. https://doi.org/10.1111/jipb.12428.

Chan, C., Paul, R., Samoray, D., Amiot, N.C., Giese, B., Jenal, U., and Schirmer, T. (2004). Structural basis of activity and allosteric control of diguanylate cyclase. Proc. Natl. Acad. Sci. U.S.A. *101*, 17084–17089.

Chatterjee, S., Killiny, N., Almeida, R.P., and Lindow, S.E. (2010). Role of cyclic di-GMP in *Xylella fastidiosa* biofilm formation, plant virulence, and insect transmission. Mol. Plant Microbe Interact. *23*, 1356–1363. https://doi.org/10.1094/MPMI-03-10-0057.

Chen, G., Zhao, Q., Zhu, F., Chen, R., Jin, Y., Liu, C., Pan, X., Jin, S., Wu, W., and Cheng, Z. (2016). Oligoribonuclease is required for the type III secretion system and pathogenesis of *Pseudomonas aeruginosa*. Microbiol. Res. 188–189, 90–96. https://doi.org/10.1016/j.micres.2016.05.002.

Chin, K.H., Lee, Y.C., Tu, Z.L., Chen, C.H., Tseng, Y.H., Yang, J.M., Ryan, R.P., McCarthy, Y., Dow, J.M., Wang, A.H., *et al.* (2010). The cAMP receptor-like protein CLP is a novel c-di-GMP receptor linking cell-cell signaling to virulence gene expression in *Xanthomonas campestris*. J. Mol. Biol. *396*, 646–662. https://doi.org/10.1016/j.jmb.2009.11.076.

Christen, M., Christen, B., Allan, M.G., Folcher, M., Jenö, P., Grzesiek, S., and Jenal, U. (2007). DgrA is a member of a new family of cyclic diguanosine monophosphate receptors and controls flagellar motor function in *Caulobacter crescentus*. PNAS 104, 4112–4117.

Christen, M., Christen, B., Folcher, M., Schauerte, A., and Jenal, U. (2005). Identification and characterization of a cyclic di-GMP-specific phosphodiesterase and its allosteric control by GTP. J. Biol. Chem. *280*, 30829–30837.

Cursino, L., Athinuwat, D., Patel, K.R., Galvani, C.D., Zaini, P.A., Li, Y., De La Fuente, L., Hoch, H.C., Burr, T.J., and Mowery, P. (2015). Characterization of the *Xylella fastidiosa* PD1671 gene encoding degenerate c-di-GMP GGDEF/EAL domains, and its role in the development of Pierce's disease. PLOS ONE *10*, e0121851. https://doi.org/10.1371/journal.pone.0121851.

D'Haeze, W., and Holsters, M. (2004). Surface polysaccharides enable bacteria to evade plant immunity. Trends Microbiol. *12*, 555–561.

D'Haeze, W., Glushka, J., De Rycke, R., Holsters, M., and Carlson, R.W. (2004). Structural characterization of extracellular polysaccharides of *Azorhizobium caulinodans* and importance for nodule initiation on *Sesbania rostrata*. Mol. Microbiol. *52*, 485–500.

Danhorn, T., and Fuqua, C. (2007). Biofilm formation by plant-associated bacteria. Annu. Rev. Microbiol. *61*, 401–422. https://doi.org/10.1146/annurev.micro.61.080706.093316.

de Souza, A.A., Ionescu, M., Baccari, C., da Silva, A.M., and Lindow, S.E. (2013). Phenotype overlap in *Xylella fastidiosa* is controlled by the cyclic di-GMP phosphodiesterase eal in response to antibiotic exposure and diffusible signal factor-mediated cell-cell signaling. Appl. Environ. Microbiol. *79*, 3444–3454.

Denny, T.P. (1995). Involvement of bacterial polysaccharides in plant pathogenesis. Annu. Rev. Phytopathol. *33*, 173–197. https://doi.org/10.1146/annurev.py.33.090195.001133.

Dodds, P.N., and Rathjen, J.P. (2010). Plant immunity: towards an integrated view of plant-pathogen interactions. Nat. Rev. Genet. *11*, 539–548. https://doi.org/10.1038/nrg2812.

Dow, J.M., Fouhy, Y., Lucey, J.F., and Ryan, R.P. (2006). The HD-GYP domain, cyclic di-GMP signaling, and bacterial virulence to plants. Mol. Plant Microbe Interact. *19*, 1378–1384. https://doi.org/10.1094/MPMI-19-1378.

Dubey, B.N., Lori, C., Ozaki, S., Fucile, G., Plaza-Menacho, I., Jenal, U., and Schirmer, T. (2016). Cyclic di-GMP mediates a histidine kinase/phosphatase switch by noncovalent domain cross-linking. Sci. Adv. *2*, e1600823. https://doi.org/10.1126/sciadv.1600823.

Edmunds, A.C., Castiblanco, L.F., Sundin, G.W., and Waters, C.M. (2013). Cyclic Di-GMP modulates the disease progression of *Erwinia amylovora*. J. Bacteriol. *195*, 2155–2165. https://doi.org/10.1128/JB.02068-12.

Engl, C., Waite, C.J., McKenna, J.F., Bennett, M.H., Hamann, T., and Buck, M. (2014). Chp8, a Diguanylate Cyclase from *Pseudomonas syringae* pv. Tomato DC3000, Suppresses the Pathogen-Associated Molecular Pattern Flagellin, Increases Extracellular Polysaccharides, and Promotes Plant Immune Evasion. mBio 20;5(3), e01168-14.

Flemming, H.C., and Wingender, J. (2010). The biofilm matrix. Nat. Rev. Microbiol. *8*, 623–633. https://doi.org/10.1038/nrmicro2415.

Food and Agriculture Organization of the United Nations (2006). World agriculture: towards 2030/2050. Interim report. Prospects for food, nutrition, agriculture and major commodity groups. Global perspectives Studies Unit, Food and Agriculture Organization of the United Nations, Rome, Italy.

Freeman, B.C., and Beattie, G.A. (2008). An Overview of Plant Defenses against Pathogens and Herbivores. The Plant Health Instructor. https://doi.org/10.1094/PHI-I-2008-0226-01.

Fuller, K. (2012). The Economic Consequences of Pierce's Disease and Related Policy in the California Winegrape Industry. Journal of Agricultural and Resource Economic 38 (2), 1–42.

Galán, J.E., Lara-Tejero, M., Marlovits, T.C., and Wagner, S. (2014). Bacterial type III secretion systems: specialized nanomachines for protein delivery into target cells. Annu. Rev. Microbiol. *68*, 415–438. https://doi.org/10.1146/annurev-micro-092412-155725.

Galperin, M.Y., Nikolskaya, A.N., and Koonin, E.V. (2001). Novel domains of the prokaryotic two-component signal transduction systems. FEMS Microbiol. Lett. *203*, 11–21.

Gómez-Gómez, L., and Boller, T. (2000). FLS2: an LRR receptor-like kinase involved in the perception of the bacterial elicitor flagellin in *Arabidopsis*. Mol. Cell 5, 1003–1011.

Guan, W., Shao, J., Elbeaino, T., Davis, R.E., Zhao, T., and Huang, Q. (2015). Specific Detection and Identification of American Mulberry-Infecting and Italian Olive-Associated Strains of *Xylella fastidiosa* by Polymerase Chain Reaction. PLOS ONE 10, e0129330.

Gust, A.A., Biswas, R., Lenz, H.D., Rauhut, T., Ranf, S., Kemmerling, B., *et al.* (2007). Bacteria-derived peptidoglycans constitute pathogen-associated molecular patterns triggering innate immunity in *Arabidopsis*. J. Biol. Chem. 282, 32338–32348.

Guttman, D.S., McHardy, A.C., and Schulze-Lefert, P. (2014). Microbial genome-enabled insights into plant-microorganism interactions. Nat. Rev. Genet. *15*, 797–813. https://doi.org/10.1038/nrg3748.

Hayashi, F., Smith, K.D., Ozinsky, A., Hawn, T.R., Yi, E.C., Goodlett, D.R., Eng, J.K., Akira, S., Underhill, D.M., and Aderem, A. (2001). The innate immune response to bacterial flagellin is mediated by Toll-like receptor 5. Nature *410*, 1099–1103. https://doi.org/10.1038/35074106.

He, Y.-W., and Chou S.-H. (2015). Cyclic di-GMP Regulation in Plant-Pathogenic Bacteria. In Virulence Mechanisms of Plant-Pathogenic Bacteria, N. Wang, J. B. Jones, G. W. Sundin, F. White, S. Hogenhout, C. Roper, L. De La Fuente, and J. H. Ham, ed. (The American Phytopathological Society, Minnesota) pp. 107–124.

He, Y.W., and Zhang, L.H. (2008). Quorum sensing and virulence regulation in *Xanthomonas campestris*. FEMS Microbiol. Rev. 32, 842–857. https://doi.org/10.1111/j.1574-6976.2008.00120.x.

He, Y.-W., Boon, C., Zhou, L., and Zhang, L.-H. (2009). Co-regulation of *Xanthomonas campestris* virulence by quorum sensing and a novel two-component regulatory system RavS/RavR. Mol. Microbiol. 71, 1464–1476.

He, Y.-W., Wu, J., Cha, J.-S., and Zhang, L.-H. (2010). Rice bacterial blight pathogen *Xanthomonas oryzae* pv. oryzae produces multiple DSF-family signals in regulation of virulence factor production. BMC Microbiology 10, 1–9.

Hengge, R. (2009). Principles of c-di-GMP signalling in bacteria. Nat. Rev. Microbiol. 7, 263–273. https://doi.org/10.1038/nrmicro2109.

Henry, J.T., and Crosson, S. (2011). Ligand-binding PAS domains in a genomic, cellular, and structural context. Annu. Rev. Microbiol. *65*, 261–286. https://doi.org/10.1146/annurev-micro-121809-151631.

Hickman, J.W., and Harwood, C.S. (2008). Identification of FleQ from *Pseudomonas aeruginosa* as a c-di-GMP-responsive transcription factor. Mol. Microbiol. 69, 376–389.

Hopkins, P.A., and Sriskandan, S. (2005). Mammalian Toll-like receptors: to immunity and beyond. Clin. Exp. Immunol. *140*, 395–407.

Huang, C.J., Wang, Z.C., Huang, H.Y., Huang, H.D., and Peng, H.L. (2013). YjcC, a c-di-GMP phosphodiesterase protein, regulates the oxidative stress response and virulence of *Klebsiella pneumoniae* CG43. PLOS ONE *8*, e66740. https://doi.org/10.1371/journal.pone.0066740.

Iriti, M., and Faoro, F. (2007). Review of innate and specific immunity in plants and animals. Mycopathologia *164*, 57–64. https://doi.org/10.1007/s11046-007-9026-7.

Ishikawa, K., Yamaguchi, K., Sakamoto, K., Yoshimura, S., Inoue, K., Tsuge, S., Kojima, C., and Kawasaki, T. (2014). Bacterial effector modulation of host E3 ligase activity suppresses PAMP-triggered immunity in rice. Nat. Commun. *5*, 5430. https://doi.org/10.1038/ncomms6430.

Jenal, U., and Malone, J. (2006). Mechanisms of cyclic-di-GMP signaling in bacteria. Annu. Rev. Genet. *40*, 385–407. https://doi.org/10.1146/annurev.genet.40.110405.090423.

Jones, J.D., and Dangl, J.L. (2006). The plant immune system. Nature *444*, 323–329.

Kang, Y., Jelenska, J., Cecchini, N.M., Li, Y., Lee, M.W., Kovar, D.R., and Greenberg, J.T. (2014). HopW1 from *Pseudomonas syringae* disrupts the actin cytoskeleton to promote virulence in *Arabidopsis*. PLOS Pathog. *10*, e1004232. https://doi.org/10.1371/journal.ppat.1004232.

Kawaharada, Y., Kelly, S., Nielsen, M.W., Hjuler, C.T., Gysel, K., Muszyński, A., *et al.* (2015). Receptor-mediated exopolysaccharide perception controls bacterial infection. Nature 523, 308–312.

Kazan, K., and Lyons, R. (2014). Intervention of Phytohormone Pathways by Pathogen Effectors. Plant Cell *26*, 2285–2309.

Koczan, J.M., McGrath, M.J., Zhao, Y., and Sundin, G.W. (2009). Contribution of *Erwinia amylovora* Exopolysaccharides Amylovoran and Levan to Biofilm Formation: Implications in Pathogenicity. Phytopathology 99, 1237–1244.

Köseoğlu, V.K., Heiss, C., Azadi, P., Topchiy, E., Güvener, Z.T., Lehmann, T.E., *et al.* (2015). *Listeria monocytogenes* exopolysaccharide: origin, structure, biosynthetic machinery and c-di-GMP-dependent regulation. Mol. Microbiol. 96, 728–743.

Kulasakara, H., Lee, V., Brencic, A., Liberati, N., Urbach, J., Miyata, S., Lee, D.G., Neely, A.N., Hyodo, M., Hayakawa, Y., *et al.* (2006). Analysis of *Pseudomonas aeruginosa* diguanylate cyclases and phosphodiesterases reveals a role for bis-(3′-5′)-cyclic-GMP in virulence. Proc. Natl. Acad. Sci. U.S.A. *103*, 2839–2844.

Kumar, A., Sunish Kumar, R., and Sakthivel, N. (2003). Compositional difference of the exopolysaccharides produced by the virulent and virulence-deficient strains of *Xanthomonas oryzae* pv. oryzae. Curr. Microbiol. 46, 251–255.

Krasteva, P.V., Giglio, K.M., and Sondermann, H. (2012). Sensing the messenger: the diverse ways that bacteria signal through c-di-GMP. Protein Sci. *21*, 929–948. https://doi.org/10.1002/pro.2093.

Lam, H.N., Chakravarthy, S., Wei, H.L., BuiNguyen, H., Stodghill, P.V., Collmer, A., Swingle, B.M., and Cartinhour, S.W. (2014). Global analysis of the HrpL regulon in the plant pathogen *Pseudomonas syringae* pv. tomato DC3000 reveals new regulon members with diverse functions. PLOS ONE 9, e106115. https://doi.org/10.1371/journal.pone.0106115.

Lamprokostopoulou, A., Monteiro, C., Rhen, M., and Römling, U. (2010). Cyclic di-GMP signalling controls virulence properties of *Salmonella enterica* serovar Typhimurium at the mucosal lining. Environmental Microbiology 12, 40–53.

Lan, L., Deng, X., Zhou, J., and Tang, X. (2006). Genome-wide gene expression analysis of *Pseudomonas syringae* pv. tomato DC3000 reveals overlapping and distinct pathways regulated by hrpL and hrpRS. Mol. Plant Microbe Interact. *19*, 976–987. https://doi.org/10.1094/MPMI-19-0976.

Laue, H. (2006). Contribution of alginate and levan production to biofilm formation by *Pseudomonas syringae*. Microbiology 152, 2909–2918.

Le Fevre, R., Evangelisti, E., Rey, T., and Schornack, S. (2015). Modulation of host cell biology by plant pathogenic microbes. Annu. Rev. Cell Dev. Biol. *31*, 201–229. https://doi.org/10.1146/annurev-cellbio-102314-112502.

Lee, H.S., Gu, F., Ching, S.M., Lam, Y., and Chua, K.L. (2010). CdpA is a *Burkholderia pseudomallei* cyclic di-GMP phosphodiesterase involved in autoaggregation, flagellum synthesis, motility, biofilm formation, cell invasion, and cytotoxicity. Infection and Immunity 78, 1832–1840.

Lee, V.T., Matewish, J.M., Kessler, J.L., Hyodo, M., Hayakawa, Y., and Lory, S. (2007). A cyclic-di-GMP receptor required for bacterial exopolysaccharide production. Mol. Microbiol. 65, 1474–1484.

Levi, A., Folcher, M., Jenal, U., and Shuman, H.A. (2011). Cyclic diguanylate signaling proteins control intracellular growth of *Legionella pneumophila*. MBio 2, e00316–10. https://doi.org/10.1128/mBio.00316-10.

Li, G., Froehlich, J.E., Elowsky, C., Msanne, J., Ostosh, A.C., Zhang, C., Awada, T., and Alfano, J.R. (2014). Distinct Pseudomonas type-III effectors use a cleavable transit peptide to target chloroplasts. Plant J. 77, 310–321. https://doi.org/10.1111/tpj.12396.

Li, W., Ancona, V., and Zhao, Y. (2014). Co-regulation of polysaccharide production, motility, and expression of type III secretion genes by EnvZ/OmpR and GrrS/GrrA systems in *Erwinia amylovora*. Mol. Genet. Genomics *289*, 63–75. https://doi.org/10.1007/s00438-013-0790-4.

Liang, Z.X. (2015). The expanding roles of c-di-GMP in the biosynthesis of exopolysaccharides and secondary metabolites. Nat. Prod. Rep. *32*, 663–683. https://doi.org/10.1039/c4np00086b.

Little, R.H., Grenga, L., Saalbach, G., *et al.* (2016). Adaptive remodeling of the bacterial proteome by specific ribosomal modification regulates *Pseudomonas* infection and niche colonisation. PLOS Genet. *12*(2), e1005837.

Lu, X.H., An, S.Q., Tang, D.J., McCarthy, Y., Tang, J.L., Dow, J.M., *et al.* (2012). RsmA Regulates Biofilm Formation in *Xanthomonas campestris* through a Regulatory Network Involving Cyclic di-GMP and the Clp Transcription Factor. PLOS ONE 7, e52646.

Macho, A.P. (2016). Subversion of plant cellular functions by bacterial type-III effectors: beyond suppression of immunity. New. Phytol. *210*, 51–57. https://doi.org/10.1111/nph.13605.

Malone, J.G., Williams, R., Christen, M., Jenal, U., Spiers, A.J., and Rainey, P.B. (2007). The structure–function relationship of WspR, a *Pseudomonas fluorescens* response regulator with a GGDEF output domain. Microbiology 153, 980–994.

Mansfield, J., Genin, S., Magori, S., Citovsky, V., Sriariyanum, M., Ronald, P., Dow, M., Verdier, V., Beer, S.V., Machado, M.A., *et al.* (2012). Top 10 plant pathogenic bacteria in molecular plant pathology. Mol. Plant Pathol. *13*, 614–629. https://doi.org/10.1111/j.1364-3703.2012.00804.x.

Martínez-Gil, M., Ramos-González, M.I., and Espinosa-Urgel, M. (2014). Roles of cyclic Di-GMP and the Gac system in transcriptional control of the genes coding for the *Pseudomonas putida* adhesins LapA and LapF. J. Bacteriol. *196*, 1484–1495. https://doi.org/10.1128/JB.01287-13.

Martínez-Granero, F., Navazo, A., Barahona, E., Redondo-Nieto, M., González de Heredia, E., Baena, I., Martín-Martín, I., Rivilla, R., and Martín, M. (2014). Identification of flgZ as a flagellar gene encoding a PilZ domain protein that regulates swimming motility and biofilm formation in *Pseudomonas*. PLOS ONE *9*, e87608. https://doi.org/10.1371/journal.pone.0087608.

Matas, I.M., Castañeda-Ojeda, M.P., Aragon, I.M., Antúnez-Lamas, M., Murillo, J., Rodríguez-Palenzuela, P., *et al.* (2014). Translocation and functional analysis of *Pseudomonas savastanoi* pv. *savastanoi* NCPPB 3335 Type III secretion system effectors reveals two novel effector families of the *Pseudomonas syringae* complex. MPMI 27, 424–436.

Matas, I.M., Lambertsen, L., Rodríguez-Moreno, L., and Ramos, C. (2012). Identification of novel virulence genes and metabolic pathways required for full fitness of *Pseudomonas savastanoi* pv. savastanoi in olive (Olea europaea) knots. New Phytol. 196, 1182–1196.

McKinney, J.S. (2004). Disruption of type III secretion in *Salmonella enterica* serovar Typhimurium by external guide sequences. Nucl.c Acids Res. 32, 848–854.

Milling, A., Babujee, L., and Allen, C. (2011). *Ralstonia solanacearum* extracellular polysaccharide is a specific elicitor of defense responses in wilt-resistant tomato plants. PLOS ONE 6, e15853.

Moscoso, J.A., Jaeger, T., Valentini, M., Hui, K., Jenal, U., and Filloux, A. (2014). The diguanylate cyclase SadC is a central player in Gac/Rsm-mediated biofilm formation in *Pseudomonas aeruginosa*. J. Bacteriol. *196*, 4081–4088. https://doi.org/10.1128/JB.01850-14.

Moscoso, J.A., Mikkelsen, H., Heeb, S., Williams, P., and Filloux, A. (2011). The *Pseudomonas aeruginosa* sensor RetS switches type III and type VI secretion via c-di-GMP signalling. Environ. Microbiol. *13*, 3128–3138. https://doi.org/10.1111/j.1462-2920.2011.02595.x.

Newell, P.D., Boyd, C.D., Sondermann, H., and O'Toole, G.A. (2011). A c-di-GMP effector system controls cell adhesion by inside-out signaling and surface protein cleavage. PLOS Biol. *9*, e1000587. https://doi.org/10.1371/journal.pbio.1000587.

Newman, M.A., Sundelin, T., Nielsen, J.T., and Erbs, G. (2013). MAMP (microbe-associated molecular pattern) triggered immunity in plants. Front. Plant Sci. *4*, 139. https://doi.org/10.3389/fpls.2013.00139.

Norelli, J.L., Jones, A.L. & Aldwinckle, H.S. (2003). Fire blight management in the twenty-first century: using new technologies that enhance host resistance in apple. Plant Disease 87, 756-765.

Nürnberger, T., Brunner, F., Kemmerling, B., and Piater, L. (2004). Innate immunity in plants and animals: striking similarities and obvious differences. Immunological Reviews 198, 249–266.

Ordax, M., Marco-Noales, E., López, M.M., and Biosca, E.G. (2010). Exopolysaccharides favor the survival of *Erwinia amylovora* under copper stress through different strategies. Res. Microbiol. 161, 549–555.

Paul, R., Abel, S., Wassmann, P., Beck, A., Heerklotz, H., and Jenal, U. (2007). Activation of the diguanylate cyclase PleD by phosphorylation-mediated dimerization. J. Biol. Chem. *282*, 29170–29177.

Pearson, J.S., Zhang, Y., Newton, H.J., and Hartland, E.L. (2015). Post-modern pathogens: surprising activities of translocated effectors from *E. coli* and *Legionella*. Curr. Opin. Microbiol. *23*, 73–79. https://doi.org/10.1016/j.mib.2014.11.005.

Pel, M.J., and Pieterse, C.M. (2013). Microbial recognition and evasion of host immunity. J. Exp. Bot. *64*, 1237–1248. https://doi.org/10.1093/jxb/ers262.

Pellock, B.J., Cheng, H.P., and Walker, G.C. (2000). Alfalfa root nodule invasion efficiency is dependent on *Sinorhizobium meliloti* polysaccharides. J. Bacteriol. *182*, 4310–4318.

Perez-Martinez, I., Rodriguez-Moreno, L., Lambertsen, L., Matas, I.M., Murillo, J., Tegli, S., *et al.* (2010). Fate of a *Pseudomonas savastanoi* pv. *savastanoi* Type III secretion system mutant in olive plants (*Olea europaea* L.). Appl. Environ. Microbiol. *76*, 3611–3619.

Pesavento, C., Becker, G., Sommerfeldt, N., Possling, A., Tschowri, N., Mehlis, A., and Hengge, R. (2008). Inverse regulatory coordination of motility and curli-mediated adhesion in *Escherichia coli*. Genes Dev. *22*, 2434–2446. https://doi.org/10.1101/gad.475808.

Peterson, L.W., and Artis, D. (2014). Intestinal epithelial cells: regulators of barrier function and immune homeostasis. Nat. Rev. Immunol. *14*, 141–153. https://doi.org/10.1038/nri3608.

Pérez-Mendoza, D., and Sanjuán, J. (2016). Exploiting the commons: cyclic diguanylate regulation of bacterial exopolysaccharide production. Curr. Opin. Microbiol. *30*, 36–43.

Pérez-Mendoza, D., Aragón, I.M., Prada-Ramírez, H.A., Romero-Jiménez, L., Ramos, C., Gallegos, M.T., and Sanjuán, J. (2014). Responses to elevated c-di-GMP levels in mutualistic and pathogenic plant-interacting bacteria. PLOS ONE *9*, e91645. https://doi.org/10.1371/journal.pone.0091645.

Pérez-Mendoza, D., Coulthurst, S.J., Sanjuán, J., and Salmond, G.P. (2011). N-acetyl-glucosamine-dependent biofilm formation in *Pectobacterium atrosepticum* is cryptic and activated by elevated c-di-GMP levels. Microbiology *157*, 3340-3348.

Pérez-Mendoza, D., Rodríguez-Carvajal, M.Á., Romero-Jiménez, L., Farias, G.d.e.A., Lloret, J., Gallegos, M.T., and Sanjuán, J. (2015). Novel mixed-linkage β-glucan activated by c-di-GMP in *Sinorhizobium meliloti*. Proc. Natl. Acad. Sci. U.S.A. *112*, E757–765. https://doi.org/10.1073/pnas.1421748112.

Pfeilmeier, S., Caly, D.L., and Malone, J.G. (2016a). Bacterial pathogenesis of plants: future challenges from a microbial perspective: Challenges in Bacterial Molecular Plant Pathology. Mol. Plant Pathol. *17*, 1298–1313. https://doi.org/10.1111/mpp.12427.

Pfeilmeier, S., Saur, I.M.-L., Rathjen, J.P., Zipfel, C., and Malone, J.G. (2016b). High levels of cyclic-di-GMP in plant-associated *Pseudomonas* correlate with evasion of plant immunity. Molecular Plant Pathology 17, 521–531.

Piqué, N., Miñana-Galbis, D., Merino, S., and Tomás, J.M. (2015). Virulence Factors of *Erwinia amylovora*: A Review. Int. J. Mol. Sci. *16*, 12836–12854. https://doi.org/10.3390/ijms160612836.

Popov, G., Fraiture, M., Brunner, F., and Sessa, G. (2016). Multiple *Xanthomonas euvesicatoria* Type III Effectors Inhibit flg22-Triggered Immunity. Mol. Plant Microbe Interact. *29*, 651–660. https://doi.org/10.1094/MPMI-07-16-0137-R.

Pratt, J.T., Tamayo, R., Tischler, A.D., and Camilli, A. (2007). PilZ domain proteins bind cyclic diguanylate and regulate diverse processes in *Vibrio cholerae*. J. Biol. Chem. *282*, 12860–12870.

Purcell, A. (2013). Paradigms: examples from the bacterium *Xylella fastidiosa*. Annu. Rev. Phytopathol. *51*, 339–356. https://doi.org/10.1146/annurev-phyto-082712-102325.

Rahme, L.G., Mindrinos, M.N., and Panopoulos, N.J. (1992). Plant and environmental sensory signals control the expression of hrp genes in *Pseudomonas syringae* pv. phaseolicola. J. Bacteriol. *174*, 3499–3507.

Ramey, B.E., Koutsoudis, M., von Bodman, S.B., and Fuqua, C. (2004). Biofilm formation in plant-microbe associations. Curr. Opin. Microbiol. *7*, 602–609.

Ramos, C., Matas, I.M., Bardaji, L., Aragón, I.M., and Murillo, J. (2012). Pseudomonas *savastanoi* pv. *savastanoi*: some like it knot. Mol. Plant Pathol. *13*, 998–1009. https://doi.org/10.1111/j.1364-3703.2012.00816.x.

Rigano, L.A., Payette, C., Brouillard, G., Marano, M.R., Abramowicz, L., Torres, P.S., Yun, M., Castagnaro, A.P., Oirdi, M.E., Dufour, V., *et al.* (2007). Bacterial cyclic beta-(1,2)-glucan acts in systemic suppression of plant immune responses. Plant Cell *19*, 2077–2089.

Rodríguez-Palenzuela P., Matas I.M., Murillo J., López-Solanilla E., Bardaji L., Pérez-Martínez I., Rodríguez-Moskera M.E., Penyalver R., López M.M., Quesada J.M., et al. (2010). Annotation and overview of the Pseudomonas savastanoi pv. savastanoi NCPPB3335 draft genome reveals the virulence gene complement of a tumour-inducing pathogen of woody hosts. Environ Microbiol. 12(6), 1604–1620. https://doi.org/10.1111/j.1462-2920.2010.02207.x.

Roelofs, K.G., Jones, C.J., Helman, S.R., Shang, X., Orr, M.W., Goodson, J.R., *et al.* (2015). Systematic identification of cyclic-di-GMP binding proteins in *Vibrio cholerae* reveals a novel class of Cyclic-di-GMP-Binding ATPases associated with Type II secretion systems. PLOS Pathogens 11, e1005232.

Romeiro, R.S., and Kimura, O. (1997). Induced resistance in pepper leaves infiltrated with purified bacterial elicitors from *Xanthomonas campestris* pv. vesicatoria. J. Phytopathol. 145, 495-498.

Ross, P., Mayer, R., and Benziman, M. (1991). Cellulose biosynthesis and function in bacteria. Microbiol. Rev. 55, 35–58.

Ross, P., Weinhouse, H., Aloni, Y., Michaeli, D., Weinberger-Ohana, P., Mayer, R., Braun, S., de Vroom, E., van der Marel, G.A., van Boom, J.H., *et al.* (1987). Regulation of cellulose synthesis in *Acetobacter xylinum* by cyclic diguanylic acid. Nature 325, 279–281.

Römling, U., Galperin, M.Y., and Gomelsky, M. (2013). Cyclic di-GMP: the first 25 years of a universal bacterial second messenger. Microbiol. Mol. Biol. Rev. *77*, 1–52. https://doi.org/10.1128/MMBR.00043-12.

Römling, U., Gomelsky, M., and Galperin, M.Y. (2005). C-di-GMP: the dawning of a novel bacterial signalling system. Mol. Microbiol. *57*, 629–639.

Ryan, R.P., Fouhy, Y., Lucey, J.F., Crossman, L.C., Spiro, S., He, Y.W., Zhang, L.H., Heeb, S., Cámara, M., Williams, P., *et al.* (2006). Cell-cell signaling in *Xanthomonas campestris* involves an HD-GYP domain protein that functions in cyclic di-GMP turnover. Proc. Natl. Acad. Sci. U.S.A. *103*, 6712–6717.

Ryan, R.P., Fouhy, Y., Lucey, J.F., Jiang, B.L., He, Y.Q., Feng, J.X., Tang, J.L., and Dow, J.M. (2007). Cyclic di-GMP signalling in the virulence and environmental adaptation of *Xanthomonas campestris*. Mol. Microbiol. *63*, 429–442.

Ryjenkov, D.A., Simm, R., Römling, U., and Gomelsky, M. (2006). The PilZ domain is a receptor for the second messenger c-di-GMP: the PilZ domain protein YcgR controls motility in enterobacteria. J. Biol. Chem. *281*, 30310–30314.

Simm, R., Morr, M., Kader, A., Nimtz, M., and Römling, U. (2004). GGDEF and EAL domains inversely regulate cyclic di-GMP levels and transition from sessility to motility. Mol. Microbiol. 53, 1123–1134. https://doi.org/10.1111/j.1365-2958.2004.04206.x.

Simpson, A.J., Reinach, F.C., Arruda, P., Abreu, F.A., Acencio, M., Alvarenga, R., Alves, L.M., Araya, J.E., Baia, G.S., Baptista, C.S., *et al.* (2000). The genome sequence of the plant pathogen *Xylella fastidiosa*. The *Xylella fastidiosa* Consortium of the Organization for Nucleotide Sequencing and Analysis. Nature *406*, 151–159. https://doi.org/10.1038/35018003.

Smith, K.D., Andersen-Nissen, E., Hayashi, F., Strobe, K., Bergman, M.A., Barrett, S.L., Cookson, B.T., and Aderem, A. (2003). Toll-like receptor 5 recognizes a conserved site on flagellin required for protofilament formation and bacterial motility. Nat. Immunol. *4*, 1247–1253. https://doi.org/10.1038/ni1011.

Sondermann, H., Shikuma, N.J., and Yildiz, F.H. (2012). You've come a long way: c-di-GMP signaling. Curr. Opin. Microbiol. *15*, 140–146. https://doi.org/10.1016/j.mib.2011.12.008.

Spoel, S.H., and Dong, X. (2012). How do plants achieve immunity? Defence without specialized immune cells. Nat. Rev. Immunol. *12*, 89–100. https://doi.org/10.1038/nri3141.

Srivastava, D., Harris, R.C., and Waters, C.M. (2011). Integration of cyclic di-GMP and quorum sensing in the control of *vpsT* and *aphA* in *Vibrio cholerae*. J. Bacteriol. *193*, 6331–6341. https://doi.org/10.1128/JB.05167-11.

Stainier, I., Iriarte, M., and Cornelis, G.R. (1997). YscM1 and YscM2, two *Yersinia enterocolitica* proteins causing downregulation of *yoP* transcription. Mol. Microbiol. 26, 833–843.

Stefani, E. (2010). Economic significance and control of bacterial spot/canker of stone fruits caused by *Xanthomonas arboricola* pv. pruni. Journal of Plant Pathology 92-1, 99-103.

Steiner, S., Lori, C., Boehm, A., and Jenal, U. (2013). Allosteric activation of exopolysaccharide synthesis through cyclic di-GMP-stimulated protein–protein interaction. EMBO J. 32, 354–368.

Su, J., Zou, X., Huang, L., Bai, T., Liu, S., Yuan, M., Chou, S.H., He, Y.W., Wang, H., and He, J. (2016). DgcA, a diguanylate cyclase from *Xanthomonas oryzae* pv. *oryzae* regulates bacterial pathogenicity on rice. Sci. Rep. *6*, 25978. https://doi.org/10.1038/srep25978.

Sun, M., He, C., Cong, Y., and Liu, Z. (2015). Regulatory immune cells in regulation of intestinal inflammatory response to microbiota. Mucosal. Immunol. *8*, 969–978. https://doi.org/10.1038/mi.2015.49.

Sun, W., Dunning, F.M., Pfund, C., Weingarten, R., and Bent, A.F. (2006). Within-species flagellin polymorphism in *Xanthomonas campestris* pv. campestris and its impact on elicitation of *Arabidopsis* flagellin sensing-dependent defenses. Plant Cell 18, 764–779.

Sunish Kumar, R., and Sakthivel, N. (2001). Exopolysaccharides of *Xanthomonas* pathovar strains that infect rice and wheat crops. Appl. Microbiol. Biotechnol. 55, 782–786.

Tal, R., Wong, H.C., Calhoon, R., Gelfand, D., Fear, A.L., Volman, G., *et al.* (1998). Three cdg Operons Control Cellular Turnover of Cyclic Di-GMP in *Acetobacter xylinum*: Genetic Organization and Occurrence of Conserved Domains in Isoenzymes. Journal of Bacteriology 180, 4416–4425.

Tamayo, R., Pratt, J.T., and Camilli, A. (2007). Roles of cyclic diguanylate in the regulation of bacterial pathogenesis. Annu. Rev. Microbiol. *61*, 131–148. https://doi.org/10.1146/annurev.micro.61.080706.093426.

Tang, X., Xiao, Y., and Zhou, J.M. (2006). Regulation of the type III secretion system in phytopathogenic bacteria. Mol. Plant Microbe Interact. *19*, 1159–1166. https://doi.org/10.1094/MPMI-19-1159.

Tao, F., He, Y.W., Wu, D.H., and Swarup, S. (2010). The cyclic nucleotide monophosphate domain of *Xanthomonas campestris* global regulator Clp defines a new class of cyclic di-GMP effectors. Journal of Bacteriology 192, 1020-1029.

Tischler, A.D., and Camilli, A. (2005). Cyclic diguanylate regulates *Vibrio cholerae* virulence gene expression. Infect. Immun. 73, 5873–5882.

Trampari, E., Stevenson, C.E.M., Little, R.H., Wilhelm, T., Lawson, D.M., and Malone, J.G. (2015). Bacterial rotary export ATPases are allosterically regulated by the nucleotide second messenger cyclic-di-GMP. Journal of Biological Chemistry. 290(40), 24470–24483.

Trda, L., Boutrot, F., Claverie, J., BrulÃ, D.E., Dorey, S., and Poinssot, B. (2015). Perception of pathogenic or beneficial bacteria and their evasion of host immunity: pattern recognition receptors in the frontline. Frontiers in Plant Science 6, 488.

Tuckerman, J.R., Gonzalez, G., Sousa, E.H., Wan, X., Saito, J.A., Alam, M., and Gilles-Gonzalez, M.A. (2009). An oxygen-sensing diguanylate cyclase and phosphodiesterase couple for c-di-GMP control. Biochemistry *48*, 9764–9774. https://doi.org/10.1021/bi901409g.

Vojnov, A.A., Slater, H., Daniels, M.J., and Dow, J.M. (2001). Expression of the gum operon directing xanthan biosynthesis in *Xanthomonas campestris* and its regulation in planta. Mol. Plant Microbe Interact. *14*, 768–774. https://doi.org/10.1094/MPMI.2001.14.6.768.

Vrancken, K., Holtappels, M., Schoofs, H., Deckers, T., and Valcke, R. (2013). Pathogenicity and infection strategies of the fire blight pathogen *Erwinia amylovora* in Rosaceae: State of the art. Microbiology 159, 823–832.

Wang, S., Sun, Z., Wang, H., Liu, L., Lu, F., Yang, J., Zhang, M., Zhang, S., Guo, Z., Bent, A.F., *et al.* (2015). Rice OsFLS2-Mediated Perception of Bacterial Flagellins Is Evaded by *Xanthomonas oryzae* pvs. *oryzae* and oryzicola. Mol. Plant *8*, 1024–1037. https://doi.org/10.1016/j.molp.2015.01.012.

Washington, E.J., Mukhtar, M.S., Finkel, O.M., Wan, L., Banfield, M.J., Kieber, J.J., and Dangl, J.L. (2016). *Pseudomonas syringae* type III effector HopAF1 suppresses plant immunity by targeting methionine recycling to block ethylene induction. PNAS 113(25), 3577-86.

Wassmann, P., Chan, C., Paul, R., Beck, A., Heerklotz, H., Jenal, U., and Schirmer, T. (2007). Structure of BeF3 Modified Response Regulator PleD: Implications for Diguanylate Cyclase Activation, Catalysis, and Feedback Inhibition. Structure 15, 915–927.

Weinhouse, H., Sapir, S., Amikam, D., Shilo, Y., Volman, G., Ohana, P., and Benziman, M. (1997). c-di-GMP-binding protein, a new factor regulating cellulose synthesis in *Acetobacter xylinum*. FEBS Lett. *416*, 207–211.

Willmann, R., Lajunen, H.M., Erbs, G., Newman, M.A., Kolb, D., Tsuda, K., Katagiri, F., Fliegmann, J., Bono, J.J., Cullimore, J.V., *et al.* (2011). Arabidopsis lysin-motif proteins LYM1 LYM3 CERK1 mediate bacterial peptidoglycan sensing and immunity to bacterial infection. Proc. Natl. Acad. Sci. U.S.A. *108*, 19824–19829. https://doi.org/10.1073/pnas.1112862108.

Win, J., Chaparro-Garcia, A., Belhaj, K., Saunders, D.G.O., Yoshida, K., Dong, S., *et al.* (2013). Effector Biology of Plant-Associated Organisms: Concepts and Perspectives. Cold Spring Harbor Symposia on Quantitative Biology 77, 235–247.

Wu, D., Ding, W., Zhang, Y., Liu, X., and Yang, L. (2015). Oleanolic Acid Induces the Type III Secretion System of *Ralstonia solanacearum*. Front. Microbiol. *6*, 1466. https://doi.org/10.3389/fmicb.2015.01466.

Wu, X., Zeng, Q., Koestler, B.J., Waters, C.M., Sundin, G.W., Hutchins, W., and Yang, C.-H. (2014). Deciphering the Components That Coordinately Regulate Virulence Factors of the Soft Rot Pathogen Dickeya dadantii. Mol. Plant Microbe Interact. 27, 1119–1131. https://doi.org/10.1094/MPMI-01-14-0026-R.

Yang, F., Qian, S., Tian, F., Chen, H., Hutchins, W., Yang, C.H., and He, C. (2016). The GGDEF-domain protein GdpX1 attenuates motility, exopolysaccharide production and virulence in *Xanthomonas oryzae* pv. *oryzae*. J. Appl. Microbiol. *120*, 1646–1657. https://doi.org/10.1111/jam.13115.

Yang, S., Perna, N.T., Cooksey, D.A., Okinaka, Y., Lindow, S.E., Ibekwe, A.M., Keen, N.T., and Yang, C.H. (2004). Genome-wide identification of plant-upregulated genes of *Erwinia chrysanthemi* 3937 using a GFP-based IVET leaf array. Mol. Plant Microbe Interact. *17*, 999–1008. https://doi.org/10.1094/MPMI.2004.17.9.999.

Yi, X., Yamazaki, A., Biddle, E., Zeng, Q., and Yang, C.-H. (2010). Genetic analysis of two phosphodiesterases reveals cyclic diguanylate regulation of virulence factors in *Dickeya dadantii*. Mol. Microbiol. 77, 787–800.

Young, J.M. (2004). Olive knot and its pathogens. Australasian Plant Pathol. 33, 33-39.

Yuan, X., Khokhani, D., Wu, X., Yang, F., Biener, G., Koestler, B.J., *et al.* (2015). Cross-talk between a regulatory small RNA, cyclic-di-GMP signalling and flagellar regulator FlhDC for virulence and bacterial behaviours. Environ. Microbiol.17, 4745–4763.

Yun, M.H., Torres, P.S., El Oirdi, M., Rigano, L.A., Gonzalez-Lamothe, R., Marano, M.R., Castagnaro, A.P., Dankert, M.A., Bouarab, K., and Vojnov, A.A. (2006). Xanthan induces plant susceptibility by suppressing callose deposition. Plant Physiol. *141*, 178–187.

Zipfel, C., and Felix, G. (2005). Plants and animals: a different taste for microbes? Curr. Opin. Plant Biol. *8*, 353–360.

Zorraquino, V., García, B., Latasa, C., Echeverz, M., Toledo-Arana, A., Valle, J., Lasa, I., and Solano, C. (2013). Coordinated cyclic-di-GMP repression of *Salmonella* motility through YcgR and cellulose. J. Bacteriol. *195*, 417–428. https://doi.org/10.1128/JB.01789-12.

Index

A

Adaptive immunity 63, 88, 136–137, 145–147, 184, 200, 201
Aeromonas hydrophila 23
Aeromonas salmonicida 19
Agrobacterium tumefacens 11
Anaplasma phagocytophilum 92
Apoptosis 17, 18, 21, 22, 24, 44, 71–73, 89, 107, 113, 136, 148, 176
Autophagosome formation 69–71, 82–84, 87, 91, 92, 145, 174
Autophagy 17, 20–22, 46, 63, 69–71, 81–95, 135, 144, 145, 170, 171, 173–175

B

Bacillus Calmette–Guérin (BCG) 86, 89, 171–173, 177, 180
Bacillus species 15
Bacillus subtilis 46, 51
Bartonella species 11, 21
Bordetella bronchispetica 48
Bordetella pertussis 5, 13, 21, 49
Borrelia burgdorferi 91
Brucella species 11, 21, 208
Burkholderia cenocepacia 48
Burkholderia pseudomallei 20, 23, 208
Burkholderia species 23
Burkholderia thailandensis 23

C

c-di-GMP 200–213
Cell-autonomous immunity 82, 86, 93
cGAMP synthase 86
Chlamydia 20
Citrobacter rodentium 47, 110
Corynebacterium glutamicum 51
Coxiella burnetii 63, 67, 74, 92
Coxiella species 22
CRISPR-Cas system 55
Cytokine production 86, 89–95, 100, 111, 147, 148, 181, 186

E

EPEC/EHEC 19, 20, 63, 68, 69, 73
Erwinia chrysanthemi 4
Escherichia coli 4, 9, 17, 19, 22, 48, 49, 51, 91, 110, 137, 141, 204, 205, 208, 209,

F

Francisella novicida 55, 109–111
Francisella tularensis 84, 92, 109–111, 179

H

Haemophilus influenzae 49
Helicobacter pylori 11, 21, 47–49, 52, 203

I

Inflammasome 47, 54, 55, 62, 72, 87, 88, 94, 99–116, 135, 136, 142, 173, 182
Interleukin-1β (IL-1β) 20, 36, 47, 62, 85–87, 91–94, 99–115, 136, 178, 182

K

Klebsiella pneumoniae 4, 48, 50, 55, 204

L

Legionella pneumophila 11, 18, 22, 55, 63, 67, 69, 72–74, 86–88, 92, 93, 106, 107, 110, 205
Lipid A 45, 48–50, 53, 55, 111
Listeria monocytogenes 46, 51, 52, 63, 65, 84, 86, 111, 112, 179

M

Macrophages 18, 21–24, 44, 46, 49, 61–74, 86–88, 90–93, 102, 106–113, 115, 134–136, 141–145, 150, 151, 169–182, 185, 186
Mycobacterium leprae 52, 112, 170, 173, 178
Mycobacterium tuberculosis 15, 24, 51, 63, 66, 67, 69–74, 86, 91–94, 112, 169–188

N

NAIP5 47, 62, 87, 88, 104, 106, 108,
Neisseria gonorrhoeae 5, 51

Neisseria meningitidis 5, 49, 51
Neisseria species 23
NLRC4 47, 62, 72, 87, 88, 100, 104, 106, 108, 109, 111–114, 136, 142
NLRP3 47, 72, 87, 100–103, 105, 106, 108–113, 115, 136, 142, 182
NOD-like receptors (NLR) family 45, 62, 85, 87, 100, 135, 172
NOD1/NOD2 46, 47, 50–52, 54, 87, 135, 142

P

Paracoccus denitrificans 22
Peptidoglycan 4, 17, 46, 48, 50–54, 85, 87, 134, 201
Phagocytosis 18, 19, 22–24, 44, 62–67, 71, 92, 109, 111, 113, 136, 141, 146, 170, 171
Porphyromonas gingivalis 7, 25
Pseudomonas aeruginosa 4, 5, 9, 19, 23, 24, 48, 91, 205, 209–211
Pseudomonas syringae 63, 200, 203, 205, 208, 210–212

R

RIG-I-like receptors (RLR) family 86, 99, 100, 135

S

Salmonella enterica 19, 22, 63, 66, 71, 86, 107, 133, 134, 137, 143, 146, 147,
Salmonella enterica serovar Typhimurium 48, 49, 54, 55, 66, 69, 84, 91, 92, 107–110, 133, 135, 137, 140–142, 144, 145, 147–152, 209
Salmonella species 9, 19, 20, 47, 63, 66, 69, 73, 82, 107–109, 114, 133–154
Sec translocase 3, 4
Shigella flexneri 9, 18, 20, 23, 50, 63, 66, 68, 70, 72, 84, 87, 91, 92, 114, 115
Staphylococcus aureus 22, 24, 44
Staphylococcus species 15
Streptococcus species 11, 15, 44

T

TLR4 45, 47–49, 61, 86, 102, 105, 108, 111, 114, 142
Toll-like receptor (TLR) family 45, 61, 85, 99, 201, 203
Twin-arginine-dependent translocation (TAT) system 3, 4
Type I secretion systems (effectors) 17
Type I secretion systems (structure) 7–9
Type II secretion systems (effectors) 18
Type II secretion systems (structure) 4, 5
Type III secretion systems (effectors) 18–20, 54, 62, 71, 72, 104, 107–109, 113, 114, 140, 142, 150, 152, 201, 203, 205, 208–211
Type III secretion systems (structure) 9–11
Type IV secretion systems (effectors) 21–23, 106–109
Type IV secretion systems (structure) 11–13
Type IX secretion systems (effectors) 24, 25
Type IX secretion systems (structure) 7
Type V secretion systems (effectors) 23
Type V secretion systems (structure) 5–7
Type VI secretion systems (effectors) 23, 24, 209, 210
Type VI secretion systems (structure) 13–15
Type VII secretion systems (effectors) 24, 117
Type VII secretion systems (structure) 15–17

V

Vibrio cholerae 4, 9, 18, 24, 49, 205, 209

X

Xanthomonas campestris 203, 205, 208, 210, 211,
Xanthomonas oryzae 199, 210, 212
Xylella fastidiosa 200, 210, 212

Y

Yersinia enterocolitica 4, 113, 114, 209
Yersinia pestis 22, 49, 50, 72, 103, 113, 114
Yersinia pseudotuberculosis 24, 72, 92, 113
Yersinia species 18–20, 48, 63, 69, 72, 113–114

www.ingramcontent.com/pod-product-compliance
Ingram Content Group UK Ltd.
Pitfield, Milton Keynes, MK11 3LW, UK
UKHW022034041026
15591UKWH00012B/671

9 781910 190692